HEAT

AF730774

MAXWELL PRESS

HEAT

J H POYNTING, Sc.D., F.R.S.

Foreign Member of the Accademia Dei Lincei, Rome;
Hon.D.Sc., Victoria University; Late Fellow of Trinity College, Cambridge;
Mason Professor of Physics in the University of Birmingham.

Sir J J Thomson, M.A., F.R.S.

Winner of Nobel Prize for Physics

Corresponding Member of the Institute of France;
Foreign Member of the Berlin Academy; Hon.Sc.D., Dublin; Hon.D.L.,
Princetown; Hon. D.Sc., Victoria; Hon.LL.D., Glasgow; Hon. Ph.D., Cracow;
Fellow of Trinity College, Cambridge; Cavendish Professor of Experimental
Physics in the University of Cambridge; Professor of Natural Philosophy in the
Royal Institution.

MAXWELL PRESS

Chennai Trichy NewDelhi

MAXWELL PRESS

ISBN 978-93-90063-50-5 Maxwell Press

No. 44, Nallathambi Street,
Printed and bound in India Triplicane, Chennai 600 005

MJP 896

Publisher : C. Janarthanan

Publisher's Note

The legacy of a country is in its varied cultural heritage, historical literature, developments in the field of economy and science. The top nations in the world are competing in the field of science, economy and literature. This vast legacy has to be conserved and documented so that it can be bestowed to the future generation. The knowledge of this legacy is slowly getting perished in the present generation due to lack of documentation.

Keeping this in mind, the concern with retrospective acquiring of rare books has been accented recently by the burgeoning reprint industry. Maxwell Press is gratified to retrieve the rare collections with a view to bring back those books that were landmarks in their time.

In this effort, a series of rare books would be republished under the banner, "Maxwell Press". The books in the reprint series have been carefully selected for their contemporary usefulness as well as their historical importance within the intellectual. We reconstruct the book with slight enhancements made for better presentation, without affecting the contents of the original edition.

Most of the works selected for republishing covers a huge range of subjects, from history to anthropology. We believe this reprint edition will be a service to the numerous researchers and practitioners active in this fascinating field. We allow readers to experience the wonder of peering into a scholarly work of the highest order and seminal significance.

MAXWELL PRESS

PREFACE

TO FOURTH EDITION.

We desire to express our hearty thanks to readers of the earlier editions of this volume who have kindly sent us lists of errata. These have been corrected, and a number of alterations and additions have been made.

June 1911.

PREFACE.

THIS volume on Heat is the third of a series forming a Text-Book on Physics. The first two volumes dealt with the Properties of Matter and Sound, and the succeeding volumes will deal with Magnetism and Electricity, and Light.

The Text-Book is intended chiefly for the use of students who lay most stress on the study of the experimental part of Physics, and who have not yet reached the stage at which the reading of advanced treatises on special subjects is desirable. To bring the subject within the compass thus prescribed, an account is given only of phenomena which are of special importance, or which appear to throw light on other branches of Physics, and the mathematical methods adopted are very elementary. The student who possesses a knowledge of advanced mathematical methods, and who knows how to use them, will, no doubt, be able to work out and remember most easily a theory which uses such methods. But at present a large number of earnest students of Physics are not so equipped, and the authors aim at giving an account of the subject which will be useful to students of this class. Even for the reader who is mathematically trained, there is some advantage in the study of elementary methods, compensating for their cumbrous form. They bring before us more evidently the points at which various assumptions are made, and they render more prominent the conditions under which the theory holds good.

J. H. P.

CONTENTS.

CHAPTER I.

TEMPERATURE.

CHAPTER II.

EXPANSION OF SOLIDS WITH RISE OF TEMPERATURE.

CHAPTER III.

EXPANSION OF LIQUIDS.

CHAPTER IV.

EXPANSION OF GASES.

CHAPTER V.

PAGES

CIRCULATION AND CONVECTION IN LIQUIDS AND GASES.

CHAPTER VI.

QUANTITY OF HEAT. SPECIFIC HEAT.

CHAPTER VII.

CONDUCTIVITY.

CHAPTER VIII.

THE FORMS OF ENERGY. CONSERVATION OF ENERGY. MECHANICAL EQUIVALENT OF HEAT. FIRST LAW OF THERMODYNAMICS.

CHAPTER IX.

PAGES

THE KINETIC THEORY OF MATTER.

CHAPTER X.

CHANGE OF STATE—LIQUID—VAPOUR.

CHAPTER XI.

CHANGE OF STATE—LIQUID—VAPOUR (*continued*).

CHAPTER XII.

CHANGE OF STATE—SOLID—LIQUID.

CHAPTER XIII.

WATER IN THE ATMOSPHERE.

CONTENTS.

CHAPTER XIV.

PAGES

GENERAL ACCOUNT OF RADIATION.

CHAPTER XV.

THEORY OF EXCHANGES.

CHAPTER XVI.

RADIATION AND TEMPERATURE.

CHAPTER XVII.

THERMODYNAMICS.

CHAPTER XVIII.

PAGES

THERMODYNAMICS OF ISOTHERMAL AND ADIABATIC CHANGES.

CHAPTER XIX.

THERMODYNAMICS OF CHANGE OF STATE AND OF SOLUTIONS.

CHAPTER XX.

THERMODYNAMICS OF RADIATION.

LIST OF ILLUSTRATIONS.

HEAT.

CHAPTER I.

TEMPERATURE.

Introductory Remarks — Temperature — Thermal Equilibrium — Construction of Mercury-Glass Thermometers—Fixed Points: Centigrade, Fahrenheit, and Reaumur Scales—Marking Fixed Points—Calibration and Graduation—Precautions in use—Limits of accuracy—Range—Scales of temperature given by expansion arbitrary—The Work Scale—Air and Hydrogen Scales—Platinum Resistance Thermometers—Table of Temperatures—Maximum and Minimum Thermometer—Thermostats.

Introductory Remarks.—In the science of Heat, we investigate those phenomena which are chiefly revealed to us by our sense of warmth or cold. We use the words "hot" or "cold" to describe the condition of external bodies which corresponds to the sensation we receive through our skin on touching or approaching them, and we habitually compare bodies with respect to the sensations so received, describing one as "hotter" or "colder" than another. Given several vessels of water, we could with very little trouble arrange them in order of hotness, and we have a number of expressions in common use to describe their conditions, such as "ice-cold," "cool," "chill taken off," "tepid," "lukewarm," "warm," "hot," "boiling hot." Our primary sensations are, therefore, those of hotness or of coldness, and we are accustomed to think of hotness as varying in degree. If we put a hot body in contact with a cold one—if, for instance, we pour hot water into a cold vessel—the hot water is cooled while the cold vessel is heated. We regard this change as the passage of something which we term heat from the hotter to the colder body, its loss by the former being accompanied by cooling, its gain by the latter by heating. We do not mean to imply by "something" some kind of matter. We may fairly describe kinetic energy as "something," and say that when one body strikes another, setting it in motion, "something," viz., kinetic energy, has passed from the one to the other, yet we do not think of energy as matter.

So, here, we only describe the heat as "something," because we believe that we can identify the heat gained by the cold vessel with that lost by the hot water. We also think of heat as greater or less in amount. If the hot water cools very considerably, we think of it as giving up more heat than if it cools only slightly. Or if the quantity of hot water cooling is comparatively large, we think of it as giving up more heat than a smaller quantity of water cooling to the same extent.

These, then, are our two fundamental ideas: that bodies are comparable as to their hotness or coldness, and that in general, on becoming hotter they receive something which we call heat—the amount of heat received depending both on the quantity and nature of the matter and on the degree to which it becomes hotter.

Our first aim must be to render our conceptions more definite, by obtaining some numerical scale to express *how* hot a body is.

We shall be occupied, in the earlier chapters of the book, with a description of the mode in which such a scale is obtained, and an account of its use in the investigation of the change of dimensions of bodies when heated. Afterwards we shall show how numerical expressions may be obtained for quantities of heat, and how these quantities may be determined by experiment.

Temperature and Thermal Equilibrium.—The number which expresses on some definite scale how hot a body is, is termed its **temperature.**

Any instrument, such as the ordinary mercury and glass instrument, used to obtain the temperature of a body is termed a **thermometer.**

We know from common observation that bodies in contact with each other, and not subjected to changes of external conditions, after a time get neither hotter nor colder—*i.e.* heat does not pass from one to another. They are then said to be in **thermal equilibrium** with each other.

For example, in a room not exposed to draughts or to sunlight, and containing no fire, the objects lying on a table will after a time all be in thermal equilibrium with the table and with each other. A glass of water on the table will neither be heated nor cooled by putting into it an iron rod which has also been lying on the table. The sensation received by an observer touching the various objects may, however, be very different. The iron will feel colder than the table on which it has been placed, and the water in the glass will feel warmer than the iron and colder than the table. But if, instead of the finger, we use a thermometer, we find that it will register the same whether it is laid on the table, placed against the iron, or put in the water—the difference of sensation not being due to difference of temperature but to different rates of communication of heat from the hand to the surfaces touched.

The thermometer, then, shows us that bodies in thermal equilibrium with each other are at the same temperature, and as the thermometer is also one of the bodies it is also at the same temperature as the substance in which it is placed.

Conversely it is true that bodies at the same temperature are in thermal equilibrium with each other.

This may be proved by direct experiment; by pouring, for instance, a quantity of mercury into a vessel containing water at the same temperature, and noting that the temperature remains constant.

Thermometers.—In order to obtain a definite scale of temperature, we make use of the fact that, in general, bodies expand on being heated. Various simple experiments show this. For instance, if a metal rod (Fig. 1) be fixed at the end V in a vice and if the end A presses down on a small roller r on a flat plate, then a pointer P attached to the roller will move over a scale S when the rod is heated, say by a gas flame.

Or, if a brass rod (Fig. 2), exactly fits into a space in a brass plate when cool, it can no longer be inserted into the space after it is heated, if the plate remains cool.

We might use the expansion of a brass rod to give us a scale of temperature, if we had a sufficiently delicate and simple method of measuring its length.

We might call the temperature of the brass when placed in ice-cold water, 0°, and describe its temperature as rising 1° for every increase in length of $\frac{1}{100000}$, but the difficulty of measuring such a small increase as would occur for any ordinary rise of temperature would render such a scale of little practical value.

FIG. 1.—Expansion of a Heated Rod.

We therefore make use of the facts that the expansion of liquids by heat is usually much greater, and that of gases enormously greater, than that of solids.

If a glass flask is filled with water, and then closed by a cork with a tube of narrow-bore passing through the cork, the warmth of the hand is quite sufficient to make the water rise very appreciably in the narrow tube. For though the glass expands, thus making the internal capacity greater, the water expands still more, not only filling up the additional volume of the flask, but also rising in the tube. This is the principle which is used in the ordinary thermometer, the bulb containing either mercury or alcohol, and the expansion in excess of that required to fill the increase of volume of the bulb being indicated by the rise of liquid in the tube.

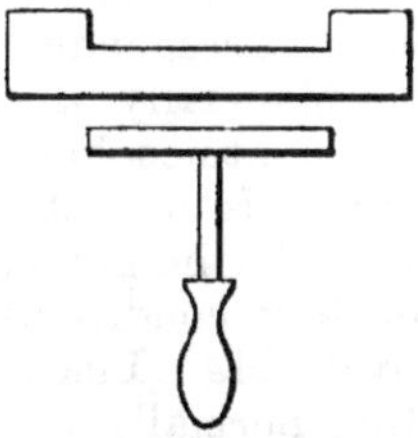
FIG. 2.—Gauge, into which a Bar fits when cold but not when hot.

Gases expand still more than liquids. If a flask partially filled with water be closed by a cork, through which passes a narrow tube with its lower end dipping under the surface of the liquid, the warmth of the hand will make the air in the upper portion of the flask expand and drive the liquid rapidly up the tube. Gas thermometers on this principle may be made far more sensitive than the liquid thermometers, but there are, as we shall see, difficulties in their use, which render them unsuitable for ordinary purposes.

The Construction of Mercury-in-Glass Thermometers.—To construct a good thermometer for scientific purposes, a tube is selected with a capillary bore as nearly uniform as possible. A bulb is blown on the end of this, the size of the bulb being adjusted by the experience of the glass-blower to the sensibility required in the thermometer—the greater the sensibility the larger the bulb or the finer the bore. In order to fill the bulb with mercury, it is heated to expel some of the air, and the open end of the tube is inserted under mercury. On cooling, the pressure of the remaining air diminishes, and the external atmospheric pressure drives some mercury up into the bulb. The tube is then held bulb downwards, and the bulb is heated till the mercury boils, the mercury vapour rising and displacing the remaining air. On again inverting the tube and placing the open end under mercury, as the temperature falls, the whole space will, as a rule, be filled with mercury. Should the air not be entirely expelled, however, the operation must be repeated. The tube is then drawn out near the end till the bore is nearly closed, the bulb is again heated until the mercury flows past the narrow part, and the latter is rapidly sealed up by means of a blowpipe flame. It is found by experience that the bulb of a thermometer undergoes contraction. The contraction is very considerable during the first few weeks or even months after it has been blown, but after two or three years it takes place very slowly and is hardly appreciable over short intervals of time. It is advisable, therefore, to wait till this stage is reached before the graduation of the instrument is proceeded with.

Fixed Points.—If we only had to use one thermometer, to indicate a rise in temperature in one particular case, it would be sufficient to mark on the stem divisions of some chosen length, say 1 mm. each, to number these from the bottom to the top of the tube, and to call each step of one "division" a degree. If, for example, the mercury stood 50 mm. above the lowest mark, we might call the temperature 50°. But we wish to use different instruments for different cases and to compare their indications. Further, we wish to compare the indications of our thermometers with those of instruments used by other experimenters on other occasions. We must, therefore, have a definite scale, as nearly as possible the same on all instruments. This is secured by the use of definite "fixed points," the same on all instruments, each corresponding to a definite fixed temperature. The two fixed points universally used are 1st, the temperature of ice when just melting under the atmospheric pressure of 760 mm.; and 2nd, the temperature of steam from water boiling normally under the same atmospheric pressure. The height of the mercury column of a thermometer placed under constant conditions in different vessels containing melting ice will remain invariable. Placed under similar conditions in the steam from water boiling normally at a pressure of 760 mm. in different vessels, it will again remain invariable, but at a point much higher in the stem. The volume of the tube—or, if it is of even bore, the length—between these two fixed points is divided into a number of equal parts, and each part indicates a degree.

*The Centigrade Scale.**—On the centigrade scale, now universally employed for scientific purposes, the temperature of melting ice is 0°, and that of boiling water under the stated conditions is 100°, and the interval is divided into 100 parts. The scale is indicated by writing C. after the temperature, as 0° C.

The Fahrenheit Scale.—On the Fahrenheit scale, the temperature of melting ice is 32°, and that of boiling water is 212°, the interval being divided into 180 equal parts. A Fahrenheit temperature is indicated by writing F. after the temperature, as 212° F.

This scale was arranged by Fahrenheit early in the eighteenth century. He found that a mixture, of which he did not state the proportions, of ice, water, and sal-ammoniac, or sea-salt instead of sal-ammoniac, gave a very low definite temperature, which he took as 0°. He found that

* Some interesting notes on the history of thermometry will be found in Bolton's *Evolution of the Thermometer.*

the normal temperature of the human body was nearly constant, and using a duodecimal scale he took this as $2 \times 12 = 24°$. He found that ice melted at 8° on this scale. These degrees being inconveniently large, he quartered them, so that he had 0°, 32°, and 96° as fixed points. Having verified Amontons' statement that water boils at a constant temperature, he found that it was at 212 of his new quartered degrees. The estimate of 96° for the human body was then found to be more nearly 98° than 96°.

The Reaumur Scale.—On this scale the fixed points are 0° for melting ice and 80° for boiling water. Reaumur used alcohol diluted with one-fifth of water as the liquid in his thermometer, and each degree marked an expansion of $\frac{1}{10000}$ of the volume of the liquid at 0°.

Conversion of Scales.—If the same temperature is indicated respectively on the three scales by C, F, and R, then evidently the equations—

$$\frac{C}{100} = \frac{F - 32}{180} = \frac{R}{80}$$

enable us at once to convert from one scale to another.

Marking the Fixed Points.—In this country the lower fixed point is marked first. For this purpose the bulb of the tube is immersed in a metal vessel containing a mixture of small pieces of melting ice and air-free distilled water. The vessel may also be surrounded by melting ice to prevent the temperature of the water in it rising above that of the ice. The thermometer is placed so that the mercury rises just to the top of the ice, and when it is steady a file mark is made to show its position.

The thermometer is then placed in a metal vessel, the construction of which is indicated by Fig. 3, the bulb and the tube being entirely surrounded by steam. The thermometer is placed so that the mercury rises just into sight and its final level, which is only attained after some time, say a quarter or half-an-hour, is again marked by a fine file.

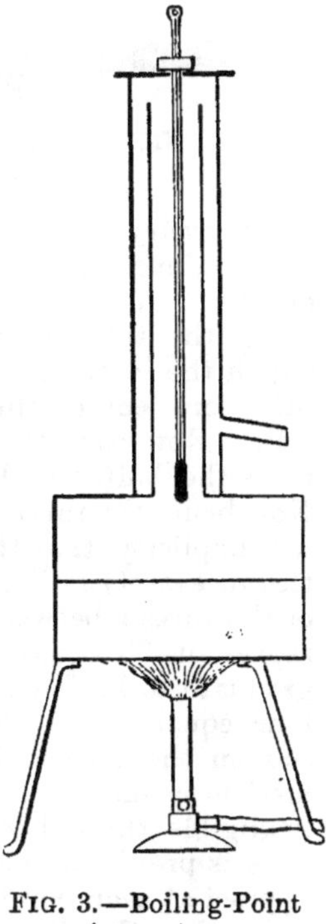

FIG. 3.—Boiling-Point Apparatus.

It is essential to read the barometer, as the boiling temperature of water varies with the atmospheric pressure, being only 99° at 733·2 mm. (slightly below 29 inches). The point marked will only be 100° on the rare occasions when the barometer is at 760 mm., and, in general, allowance must be made for the deviation.

Pressure also affects the melting point of ice, but the ordinary variations of pressure produce no appreciable effect. Practically it is only important in ordinary thermometers to use pure water, and to be sure that the ice is all at the melting temperature. Impurities lower the melting point, and large lumps of ice which have not been long melting may easily be below the melting point inside.

It may be noted that 760 mm. of mercury is not an invariable standard, but varies with the variation of gravity. At the equator, 760 mm. of mercury would only imply a pressure equal to something less than 758 mm. in England, and water would boil therefore at slightly above 99·9°. If great accuracy is required, the standard pressure is taken as 760 mm. of mercury in lat. 45°. The same barometric height at Greenwich corresponds to a pressure greater by about 56 in 100,000, which will alter the boiling-point about ·016° C.

Calibration.—If the thermometer is to be used as a standard, it must be calibrated, *i.e.* the variations in the capillary bore must be determined and allowed for. There are various modes of effecting this,* the simplest and quickest being as follows :—

The mercury in the tube is detached near the neck of the bulb, either by warming the tube in a very fine gas flame at the point where it is to be detached, or by manipulation of the trace of air still remaining, and which collects in the vacuum left in the bulb if the thermometer is inverted and the mercury is sent down to the end of the tube. The detached thread is then run down into an enlargement of the bore provided for the purpose at the end farthest from the bulb, leaving the bore clear.

A short thread, say about 20 mm. in length, is then detached from

FIG. 4.—Calibration of a Thermometer; AB, BC,—HK, = equal volumes of the bore.

the mercury still remaining in the bulb and is measured at various points along the tube by a travelling microscope, or by the dividing engine.† Let us suppose, for example, that the length AK, Fig. 4, along the tube, represents 100°, and that we wish to know how to adjust the intervening divisions to allow for variations of bore. Let us bring one end of the thread to A, the other end being at B. Measure AB. Now pass the thread along by gently tapping the end farthest from the bulb till the thread occupies the position BC. Measure BC. Now bring the thread to CD. Measure CD; and so on. Let us suppose, for simplicity, that the last position of the thread is HK, the end of the thread exactly falling at K, and suppose that there are in all 25 lengths of the thread between A and K. Then the lengths AB, BC, CD—HK represent 25 equal volumes, and each of them must contain 4°. To graduate to 100°, we must divide each of the lengths AB, BC, &c., into four equal parts. At each of the points B, C, &c., there is, therefore, a sudden though small change in the length of the degrees. Hence the method is only applicable when the variation in the bore is practically negligible through one thread-length. For still greater accuracy, the tube is previously graduated to equal lengths, and the correction to be applied at each point is determined by graphic methods. The mode of employing the dividing engine for graduation is described in Stewart and Gee's *Practical Physics*, p. 24.

* *Report of Committee on Methods Employed in the Calibration of Mercurial Thermometers*, British Association, 1882.

† See Stewart and Gee, *Practical Physics*, vol. i. p. 16. An excellent instrument of simple construction is described by Brown in *Phil. Mag.*, vol. xiv. 1882, p. 57.

When the thermometer is not to serve as a standard, the labour of calibration may be saved by comparing the readings with some standard instrument, and observing the deviation from the true reading at various points along the scale. Such comparisons are undertaken by the National Physical Laboratory, and a certificate is issued with each thermometer compared, stating its errors.

Precautions in Use.—In order to obtain consistent values for a given temperature, certain precautions must be observed in using the mercury-glass thermometer. If the instrument is first used for a low temperature, then exposed to a high one, and lastly brought back to the first low temperature, it will give a lower indication than before. The effect is entirely due to the glass, which does not on cooling at once contract to its original volume. In the course of days or weeks, however, it does return to that volume. The effect may be well observed by immersing a thermometer in melting ice till it is at 0°, then putting it into steam for twenty minutes or so and then returning it to the ice. The zero point will be found to be depressed by an amount differing with the kind of glass used, ranging from about 0·1° C to perhaps 0·5° C. With some kinds of glass the depression is nearly proportional to the high temperature reached, but with others the relation is not so simple. If time cannot be allowed to eliminate the effect, the thermometer, if of English make, should have its zero point re-determined immediately before being used for any temperature lower than the high one to which it has been subjected. Abroad it is usual to mark the fixed point 100° before the 0° point in graduation, and with such a thermometer the zero point should be re-determined immediately after the reading of an intermediate temperature.

We have already mentioned the gradual contraction of the bulb and consequent rise of zero point. Though after a few years this becomes very small, it may still be sensible for delicate instruments, and it is necessary, therefore, to find the zero point at intervals and subtract the rise from the indication on the scale. Dr. Joule observed the rise of zero point on two delicate thermometers at intervals during forty years, and the results obtained for one of them (*Scientific Papers*, vol. i. p. 358) may be represented by the curve in Fig. 5. It practically coincides with the curve,

$$y = 13{\cdot}94 - 9{\cdot}51e^{\frac{-t}{15\cdot3}}$$

the height y being in arbitrary divisions of the stem, 13 divisions to 1° F., and t being the time in years from 1844.

The total rise in thirty-eight years was 1° F., and if the curve truly represents the results, it appears to show that it had still, in 1882, about $\frac{1}{20}$th of a degree Fah. to rise, and that it will halve its distance from the final value about every ten years.

In taking a temperature the whole instrument should if possible be at that temperature. If, for instance, the bulb alone is in a hot liquid while the stem emerges into the colder air, not only is the stem conducting heat from the bulb and keeping it at a lower temperature than the liquid, but the glass and the part of the mercury in the air have too small volumes, and on both accounts the temperature indicated is too low. There are formulæ for correcting for the emergence of the stem, but they are unsatisfactory, and where possible their use should be avoided.

A thermometer should be used when possible in the position in which its fixed points were marked. If, for instance, they were marked with the stem vertical, then in the horizontal position the internal pressure on the bulb due to the column of mercury in the stem is removed and the bulb contracts slightly, indicating too high a temperature.

Corrections can, however, be determined by direct experiment and can be applied to the observed reading. With the most sensitive instruments it is also necessary to take into account the varying pressure on the outside of the bulb due to change of atmospheric pressure or depth of immersion in a liquid.

Limits of Accuracy.—With different thermometers made of the same kind of glass and carefully graduated, the indications of a given temperature should agree to within about 0·01° C. Different kinds of glass have

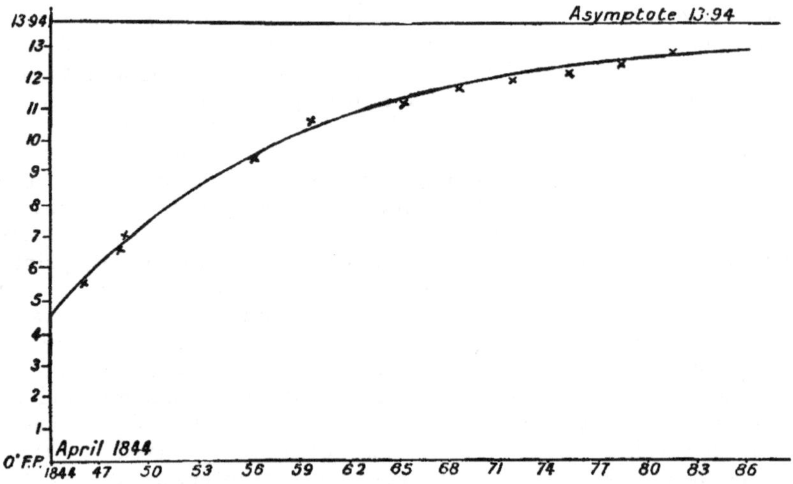

FIG. 5.—Joule's Observations on the Alteration of the Freezing Point in Thermometers, and the Comparison with the Curve $y = 13{\cdot}94^\circ - 9{\cdot}51^\circ e^{\frac{-t}{15\cdot3}}$

different expansions and, though their indications at 0° and 100° will agree, intermediate indications may differ by, perhaps, more than 0·1° C. Above 100° their indications may differ much more widely than this. In recent years very careful attention has been paid to the qualities of different glasses for thermometer purposes and to the methods of correction to be employed to make the readings of different thermometers give the same value of the same temperature.*

It is to be hoped that, as a result of these investigations, the quality of glass used will be improved, and that scientific thermometer makers will everywhere use the same glass, the same mode of marking the fixed points, and the same mode of graduating, so that the indications of different instruments may be immediately compared.

Range of the Mercury-Glass Thermometer.—There are limits

* In Dr. Chree's "Notes on Thermometers," in the *Philosophical Magazine*, xlv. 1898 p. 205, will be found a description and full discussion of the various corrections.

to the use of the mercury thermometer in both directions. The freezing point of mercury being about $-39°$ C. it cannot be used for lower temperatures, and hence, for meteorological purposes, it is usual in cold climates to replace it by alcohol, since alcohol has a much lower freezing point. Nor is it safe to use an ordinary mercury thermometer much above 350° C., the pressure of mercury rising from about $\frac{1}{3}$ atmosphere at 300° to 1 atmosphere at 356°, the normal boiling-point of mercury, and then increasing still more rapidly to 2 atmospheres at 400° and 4 atmospheres at 450°. This great increase of internal pressure may very seriously alter the capacity of the bulb. Mercury thermometers made of specially hard glass and containing nitrogen above the mercury are now, however, made with a range up to 500° C., but they are hardly suitable for exact work.

Scales of Temperature given by Expansion depend on the Substances used.—Beginners in the study of heat sometimes suppose that mercury and glass are chosen in the construction of thermometers, because their expansion is regular and equal for each successive degree. But this regularity is simply due to our definition, that equal degrees shall be such equal expansions. Within the short range from 0° C. to 100° C. most substances which remain otherwise in the same physical condition between these points expand nearly regularly with rise of temperature as indicated by the mercury-glass thermometer. And so most substances, with the same fixed points would give nearly the same scale. Thus, if the expansion of a brass rod between 0° and 100° were used and divided into 100 equal steps, each step would have very nearly the same value as the corresponding steps on the mercury-glass scale. But not exactly, for even between 0° and 100° there are measurable deviations from expansion in the same rates, and outside that range the deviations become more considerable. As we have seen, even different kinds of glass expand differently, so that two thermometers of different glasses agreeing at 0° and 100° will not agree exactly at all intermediate points. The disagreement fortunately is very small within that range.

The Work Scale of Temperature.—There is one scale of temperature, due to Lord Kelvin, which is quite independent of the particular substance used to indicate it. We shall give a full account of this scale in chapter xvii. Here we can only attempt a brief sketch of its nature in order that the reader may know that such a scale exists.

The work scale depends on the amount of work obtained from a given supply of heat to a heat engine.

We may roughly describe an ordinary steam engine as a heat engine which takes in heat at the temperature of the boiler, and turns some of this heat into work by the expansion of the steam in the cylinder. Though the steam cools as it expands it does not turn all the heat received into work but retains some of it till it passes into the cooler or condenser where it returns to the liquid form. It can therefore at the very most only convert into work the difference between the heat taken into the boiler and the heat put out in the condenser.

We can imagine an ideal engine, in which any substance is used, like steam in the ordinary engine, to do work by expansion. The substance works between a source of heat, like the boiler in an ordinary engine, and a cooler receiver like the condenser. The working substance takes in heat from the source, converts some of it to work, and gives out the balance to the receiver. When the engine is imagined to work under certain ideal conditions first prescribed by Carnot (whence it is known as a Carnot engine), the fraction of the heat received which is converted into work depends solely on the temperatures of the source and receiver, and for two given temperatures is the same whatever working substance is used. Or, putting the statement in another way, the ratio of the heat put in at the higher temperature to the heat put out at the lower temperature depends solely on these temperatures. We may therefore use the Carnot engine to give us a scale of temperature in the following way. Let a quantity of heat Q_1 be put in at the higher temperature which we will denote by θ_1 and let a quantity Q_2 be put out at the lower temperature θ_2. Then we fix the ratio of these temperatures by putting

$$\frac{\theta_2}{\theta_1}=\frac{Q_2}{Q_1}$$

If we keep θ_1 and Q_1 constant, Q_2 is less the lower θ_2. If all the heat Q_1 is turned into work, none remains to be put out at θ_2. In this case Q_2 is zero and therefore θ_2 is zero. This implies that the new scale dates from a point such that a Carnot engine working down to that point will turn all the heat which it receives from the source into work. We can imagine no greater degree of cold than that of such a receiver, and its temperature is therefore termed the absolute zero.

It can be shown that if a Carnot engine works between the temperatures of boiling water as source, and melting ice as receiver, then for every 373 parts of heat put in at 100° C. it will turn out about 273 parts at 0° C. and convert 100 parts into work. The ratio of these temperatures on the work scale is therefore 373 : 273 very nearly. If we decide to make the length of degree on the scale such that there are 100 of them between melting ice and boiling water, then melting ice is at 273° A. (where A denotes the work, or, as it is often termed, the absolute scale), and boiling water is at 373° A. The absolute zero then is at $-273°$ C. or 273 absolute degrees below the temperature of melting ice. Though the Carnot engine is ideal merely, and though we can only approximate to it in practice, we shall see in chapter xvii. that we can tell how it would work if realisable. The new scale is, therefore, a perfectly definite one, and it is possible to determine the relation between the work expression of a temperature and its expression on other scales.

Air and Hydrogen Scales.—By the experiments which we shall describe in chapter iv. it has been found that different gases of sufficient tenuity, and sufficiently above their condensing points, expand nearly equally for equal rises of temperature when kept at the same pressure, and that if their density is kept constant their pressure increases nearly equally. Two gases have been chiefly used for thermometric purposes, dry air, and hydrogen, and it is usual to employ the increase of pressure at constant density to give a scale of temperature. Taking $\frac{1}{100}$th of the increase between 0° C. and 100° C. to indicate a degree the scale agrees very nearly with the mercury glass scale within that range. A gas thermometer is applicable through a far wider range than the mercury-glass thermometer. Its scale has the further advantage of being nearly coincident with the work scale. Formerly the air scale was the standard, but now the hydrogen scale, as used at the Bureau International des Poids et Mesures has taken its place. At the Bureau there is a hydrogen thermometer in which the pressure at 0° C. is 1 metre of mercury, and in which the density is kept constant. The degrees of the scale of this instrument are increments of pressure, each $\frac{1}{100}$th of the increase of pressure between 0° C. and 100° C. We shall return to the subject of gas thermometry in chapter iv.

Platinum Resistance Thermometer.—The electrical resistance of pure metals increases almost in direct proportion to the rise of temperature as indicated by the mercury-glass and gas scales. Siemens was the first to employ the resistance of a platinum wire to indicate temperature, and the method has been thoroughly investigated by Callendar ("On the Practical Measurement of Temperature," *Phil. Trans.*, 1887, A., p. 161; *Phil. Mag.*, 1899, xlviii., p. 519). He has shown that it gives an instrument convenient in form, easy to use, and applicable through a far wider range than any other. The platinum wire he used in his original experiments was ·017 cm. diameter, about a metre long, and about 5 ohms resistance. This was wrapped as a spiral on a glass tube, and the ends soldered into thicker platinum leads ·073 cm. diameter, the tube being inserted in the enclosure of which the temperature was to be measured. The resistance of the leads could easily be allowed for, and the resistance of the platinum spiral itself at any temperature could be found. The platinum scale is defined as giving equal degrees by equal increments of resistance, 100 of such degrees making the interval from 0° C to 100° C. Let R be the resistance at any temperature, R_0 and R_{100} the resistances at 0° C. and 100° C. If we denote a temperature on the platinum scale by *pt*, then

$$pt = \frac{R - R_0}{R_{100} - R_0} \times 100.$$

Callendar compared this scale with the air scale, and found that if t is the temperature on the latter, then to a close approximation

$$t - pt = \delta\left(\frac{t^2}{10000} - \frac{t}{100}\right)$$

where δ is constant for a given wire and has nearly the same value, 1·57 for all specimens of pure platinum.

It will be seen that the difference between the two scales must by definition vanish at 0° C. and 100° C. At 50° C. it is a maximum—t being less than pt by about 0·4°. Above 100° t is the greater, at 200° by about 3°, at 300° by about 9°, at 500° by about 31°.

One great advantage of the platinum thermometer lies in its easy use for the determination of low temperatures.

Thermo-Electric Thermometer.—This thermometer makes use of the fact that when a circuit consists of two different metals, A and B, an electric current in general flows round the circuit when the two junctions are at different temperatures. The driving E.M.F. depends solely on the nature of the metals A and B and on the temperatures of the two junctions. Further, if a galvanometer be included in the circuit, with wire of another metal C, inserted, say, in the course of the wire B, then so long as the temperatures of the junctions of C with B are equal, the E.M.F. is the same as if the circuit consisted of A and B only. The thermo-electric thermometer is made in many forms, and with many pairs of metals, according to the purpose for which it is used. It will be sufficient here to describe one form, devised by Le Chatelier, and used by Roberts-Austen for the determination of certain high temperatures (*Nature*, xlv., 1891–2, p. 534).

The active metals in this form are platinum, and an alloy of platinum with 10 per cent. of rhodium. The junction to be inserted in the vessel of which the high temperature is to be measured, consists of a platinum wire round which the platinum-rhodium wire is twisted. The two wires are brought out of the vessel and connected up to a D'Arsonval galvanometer, the junctions with the galvanometer being kept at the same lower temperature, that of the room. The E.M.F. drives a current deflecting the galvanometer, by an amount depending on the difference of temperatures. The instrument is calibrated by inserting the testing junction into vessels of known temperatures, containing in succession, say, boiling water 100° C., melting lead 326° C., and boiling zinc 940°, the deflection of the galvanometer being observed for each of these, and other temperatures being determined by interpolation.

Some Important Temperatures.—The following table gives a few important temperatures determined in various ways. They are put here merely to enable the student to realise the range of measurement possible with the instruments and methods now available :—

TABLE OF TEMPERATURES.

(Chiefly from "Travers' Study of Gases," and from Callendar, *Phil. Mag.*, xlviii., 1899, p. 519.)

	Temperature on Centigrade Scale.
Absolute zero, work scale	– 273
Melting point of hydrogen	– 256 to – 257
Boiling " "	– 252 to – 253
Boiling " oxygen	– 183
Melting " mercury	– 38·8
" " ice	0
Boiling " alcohol	78·3
" " water	100
" " aniline	184·1
" " mercury	356·7
Melting " lead	327·7
" " zinc	419·0
Boiling " sulphur	444·5
Melting " silver	961
" " gold	1061
" " platinum	1820
Crater of electric arc, of the order	3500
Sun's radiating surface, of the order	6000
Bodies begin to emit visible rays, about	380
Red heat, about	500 to 1000
White heat above	1000

Maximum and Minimum Thermometers.—It is often important for meteorological purposes to register the highest and lowest temperatures which have been attained in any period during the absence of the observer. For *maximum* thermometers, a common device is to have a short rod of iron in the tube above the mercury; the thermometer is then placed with its stem horizontal. As the mercury moves outwards along the stem, the rod is pushed in front of it; but when the mercury recedes, it leaves the rod behind, thus indicating the farthest point reached. The iron index may be brought back into contact with the mercury by means of a magnet.

For *minimum* thermometers, a small glass rod is put in the tube, which is horizontal. The liquid (in this case usually alcohol) flows past the glass rod in rising, but in falling it pulls the rod back with it owing to capillary adhesion. The glass index may be brought again to the end of the liquid column by inverting the thermometer.

Six's Thermometer.—This instrument, which is a maximum and minimum thermometer in one, is now very commonly used. The construction is shown in Fig. 6. It consists essentially of a U tube with a bulb at each extremity of the U. The bulb G contains alcohol or other suitable liquid, extending along the tube to *a*; *ab* is a thread of mercury extending round the bend to *b*; above *b* the tube and part of the bulb H to which it leads are filled with the same liquid as that in G, and above the liquid is a space V containing the vapour of the liquid, which can be compressed or extended, and serves as a sort of spring.

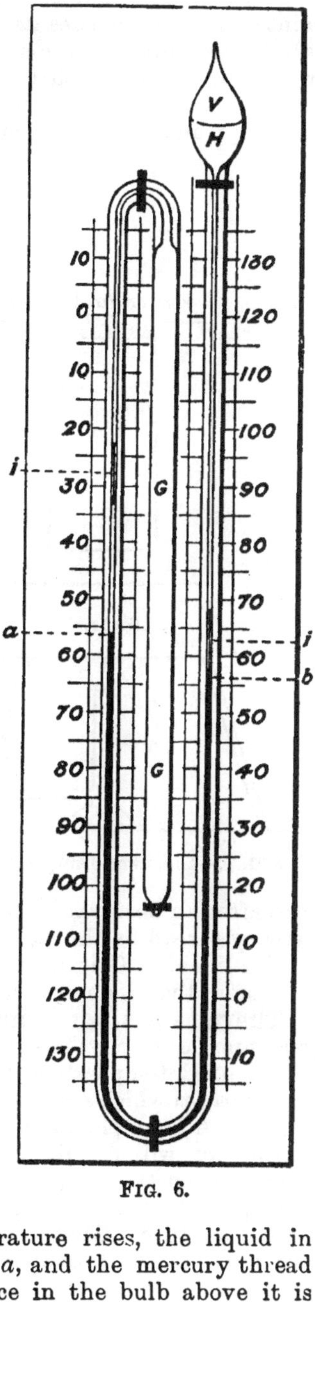

FIG. 6.

In the tubes above *a* and *b* are two small iron rods or indices with hairs attached to them, the hair giving just enough friction to keep the index in position when the mercury column retreats and leaves it. The indices may be brought into contact with the mercury at *a* and *b* by means of a small magnet applied outside the tubes.

Suppose that at a given temperature the indices *ii* are in contact with the mercury on each side. If now the temperature rises, the liquid in G expands, pushes down the mercury from *a*, and the mercury thread *ab* moves round, *b* rises and the vapour space in the bulb above it is decreased. The index above *a* is left in its initial position, while the index above *b* is pushed to the furthest position reached by the mercury. If, on the other hand, the temperature falls, the liquid in G contracts, the point *a* rises and pushes its index in front of it, while the index above *b* is left in its initial position. Two graduated scales are fixed behind the two limbs of the U, that on the left running from above downwards, that on the right from below upwards. Evidently the former gives minimum and the latter maximum temperatures. The instrument is

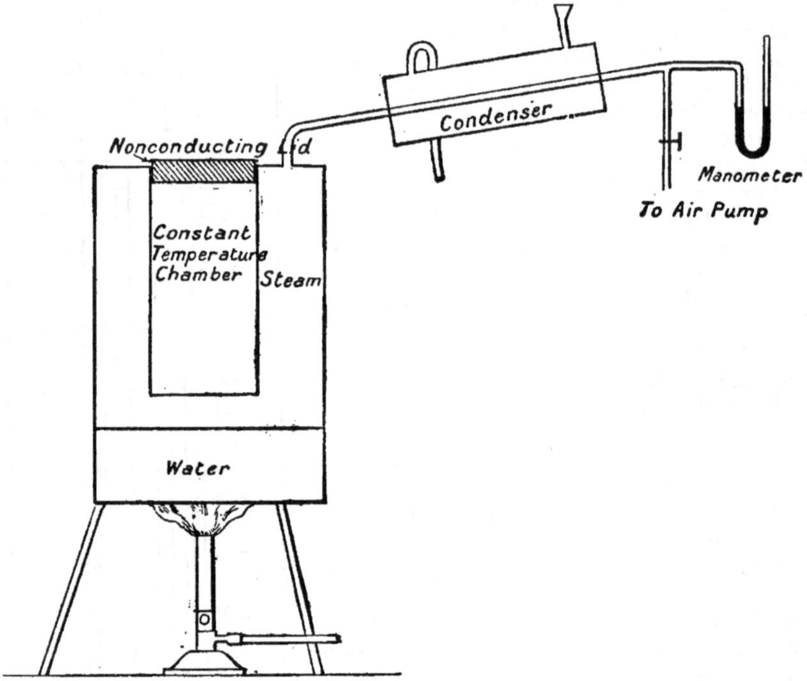

FIG. 7.—Constant-Temperature Apparatus, using the Boiling-Point Method.

reset after an observation by moving the indices back to the ends of the mercury thread by the magnet.

The liquid in G is here the chief expanding liquid. The mercury is little more than a device for moving the indices.

Thermostats, or Constant-Temperature Instruments.—In a great number of physical experiments, it is desirable to keep a body at a known fixed temperature for a considerable time. There are certain temperatures which are easily maintained, as, for instance, that of melting ice. A body placed in melting ice, or in a chamber surrounded by melting ice, will remain indefinitely at 0° C, if proper precautions are taken to keep the temperature of the ice and water uniform either by stirring or by surrounding the vessel with a second vessel containing a similar mixture. There are also definite "freezing mixtures" which give fairly constant temperatures below 0°. Or, again, the steam from water boiling in a metal vessel is very nearly 100° at ordinary altitudes, and by observation of the barometer any deviation from 100° can be accurately determined. But, as we know, the boiling-point varies in a definite manner with the variation of pressure, so that it is possible, by regulating the pressure, to keep the steam at temperatures other than 100°. This method is made use of in one class of constant-temperature apparatus. Water, or some other liquid suitably chosen, and contained in a closed vessel, is supplied with so much heat that it boils. The vapour passes into a cooling arrangement, so that it is condensed back into liquid as fast as it is formed. The pressure is so regulated by varying the amount of air in the vessel, that the boiling-point is the desired fixed temperature. To adapt this apparatus to secure a constant temperature, the vapour is made to surround the chamber in which the constant-temperature operations are being carried on.

The sketch in Fig. 7 will illustrate an application of the method.

This will be seen to be merely an adaptation of Regnault's apparatus for determining the pressure of water vapour at high temperatures, described in chapter x.

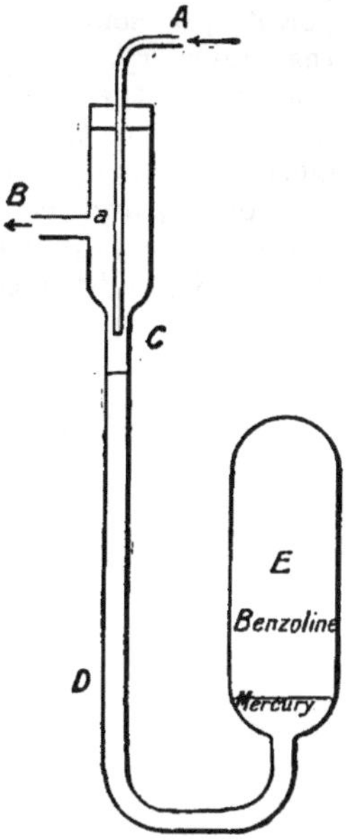

FIG. 8.—Thermostat.

Ramsay and Young have investigated the change in vapour pressure of a number of liquids in the neighbourhood of their boiling points, choosing liquids which are suitable for use in such an apparatus as this, and tables embodying their results will be found in the *Journal of the Chemical Society*, Sept. 1885, vol. xlvii. p. 640. The liquids investigated were:—

Liquid.	Approximate Boiling-Point.
Carbon bisulphide	46°
Ethyl alcohol	78°
Chlorobenzene	132°
Bromobenzene	155°
Aniline	184°
Methyl salicylate	222°
Bromonaphthalene	280°
Mercury	358°

In another class of constant-temperature apparatus, the constant temperature enclosure is heated by gas, and the supply of gas to the burner or burners is regulated by means of a "thermostat," so that if the temperature tends to rise above that required, the gas supply is checked, while if it tends to fall below it, the gas supply is increased. There are many devices for effecting this. The following (Nicol, *Phil. Mag.*, 1883, xv. p. 340) will serve as an example. The thermostat (Fig. 8) is placed with its bulb in the constant-temperature chamber. The gas passes from A to B and thence to the burner, partly through a small hole at *a* and partly up from the end C of the smaller tube. The larger tube D is filled with mercury, which extends round to the lower part of the bulb E, the upper part of which is filled with benzoline or paraffin, or some more expansible, and, therefore, more sensitive liquid than mercury. The tube A can be raised or lowered, and is so adjusted that when the thermostat is at the desired temperature, the mercury just reaches to the end C of the tube. If the temperature now rises, the mercury seals up the end C, and the gas only passes through the small sidehole *a*, and is just enough to keep the burner lighted. If the temperature falls, the mercury allows a free passage to the gas through the end C, and the supply of heat increases.

CHAPTER II.

EXPANSION OF SOLIDS WITH RISE OF TEMPERATURE.

Linear Expansion of Solids—Ramsden's Method—Modern Use of the Method—Method of Lavoisier and Laplace—Results—Fizeau's Optical Method—Applications of Linear Expansion—Volume Expansion of Solids.

As a general rule bodies expand with rise of temperature. As a general rule, also, gases expand more than liquids, and liquids more than solids. As regards both liquids and gases, we have only to consider change of volume, for fluid substances have no shape of their own; but with solids we have also to consider change of length as well as change of volume. We shall first deal with the change of length of solids.

Linear Expansion of Solids.—Though the increase of length of solids with an ordinary rise of temperature is small, it is still sufficiently considerable in many cases to be of great practical importance. For instance, in the construction of railways it is necessary to leave a small interval between the rails to allow free play for expansion of the iron. Iron tubular bridges, again, have to be fitted on rollers, so that on expansion they may lengthen freely. Iron water-pipes have sometimes to be provided with telescopic joints. In tubular boilers, the fact that copper expands more than iron is made use of to secure water-tight joints. The copper tubes are fitted into the iron end-plates when cold, and on expansion they fit still more tightly, and so prevent leakage. Pendulum clocks, especially with metal pendulum-rods, go appreciably slower in summer than in winter, owing to the lengthening of the pendulum, and the rate may easily change to the extent of one minute per week. These examples all show the importance of an exact knowledge of expansion, while the last prepares us for the difficulty of the investigation by showing us how small is the quantity to be measured. A change of rate of one minute in a week is a change of 1 in 10,080, which may be shown to correspond to a change in length of the pendulum of 1 in 5000.

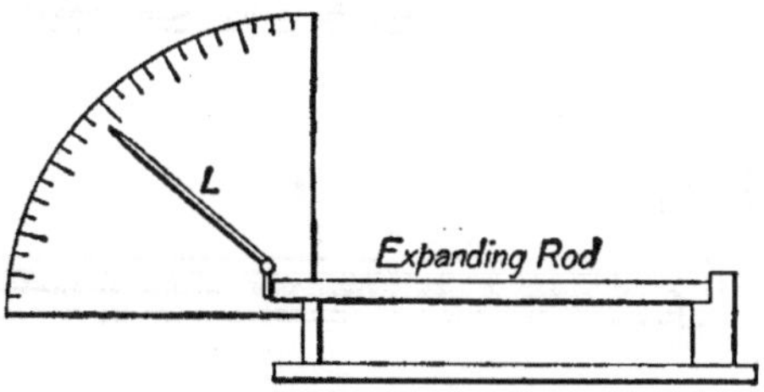

FIG. 9.—Expansion Apparatus.

The earliest attempt to measure linear expansions appears to have been made by apparatus resembling in principle the well-known instrument given in Fig. 9, the bar being placed against the short arm of the lever L when cold, and again when hot, and the movement of the long arm on the scale being noted. But there are two serious objections to this apparatus. Firstly, it is difficult to maintain the bar at any desired temperature; and secondly, during the experiment the measuring part of the apparatus may change in dimensions, as well as the body to be measured. For accuracy it is necessary that the measuring apparatus shall remain absolutely at a constant temperature, while the body to be measured shall be varied in temperature in a known manner. The latter condition may be fulfilled by taking the length of the body first in melting ice, then in boiling water. To fulfil the former condition several methods have been adopted.

Ramsden's Method.—One of the best is that of Ramsden, devised in 1785 to determine the expansion of the rods used by General Roy to measure the base line on Hounslow Heath, on which was founded the original Ordnance Survey of the United Kingdom.

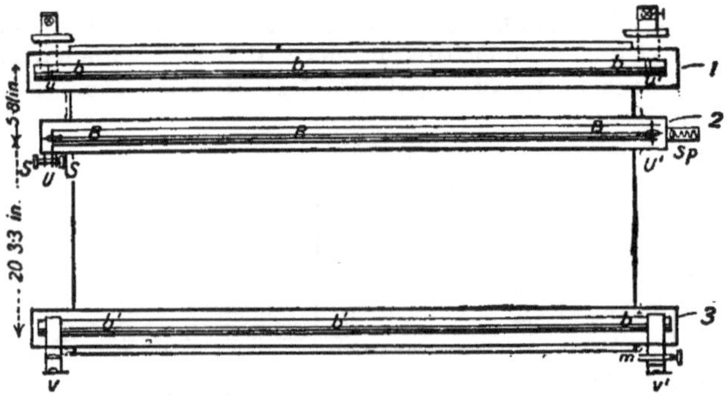

FIG. 10.—Plan of Ramsden's Expansion Apparatus.—(1) Wood trough at 0° containing cast-iron bar *bb*, with cross wires in uprights *uu'* fixed near ends; *uu'* shown also in elevation; (2) copper trough with lamps underneath to raise temperature, and containing bar BB to be tested, upright U pressed against left end by adjusting screws SS, upright U' pressed against right end by spring *sp*. UU' carry microscope object-glasses; (3) wood trough at 0° containing cast-iron bar *b'b'* with uprights *vv'* fixed near ends carrying microscope eyepieces, that on *v'* being provided with a micrometer *m*.

The apparatus, of which the general arrangement may be gathered from Fig. 10 and the accompanying description, consisted of three troughs, each over 5 feet long, placed parallel on a table. Cast-iron bars were fixed in the two end troughs, which were always filled with melting ice, so that the bars were of invariable length. Near the ends of the bar, *bb*, were fixed uprights carrying cross wires. In the middle trough were two sliders, moving only horizontally along the trough with uprights UU' at their more distant ends, carrying the object-glasses of two microscopes. The bar to be tested was placed on rollers in this trough, and its ends were made to bear always against the uprights UU'. The upright U was kept fixed so that any expansion pushed out the upright U'. Near the ends of the bar *b'b'* were fixed uprights, carrying the eyepieces of the microscopes, each with a vertical cross wire, the wire at *v* being fixed, that at *v'* being movable by a micrometer screw *m*. The middle trough could be heated by means of lamps.

The general nature of an experiment was as follows: All three troughs being filled with melting ice, the uprights were so adjusted that the cross wires on *uu'*, seen in the microscopes, were in the centre of the field, and coincided with the eyepiece wires. The middle trough was then filled with hot water, which was further heated by the lamps to boiling. One observer at the left-hand microscope took care to keep the upright U (and therefore the left end of the bar) in its original position, by means of the adjusting screws SS; while another observer at the right-hand microscope, measured the displacement of the image of the right hand cross wires at *u'* across his field of view, by means of the micrometer *m*. The displacement of the object-glass U' in terms of the micrometer divisions was determined thus. In a preliminary experiment, two vertical fibres fixed at *u'*, one on each side of the diagonal fibres and exactly $\frac{1}{10}$ inch apart, were viewed by the microscope, and the distance of their images apart measured by the micrometer. To this was added the value in micrometer divisions of $\frac{1}{10}$-inch motion of the micrometer thread, and the total gave the value of $\frac{1}{10}$-inch motion of the object-glass. For, if LM, Fig. 11, be the two vertical wires at *u'* and *lm* their images, when *o* is moved to *o'*, a distance equal to ML, the image of L moves to *l'*, where L*l'* is parallel to M*m*, and therefore $ml' = oo' =$ LM. Hence, $ll' = lm + ml' = lm$ + LM; or, the distance apart of the images of LM + the distance apart of the objects, is the displacement of the image of them by a motion of *o* through a distance LM.

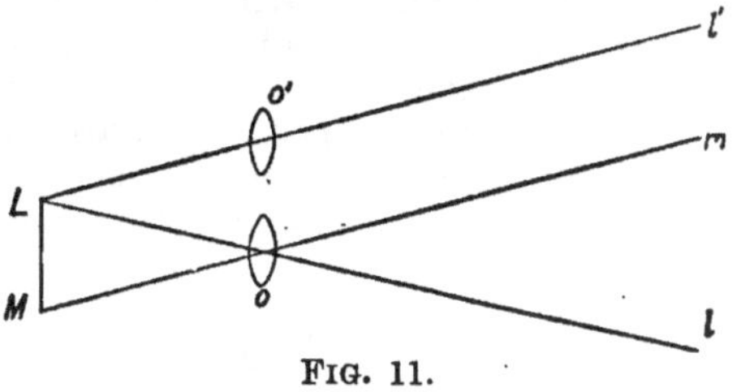

FIG. 11.

Ramsden's results may be put thus:—

Standard brass scale, 1,000,000 parts at 0° expanded to 1,001,855 at 100°.

Brass rod	1,001,893
Brass trough	1,001,895
Steel rod	1,001,145
Iron rod	1,001,109
Glass rod	1,000,808

Ramsden also found that on dividing the interval from 0° to 100° into three equal steps, the expansion for each step upwards in temperature was the same, within the limits of errors of observation. This result has been shown to be not quite true when the measurements are made with extreme accuracy; but assuming it as sufficiently exact for ordinary purposes, it follows that a rod expands by the same fraction of its length at 0° for each rise of 1°. This fraction is termed the coefficient of expansion, and is usually denoted by *k*. Hence, if l_0 l_t are the lengths of a bar at 0° and t°,

$$l_t = l_0(1 + kt).$$

The method of Ramsden has been since modified by attaching a micrometer to the object-glass *o'*, instead of to the eyepiece. The expansion is then measured directly by the distance through which the object-glass must be moved back after expansion to give coincidence again.

Modern Form of the Method.—A very similar arrangement to that of Ramsden is adopted at the Bureau International des Poids et Mesures at Paris, for the determination of the expansion of metre scales. The general plan consists in keeping one scale at a constant temperature, and therefore of invariable length, and in measuring the difference between this length and the length of the scale whose expansion is sought, the temperature of the second scale being varied. For the comparison, two microscopes are placed vertically, as nearly as possible one metre apart, passing through projections overhanging from the tops of two massive pillars (Fig. 12), the bodies of the microscopes being very firmly attached to the projections. Two parallel troughs, somewhat more than a metre in length, are fixed to a table running to and fro on rails, one of the scales being placed in each, so that either scale may be brought with its end-marks under the microscopes. The troughs are double-walled, and a stream of water, kept at a constant temperature by a thermostat, circulates in the space between the walls in each. The inner troughs are also filled with water, one being kept at a constant temperature, while the other is heated to successive higher temperatures. The difference in the lengths of the two scales is measured at each temperature by the microscopes, which are provided with micrometer eyepieces for the purpose. Special arrangements are adopted for the adjustment of the scales in position, and stirrers in the form of turbines are used to

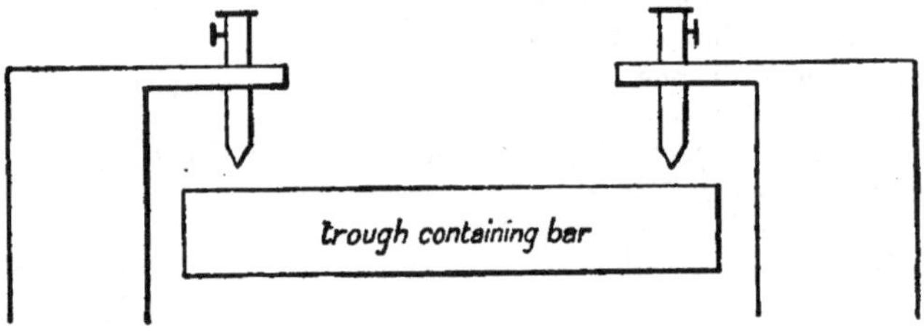

FIG. 12.—Diagram of Expansion Apparatus at the Bureau International.

mix up the water thoroughly before each reading is taken, so that the temperature throughout the trough is uniform. The thermometers used are all studied carefully, so that the temperature is ascertained with great accuracy. Full details of the method are given in *Travaux et Memoires du Bureau International des Poids et Mesures*, vol. ii.

Another method of the same class was devised by Pouillet, to measure expansions of bars at very high temperatures, as in a furnace. The bar to be tested was placed horizontally in a chamber, the temperature of which could be regulated, and its two ends were viewed through windows in the chamber by two horizontal telescopes of short focal length. Arrangements were adopted so that all expansion took place at one end and the rotation of the telescope to keep the end in the centre of the field at the high temperature was observed.

Method of Lavoisier and Laplace.—Shortly before Ramsden made his experiments, an apparatus was devised and used by Lavoisier and Laplace, though their results were not published till many years later. The arrangement will be understood by the aid of Fig. 13.

The bar BB, of which the expansion was sought, was supported horizontally on glass rollers *rr* in a trough filled with liquid. One end abutted against a vertical, fixed, glass rod FF, suspended from a cross-piece T supported by two firm pillars, of which only the back one is shown; the other end pressed against a lever CL, which could rotate round the axis C, supported in bearings on two other pillars, of which again only the back one is shown. To the axis of the lever CL was attached a telescope about 6 feet long, directed to a vertical staff divided to twelfths of a French inch, and distant 600 feet from the telescope. The temperature of the trough was raised by a furnace underneath it, but from their size and distance from the furnace the pillars were unaffected. The axis CC, therefore, remained a constant distance from the fixed piece TFF. Hence, an expansion of the bar moved the end L of the lever and rotated the telescope, the length CL being such that an expansion of the bar through one line moved the cross-wire in the eyepiece over 744 divisions of the image of the staff. This must not be taken to imply that the accuracy was increased 744 times, for the passage of the cross-wire over one division of the scale

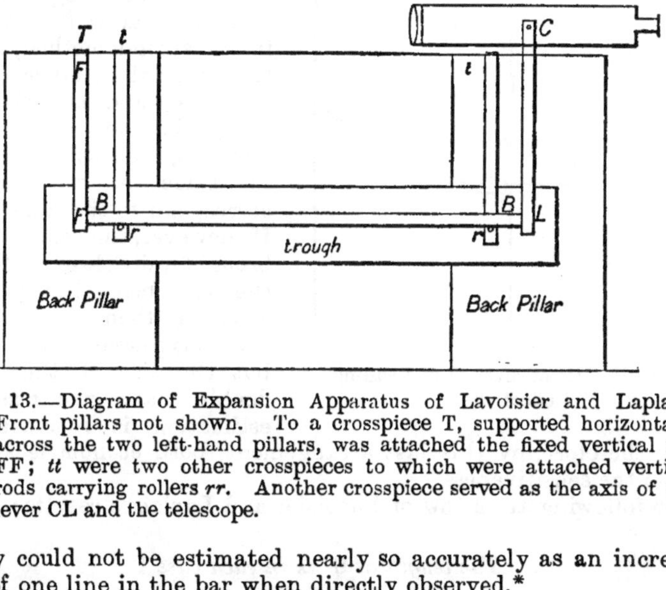

FIG. 13.—Diagram of Expansion Apparatus of Lavoisier and Laplace. Front pillars not shown. To a crosspiece T, supported horizontally across the two left-hand pillars, was attached the fixed vertical rod FF; *tt* were two other crosspieces to which were attached vertical rods carrying rollers *rr*. Another crosspiece served as the axis of the lever CL and the telescope.

probably could not be estimated nearly so accurately as an increase in length of one line in the bar when directly observed.*

Modifications of the method have since been made by Müller and others, in which the telescope is replaced by a mirror on the axis C. The reflection of a fixed scale is viewed in the mirror by a fixed telescope, and the motion of the image of the scale across the field of view through the rotation of the mirror gives the expansion of the bar. The bar abuts against small rounded projections attached to the vertical pieces at each end, so that it touches the lever CL at a fixed point. The adoption of the mirror method has two great advantages:—(1) It reduces the weight of the moving parts by the substitution of a light mirror for a heavy telescope; (2) it economises space, for the telescope and scale may be placed at half the distance of the scale in the original experiment,

* The details of Lavoisier and Laplace's work were not very fully published, and it is not known whether they took sufficient precaution to maintain the length CL of the lever invariable or not.

and the same accuracy will be obtained. This may be seen at once from Fig. 14.

If mm be the position of the mirror which reflects the central division c of the scale S into the telescope, when the mirror turns to $m'm'$ through an angle mOm' the division reflected into the telescope will be a where $aOc = 2mOm'$. Hence, what we may call the reflected line of sight, Oa, passes over twice as many divisions of the scale as the normal to the mirror On, and therefore over as many divisions as the line of sight of a telescope at O, directed along On to a', on a similar scale S′ at the full distance. It may be worth noting that on contrasting Lavoisier and Laplace's telescope method with the mirror-telescope method with scale and telescope close together, and using the same telescope and scale in each case, there is no gain in sensitiveness by the use of the mirror. For though the scale is half the distance from the mirror in the second method, it is still the original distance from the telescope, and the size of the image is the same in both methods, and, as we have seen, the same number of divisions correspond to the same rotation. If, however, the telescope can be brought quite close to the mirror, the scale being left at the half distance, then the number of divisions passed over is the same, but the size of the image is doubled, and the accuracy of estimation is increased. We should, therefore, by this last arrangement, both economise space and increase the sensitiveness.

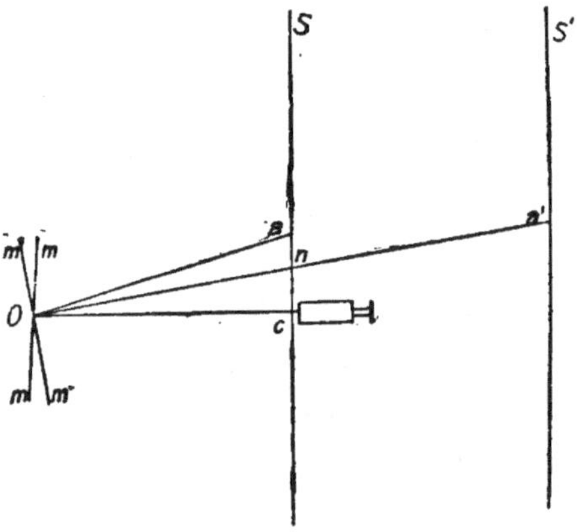

FIG. 14.—Mirror Method of Reading Deflections.

The following are a few of Lavoisier and Laplace's results :—

The length at 0° *is in each case* 1.

	Length at 100°.	
Untempered steel .	1·00107912	Coefficient of expansion between 0° and 100° obtained by dividing in each case the total increase in length by 100.
Tempered steel . .	1·00123956	
Silver	1·00190974	
Copper . . .	1·00171733	
Brass	1·00187821	
Iron (soft) . . .	1·00122045	
Iron wire drawn . .	1·00123504	
Glass with lead . .	1·00087199	
Glass without lead .	1·00087572	

The increments in length here given contain five or six *significant* figures, but the accuracy of the methods does not justify us in placing confidence in more than the first two or three figures. The determination of the temperature by the older observers hardly warrants us in going further. But even with accurately observed temperatures the numbers obtained to four places would not be the same for different specimens of a substance, so much does the expansion depend on physical condition. This is well illustrated by the difference between tempered and untempered steel in this table, and is also shown by the difference between soft iron and iron passed through a draw-plate. Comparing the results of Lavoisier and Laplace with those of Ramsden, prieviously given, for substances with the same name, the difference is still more striking. In fact, we can hardly trust the generality of the results in the tables beyond two figures ; and if greater accuracy is required, the expansion must be determined for the particular specimen concerned.

Fizeau's Method.—A very remarkable and accurate method has been devised by Fizeau, in which use is made of Newton's rings. It is well known that, if a slightly convex lens is placed on a flat glass plate, several bright rings are seen round the point of contact, these being due to the interference of the light reflected upwards at the curved glass surface with that which, passing down through the film of air, is reflected back at the lower plane glass surface. If, instead of white light, we use the nearly homogeneous light from a sodium flame, the number of rings is very greatly increased, and they extend, in general, to the edge of the lens.

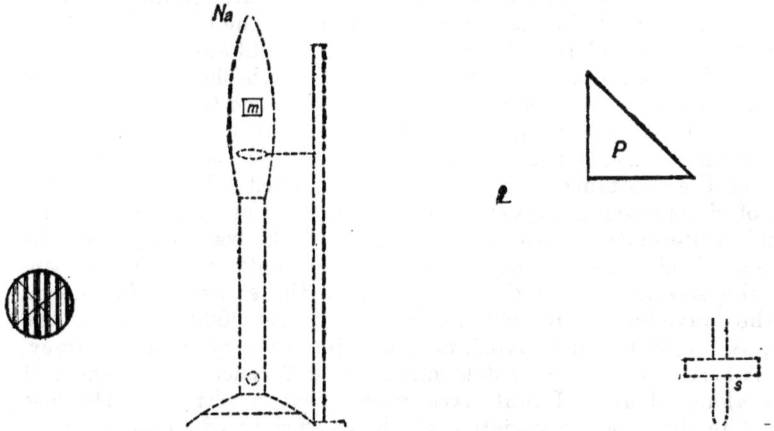

FIG. 15.—Diagram of Fizeau's Expansion Apparatus.—B, block of which expansion is to be measured placed on table T ; *ss*, levelling screws, also supporting lens, L just above B ; *f*, film of air-forming interference bands ; P, right-angled prism throwing light down and reflecting it out again ; N*a*, sodium flame behind the plane of the figure ; *m*, mirror in the plane of the figure reflecting the sodium light towards P ; *tel*, telescope directed towards P and viewing the interference bands ; F represents the appearance of the bands in the telescope, which is provided with cross-wires. The expansion table is protected by a case which can be heated. This case is not represented.

Now if the path of the rays of light is normal to the lower glass plate, each concentric bright ring from the centre outwards corresponds to a thickness of the film of air $\frac{\lambda}{2}$ greater than the preceding one, where λ is the wave-length of the light employed; so that even if we do not know the absolute thickness at any point, we know the difference in thickness of the film of air at any two points by counting the number of rings between them. Further, it is not necessary to have contact between the curved and plane surfaces. If the lens is gradually raised upwards, the rings contract, disappearing as they reach the centre; but the distances apart of the successive rings remain the same at the same points, corresponding to the same differences in the thickness of the film of air. In Fizeau's experiment the rings were still visible when the air-film was over a centimetre thick. His method, as used at the Bureau International des Poids et Mesures, is as follows*:—A flat metal plate T (Fig. 15) is supported horizontally by three screws passing upwards through three holes near its edge. On the three screws is supported a plano-convex lens LL with the plane surface downwards. The so-called plane surface is in reality slightly convex, as is the case with most plano-convex lenses, and if allowed to touch another truly plane surface, Newton's rings are seen round the point of contact. A plate B of the substance of which the expansion is required is prepared with parallel polished faces, and about 1·5 cm. thick, and this is laid on the centre of the metal plate T. The lens is then adjusted, at some small distance above it, so that when sodium light is thrown on it normally, Newton's rings are seen through interference between the rays reflected, at the lower surface of the lens and those passing through and reflected at the upper surface of the plate of the substance. The light of a sodium flame is thrown in and reflected out again by a right-angled prism, and then received by a telescope. The metal plate with the substance and lens is enclosed in a chamber maintained at a uniform temperature by a thermostat. When the temperature is raised, the thickness of air is altered by the difference between the expansion of the supporting screws and that of the substance, and the rings are shifted. By counting the number of rings passing a given point in the field of view of the telescope, this difference is measured in terms of the wave-length of the light used. Preliminary experiments are made to determine the expansion of the screws, and so the expansion of the substance is known. Since the wave-length of sodium light is about ·000589 mm., the method, as might be anticipated, is susceptible of very great accuracy, and by it Fizeau was able to determine the difference of expansion of crystals along their different axes with great exactness. He also succeeded in showing the variation of the co-efficient of expansion with the temperature, taking as the co-efficient the increase per 1° rise in temperature of a length which is equal to 1 at 0° C.

Applications of Linear Expansions.—In many cases in which metals are used in construction, account must be taken of variation in their length with variation of temperature. Railway lines must be laid with a small interval between the successive rails, otherwise in hot weather the rails would tend to force each other out of the straight so as to allow the necessary expansion. This is perhaps most easily realised by calculating the total increase on some long line. For instance, the distance from London to Edinburgh is about 400 miles. If the rails are laid in cold weather, we may easily conceive the possibility

* Vol. i. c., *Travaux et Mémoires.*

of a rise in temperature of 25° C. Now Ramsden's value of the coefficient of expansion of steel is ·00001145. Then 400 miles will increase with a rise of 25° by

$$400 \times \cdot 00001145 \times 25 \text{ miles} = \cdot 1145 \text{ mile,}$$
or over 200 yards.

Then this distance, at least, must be left in intervals between London and Edinburgh.

In iron bridges some arrangement must be adopted to allow for expansion without serious change of shape in the structure. The Menai tubular bridge, of which the total length is over 1500 feet, is mounted on rollers, and the joints between the successive tubes are telescopic, and have a play of several inches. In the Forth Bridge rocking columns are interposed between the ends of the central girders and the extremities of the central Inchgarvie cantilever arms, and the shore arms of the side cantilevers are left free to slide on their abutments.

In the long lengths of iron rod used to work railway points at a distance from the signal box, expansion or contraction through change of temperature might seriously interfere with the working. The rod is therefore divided into successive lengths connected by short cross-pieces. Thus in the simplest case, represented in Fig. 16, it may

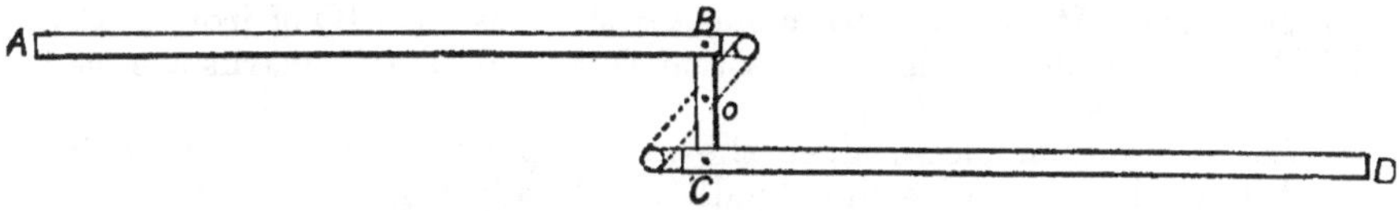

FIG. 16.—Provision for Expansion in Rods Working Railway Points.

be divided into two halves AB, CD, with a connecting piece BOC turning about O, which is fixed. If A and D are fixed, expansion is provided for by BC turning round as indicated by the dotted lines. But if A be moved back by the pointsman, D will move the same distance forward.

An interesting application of the expansion of iron was first made many years ago in order to draw together the two side-walls of a gallery at the Conservatoire des Arts et Metiers at Paris, which were bulging outwards through the weight of the roof. Long iron bars were passed through the two walls, and circular plates were screwed outside on to the two ends of each, till they came against the walls. The bars were heated inside and expanded. The plates, having been thus pushed out, were screwed farther on till they again touched the walls, and, on cooling, the bars contracted and drew the walls together. By several repetitions of the process, the alternate bars being heated in each operation, the walls were brought to their proper position.

It is well known that a thick glass vessel is very liable to crack, if hot water is poured into it. This is due to the sudden expansion of the inner layers, the outer layers not at once receiving heat, since glass is a poor conductor. The strain in the glass is relieved by the rupture.

The contraction of iron on cooling is made use of in putting hoops on casks and tires on wheels; the iron is put in position while hot, and as it cools the contraction keeps it firmly in its place.

The contraction of iron in cooling must be allowed for in iron castings, the pattern always having to be made slightly larger than the casting required.

In measuring base lines for trigonometrical surveys "compensation bars," first designed by General Colby for the Indian Survey, are now used. To understand their construction, suppose that we have two parallel bars, AB of brass and CD iron, held at their middle points PQ. Let these be of equal length at some given temperature. Let their temperature rise and let their expansions be as 3 : 2.

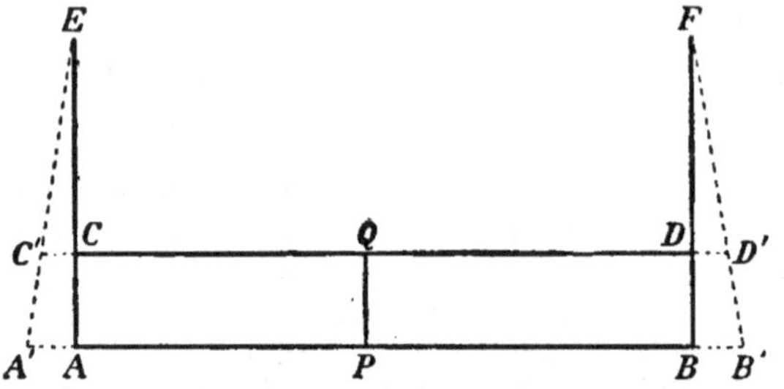

FIG. 17.—Compensation Measuring-Bar.—EF, of invariable length; AB, brass; CD, iron.

If then AB expands to A'B' and CD to C'D', AA' : CC' = BB' : DD' = 3 : 2. If A'C' be produced to cut AC produced in E then EA : EC = 3 : 2 and E is a fixed point. Similarly F is fixed and the distance EF is independent of temperature variations so long as the two bars are all at one temperature. It is usual to make AB of brass and CD of iron. ACE, BDF are steel pieces jointed at A and C, B and D, and marks are made at the extremities E and F.

In a pendulum clock, since the rate depends on the length of the pendulum, the time of swing tends, as already pointed out, to become longer in summer than in winter. An iron pendulum, for example, increases by about ·000012 of its length for a rise of 1°, altering the rate by half this proportion, and so tending to make the clock lose about half a second per day. With a variation of 20° or 30°, the change of rate becomes very serious. This effect of temperature is eliminated as far as possible by employing a "compensating" pendulum, made up of two metals, arranged in such a way that the distance between the centres of suspension and oscillation—the effective length—remains constant.

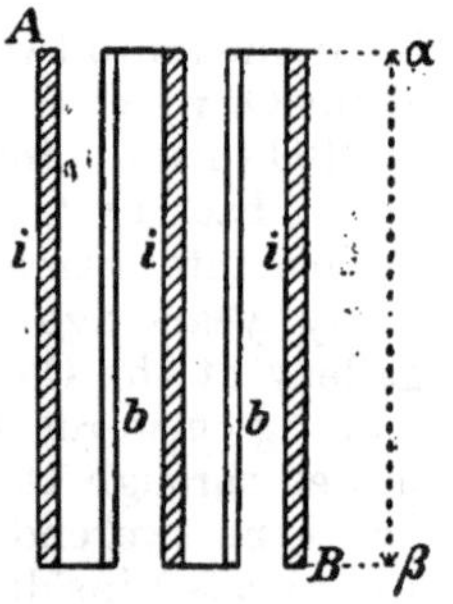

FIG. 18.—Principle of Gridiron Pendulum of Iron and Brass Bars.

The form most commonly used* is the "gridiron" pendulum, invented by a clockmaker named Harrison about 1720. If we make an arran ement of bars of iron and brass as in Fig. 18, *iig* being iron, *bb* brass, and suppose the expansions to be in the ratio 2 : 3, then the distance between A and B, measured parallel to the bars along $\alpha\beta$, is invariable. For suppose that each iron bar expands a length i, and that each brass bar expands a length b, then $3i = 2b$. If we first suppose the iron alone to expand, it is evident from the arrangement that,

* A simple form, on exactly the same principle as that employed by Colby in his compensation bar, was devised by Ellicott and described in the *Philosophical Transactions* for 1751.

if A is kept fixed, B will be lowered a distance, $3i$. If after this we allow the brass to expand, B will be now raised a distance $2b$, and since $2b = 3i$, B resumes its original position.

The pendulum as actually made is merely a double arrangement of this kind as in Fig. 19, where two metals, of expansion in the ratio 2 : 1, are taken for simplicity of construction.

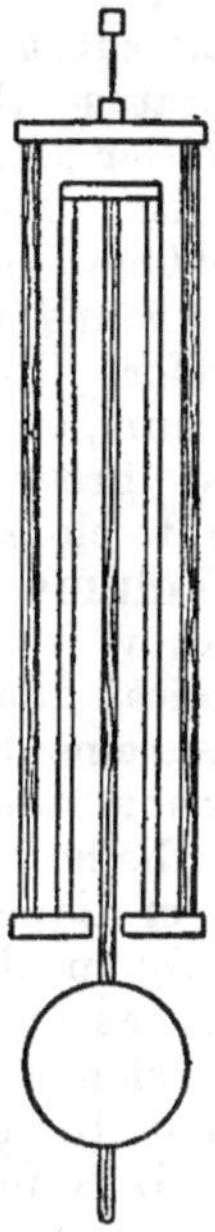

FIG. 19.—Gridiron Pendulum of 2 metals, with expansion :: 1 : 2.

A mercurial compensating pendulum is frequently employed. In this the bob consists of a glass cylindrical vessel containing mercury. As the temperature rises, the vessel is lowered through the increase in length of the pendulum rod. But at the same time the mercury expands more than the glass, and its centre of gravity rises in the vessel. The depth of the mercury is so adjusted that the effective length of the pendulum is invariable.

M. Guillaume has discovered a nickel-steel, which he terms "Invar," which has a negligible co-efficient of expansion at ordinary temperatures. This is very suitable for pendulum rods, and will probably supersede compensating pendulums. It has already superseded compensating measuring rods for survey purposes.

In compensating chronometers, the difference of expansion is made use of in another way. The rate of the chronometer depends on the resistance of the hair-spring to coiling, and on the disposition of the mass of the balance-wheel. As the temperature rises, the resistance of the hair-spring decreases, and if the wheel were of one metal it would expand and throw its mass farther from the centre. On both accounts the time of vibration would be longer and the chronometer would lose. But the rim of the balance is made of two metals, say brass and steel, of unequal expansion, that with the less expansion being inside, and the wheel is discontinuous as shown in Fig. 20. The outer rim expands more than the inner, and the two loaded ends therefore curl inwards with rise of temperature. The loads are so adjusted that the mass is thrown inwards by a sufficient amount to compensate for the weakening of the hair-spring and the expansion of the spokes. It may be noted that the effect of the weakening of the spring is far greater than that of the expansion.

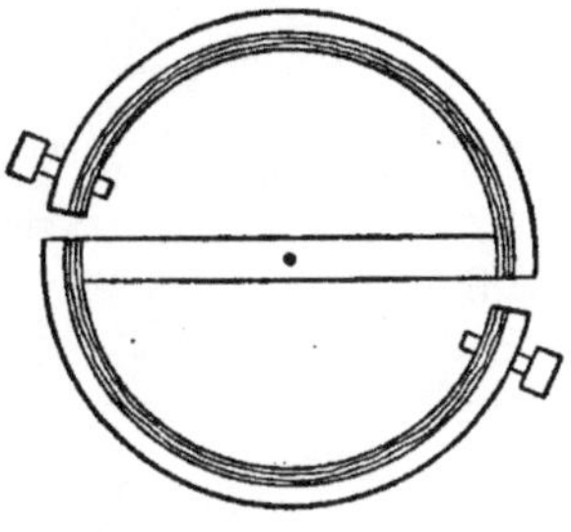

FIG. 20.—Compensating Balance for Chronometers. Outer rim brass, inner rim steel.

Metallic thermometers are made in which the unequal expansion of the two parts of a compound curved bar, formed by two strips of different metals brazed together, results in a change of curvature. But the strains to which the metals are subjected on being raised to a higher temperature are usually so great in these instruments that the metals on being brought back to the original lower temperature do not recover at once from them, and do not give consistent indications.

There is one case in which the nearly equal expansion of two different substances is of great importance in the construction of scientific apparatus—that of glass and platinum. Referring to the table of Lavoisier and Laplace, it will be seen that they found as the co-efficient for glass ·0000087, while Borda obtained for platinum a value nearly ·0000086. If then a platinum wire be inserted in glass, and the glass be fused round it, in cooling the two contract nearly equally. There is therefore little strain, and the glass does not break away from the platinum, as it would from other metals. Platinum wires can thus be fused through glass, and through these wires electric currents can be led into closed glass vessels, such as vacuum tubes and eudiometers.

Volume Expansion of Solids.—We do not very often require to know the volume expansion, or, as it is frequently termed, the "cubical dilatation" of solids, except in so far as it is necessary in the measurement of the expansion of liquids and gases, when we may require to know the volume expansion of the containing vessel. We shall describe how this expansion is determined in connection with the expansion of liquids, and shall here only state that, if the solid expands equally in all directions, its volume expansion may be found from its linear as follows: If a solid of cubical form has a co-efficient of linear expansion k, the length l of the edge at 0° becomes $l\,(1+kt)$ at $t°$. The volume, therefore, increases from l^3 to $l^3(1+kt)^3 = l^3(1+3kt+3k^2t^2+k^3t^3)$. Now kt is, for moderate temperatures, so small that we may with more or less exactness neglect k^2t^2 and k^3t^3, and the volume is very nearly $l^3(1+3kt)$. The co-efficient of volume expansion—the increase of the volume which is 1 at 0° for each degree rise of temperature—is, therefore, $3k$ or three times the linear coefficient of expansion.

CHAPTER III.

EXPANSION OF LIQUIDS.

Volume Expansion of Liquids—U-Tube Method applied to Mercury—Dulong and Petit—Regnault—Expansion of other Liquids by Specific Gravity Bottle—By Dilatometer—Matthiessen's Hydrostatic Method—The Expansion of Water—Hope's Apparatus—Apparatus of Joule and Playfair—Results.

Volume Expansion of Liquid.—Since liquids have no definite shape of their own, their dimensions depend on the containing vessel. We are, therefore, only concerned with their volume expansion as the temperature rises.

The most obvious mode of observing the expansion of a liquid is to enclose it in a vessel such as a thermometer bulb, and to note the change in level of the liquid in the stem as the temperature rises. If we know the expansion of the bulb and stem we can deduce the expansion of the liquid. For suppose, at a given temperature, the liquid reaches to a certain mark on the stem. On raising the temperature the internal capacity of bulb and stem increases, but in general the volume of the liquid increases still more, so that it not only occupies the increased space in the vessel, but also rises in the stem. Its total increase of volume is therefore:—increase of volume of bulb and stem to first level + additional volume of stem between first and second levels, and when we know these we can determine the increase in volume of the liquid.

The linear expansion of the bulb and stem may be measured directly, and the volume expansion is approximately three times as great. Chappuis (*Travaux et Mémoires du Bureau International*, 1907) has used this method with a weight thermometer to determine the expansion of mercury. But it is not satisfactory, as we cannot assume that the expansion of the bulb is the same in all directions. Practically there is only one method of accurately gauging the internal volume of a vessel, and this consists in finding the weight of a liquid of known specific gravity filling it. Thus, if ω be the weight filling it at one temperature and ρ is the specific gravity of the gauging liquid used, ω/ρ is the volume. If at a higher temperature the weight is ω' and the specific gravity of the gauging liquid is ρ', the volume is ω'/ρ'. So that ω/ρ has expanded to ω'/ρ'. The problem of determining liquid expansion in general, therefore, resolves itself into that of determining the specific gravity of some one gauging liquid through the range of temperature over which the expansion is required. This specific gravity must be determined independently of the containing vessel.

U-Tube Method.—The best method yet devised for this purpose is a special application of the U-tube hydrometer. As a simple illustration, let us suppose that two glass tubes are fixed vertically with their upper ends open and their lower ends joined by narrow horizontal tubing. A short length of indiarubber tubing, having a pinch-cock upon it (Fig. 21), is inserted in the horizontal tube.

The pinch-cock being closed, hot water is poured into one tube, and cold water, to about the same level, into the other. On now opening the pinch-cock, it is found that the level of the hot water is quite appreciably higher than the level of the cold, since it takes a longer column of lighter hot water to balance a given column of heavier cold water. In fact, if h, h' be the heights, ρ, ρ' the densities, we have, on equating the hydrostatic pressures at the bottom,

$$h\rho = h'\rho',$$

But if V, V′ be the volumes of equal masses of densities ρ, ρ',

$$V\rho = V'\rho'.$$

Hence $$\frac{V}{h} = \frac{V'}{h'}, \text{ or } V' = \frac{Vh'}{h},$$

so that we can express the expanded volume in terms of the original volume

Though water serves very well for an illustration of the method, mercury has been chosen as the liquid on which to make exact experiments for the following reasons:—it is easily obtained pure; it expands nearly regularly as the temperature increases, when this is indicated either by the mercury-glass scale or the air scale; it has a high boiling-point (about 356° C.); it does not evaporate much until the temperature is approaching 200°; it does not wet the vessel in which it is placed, if this be glass, so that its level may be read with great accuracy; and its great specific gravity renders it especially suitable for gauging vessels, when its expansion has been found.

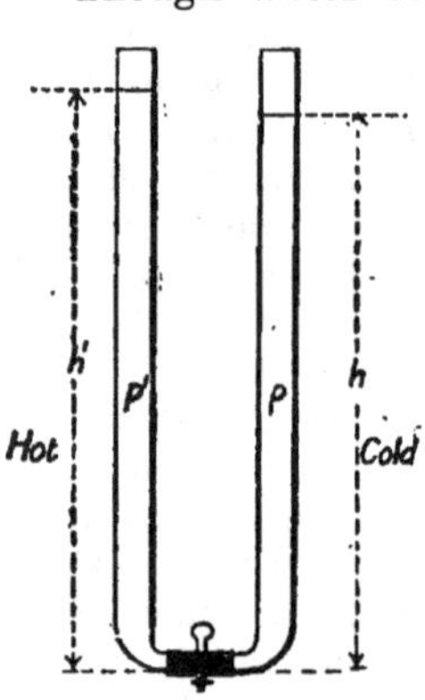

FIG. 21.—U-Tube Hydrometer, with hot and cold water.

The first experiments of this kind with mercury were made by Dulong and Petit, who employed an arrangement exactly similar in principle to that represented in Fig. 21. The two tubes were enclosed, the hot one in an oil-bath, and the cold one in a vessel filled with melting ice, the upper ends rising above the enclosing vessels, and the upper levels of the mercury being just visible when the height was measured. The connecting cross-tube at the bottom was of very fine bore, so as to prevent circulation from one tube to the other when the temperatures were steady. An air-thermometer was used to give the temperature of the hot vessel.

If H and H_0 are the two heights, hot and cold, and ρ and ρ_0 are the respective densities,

$$H\rho = H_0\rho_0, \text{ or } \frac{H}{H_0} = \frac{\rho_0}{\rho}$$

But if a volume V_0 at 0° expands to $V_0(1+\Delta_t)$ at $t°$, since the mass is the same at each temperature,

$$\rho_0 V_0 = \rho\, V_0(1+\Delta_t)$$

or $$\frac{\rho_0}{\rho} = 1 + \Delta_t$$

Hence $$\frac{H}{H_0} = \frac{\rho_0}{\rho} = 1 + \Delta_t$$

and Δ_t, or the expansion of unit volume at 0° for a rise of t° can be found from H and H_0.

As there were several details in this method open to criticism, Regnault, in making a redetermination, introduced some modifications in the apparatus. He carried out a very extensive series of experiments, which have till recently been the standard ones on this subject.

He arranged the apparatus in two forms.

The general principle of one arrangement may be understood from Fig. 22. The two vertical tubes AB, A′B′, are united by the horizontal crosspiece AA′ with a small hole at o, so that at that level the mercury is exposed to the atmospheric pressure. The lower crosspiece is broken in the middle by the insertion of the inverted U tube DCD′, connected with the reservoir M, of compressed air, the pressure being adjusted so that the mercury rises to the levels CC′ in the two arms of the U. The temperature of A′B′ is kept constant throughout by surrounding it with water running from the mains, while that of AB may be raised to any desired point by surrounding it with a bath of oil heated by a furnace. The temperature of AB is given by an air thermometer. The heights AB, A′B′, CD, and C′D′ are all measured at each temperature by a cathetometer, and from these the expansion may be found as follows:

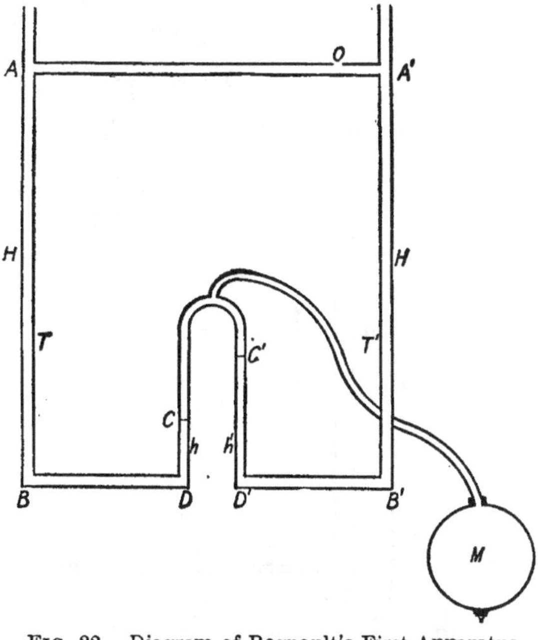

FIG. 22.—Diagram of Regnault's First Apparatus for finding the Expansion of Mercury.

The pressure at A is equal to that at A′, both being equal to the atmospheric pressure through the communication with the atmosphere at o. Also the pressure at C is equal to that at C′, both being equal to the air-pressure in the reservoir M.

$$\text{Then, pressure at C} - \text{pressure at A} = \text{pressure at C}' - \text{pressure at A}' \qquad (1)$$

Expressing these differences by the usual hydrostatic formula, we obtain an equation giving the expansion of the mercury. For, let H, H′ be the heights of AB, A′B′, let h, h' be the heights CD, C′D′, and let T, T′ be the temperatures of AB, A′B′; and let us, to simplify matters, suppose the temperature of CD, C′D′, to be T′. Let ρ, ρ' be the densities of mercury at T, T′, Δ_T, $\Delta_{T'}$ the expansions of unit volume at 0°, on raising the temperature respectively to T and T′.

Then $$H\rho - h\rho' = H'\rho' - h'\rho' \text{ from (1)}$$

or $$H\rho = (H' + h - h')\rho'$$

But if ρ_0 is the density at 0°, the volume of the same mass being inversely as the density, we have

$$\frac{\rho}{\rho_0} = \frac{1}{1+\Delta_T} \qquad \frac{\rho'}{\rho_0} = \frac{1}{1+\Delta_{T'}}$$

Then $$\frac{H}{1+\Delta_T} = \frac{H'+h-h'}{1+\Delta_{T'}}$$

or $$1+\Delta_T = \frac{H}{H'+h-h'}(1+\Delta_{T'}) \qquad (2)$$

Now $\Delta_{T'}$ is in practice only small, and we may take an approximate

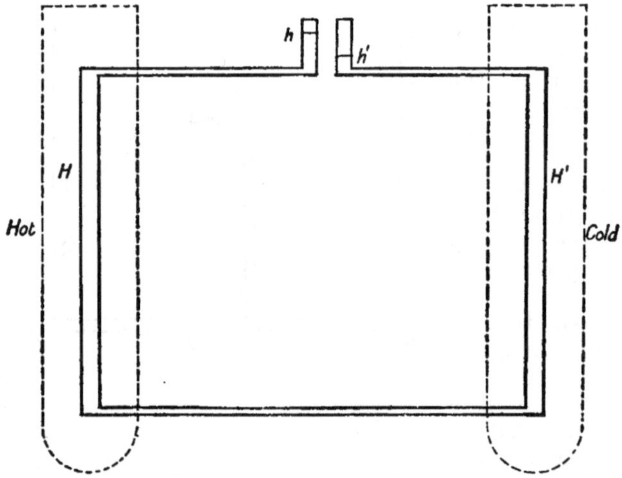

FIG. 23.—Regnault's Second Apparatus for the Expansion of Mercury.

value for it without seriously affecting the value of Δ_T. Then (2) gives us Δ_T in terms of known and measurable quantities.

Regnault also used a method very nearly the same as that of Dulong and Petit, which will be understood from Fig. 23, the U tube in the lower crosspiece being replaced by a flexible iron tube, so that the two halves could expand independently.

The great advantages over the arrangement of Dulong and Petit consisted in the maintenance of known fixed temperatures in each vertical part and in the greater accuracy of measurement of the vertical heights, attained by bringing the two levels close together.

It will be easily seen that in the arrangement of Fig. 23

$$\frac{H}{1+\Delta_T} + \frac{h}{1+\Delta_{T'}} = \frac{H'+h'}{1+\Delta_{T'}}$$

or $$1+\Delta_T = \frac{H}{H'+h'-h}(1+\Delta_{T'}).$$

Regnault made, in all, about 130 observations in the range of temperature from 25° C. to 350° C., and from these he constructed a table of expansions by a "graphic" method, *i.e.* by representing his results on a volume-temperature diagram. The principle he adopted is as follows:—Taking two lines at right angles, and marking along one temperatures and along the other the volumes assumed by the mass which has at 0° the volume 1000, each observation will be represented by a point on the diagram. If the observations were perfectly free from error, we might expect that all these points would lie on a regular curve or a straight line, the true curve which we seek to determine. But through imperfections in the measurements of length and temperature, the points will probably lie sometimes above, sometimes below the true curve.

Let us suppose that the observations are indicated by the crosses in Fig. 24. Then a curve is drawn so as to pass as nearly as possible midway through the points, *i.e.* so as to make the sum of the heights above the curve equal to the sum of the depths below it. Of course, if there is any systematic error running throughout the work and always in the same direction, this method will not detect it.

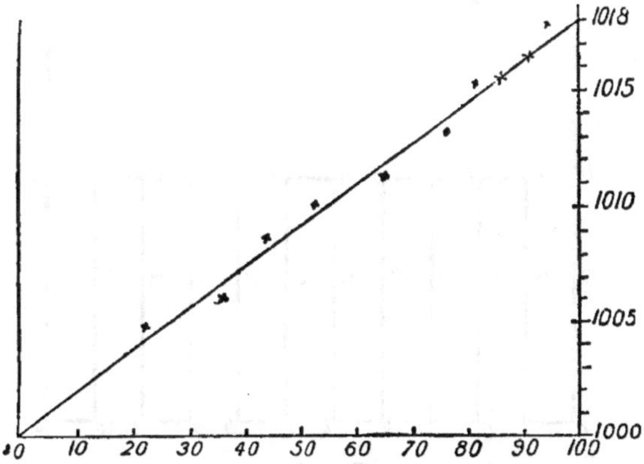

FIG. 24.—Graphic Method of Determining Results.

If the expansion per 1° were the same throughout the range, the curve would be a straight line. Regnault found, however, as Dulong and Petit had already observed, that the line bends slightly upwards, that is, that the co-efficient of expansion slightly increases.

A constant co-efficient of expansion would be represented by

$$\Delta_T = aT.$$

The slight bending upward may be represented nearly by taking

$$\Delta_T = aT + \beta T^2,$$

and Regnault, from his diagram, found that

$$a = \cdot 00017905$$
$$\beta = \cdot 0000000252.$$

Regnault's results have since been studied by others who have sought to obtain formulæ more nearly representing his observations, but the differences they have introduced are very slight. If we do not require very great accuracy, we may take

$$\alpha = \cdot 0001800$$
$$\beta = \cdot 00000002.$$

Callendar* has modified the second method used by Regnault by employing six pairs of hot and cold columns placed in series as represented diagrammatically in Fig. 25A, the successive columns being alternately cold and hot, as marked by C and H. If the mercury when in equilibrium stands at a in the gauge-tube connected to the first cold column, and at z in the gauge-tube connected to the last hot column, the difference of level to be measured, represented by a', z, will be six times that due to a single pair of hot and cold columns. As the columns were

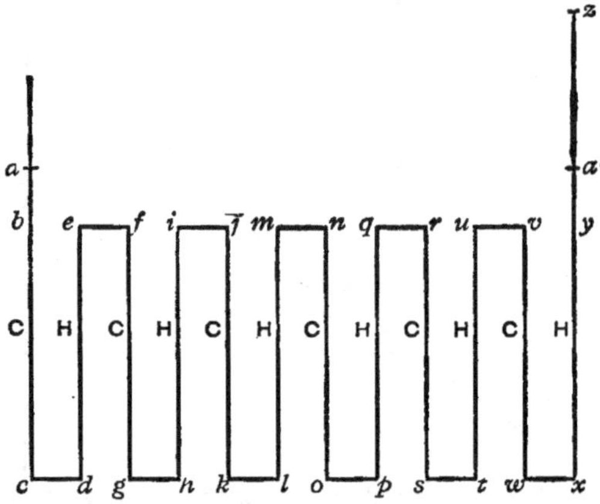

FIG. 25A.

FIG. 25B.

nearly 2 m. long in place of 1·5 m., the length of Regnault's columns, the expansion was nearly eight times as great. In the actual apparatus the cross tube *ef* was doubled back, so that *fg* lay behind *bc*, and *ih* behind *ed*, and so on, as shown in plan in Fig. 25B. All the cold columns were placed in one tube, and all the hot columns in another, the tubes containing oil rapidly stirred. One was surrounded by ice and the other was electrically heated. For further details we refer the reader to the original paper.

Callendar found that if ${}_0\alpha_t$ is the mean co-efficient of expansion between 0° C. and t° C., then

$$10^{10} \times {}_0\alpha_t = 1805553 + 12444t/100 + 2539t^2/10000.$$

For approximate work we may put

$$10^8 \times {}_0\alpha_t = 18006 + 2t.$$

* *Phil. Trans.*, A. 211, p. 1, 1911.

We may use these values to give us the specific gravity at any temperature. For if ρ_o is the specific gravity at 0° and if ρ_t is that at t° and Δ is the expansion of unit volume at 0° when raised to t°, we have

$$\rho_o = \rho_t(1 + \Delta).$$

The specific gravity of mercury at 0° is, according to Regnault, 13·596. Then the specific gravity at 100°, for example, is

$$\rho_{100} = \frac{\rho_o}{1+\Delta} = \frac{13{\cdot}596}{1{\cdot}0182} = 13{\cdot}353.$$

Determination of Liquid Expansion, using the known Expansion of Mercury.—We may use our knowledge of the specific gravity of mercury to gauge a vessel at different temperatures and then fill it with a liquid of which we require the expansion. Two methods are employed, one which we may illustrate by the use of a specific gravity bottle, the other in which a dilatometer, virtually a thermometer bulb, is used.

Specific Gravity Bottle Method.—In its original form the specific gravity bottle is a flask of thin glass with carefully ground stopper perforated by a very fine tube. When the bottle is filled with liquid and the stopper is inserted, the excess of liquid is forced out through this tube, and it is assumed that the stopper takes a definite position. There is a newer form in which the bottle is a U-tube with two fine tubes turning out horizontally from the ends of the U and provided with stoppers. A mark is made in each fine tube, and the liquid occupies the U and the fine tubes to the marks. Any excess of liquid may be easily removed.

The first aim is to determine the expansion of the bottle. For simplicity, we will suppose that we are going to heat it from 0° C. to 100° C. We first fill the bottle, when surrounded with ice at 0°, with mercury, and find the weight W_o of this mercury. Then we fill it when surrounded with steam at 100° and we find the new weight W_{100} of the mercury.

Let V_o be the internal volume of the bottle at 0° C. and
V_{100} " " 100° C.

Let ρ_o be the density of mercury at 0° and
ρ_{100} " " 100°

Let G be the expansion of 1 cc. of the internal volume of the bottle from 0° to 100°
γ " " mercury.

Then $V_{100} = V_o(1+G)$ and $\rho_o = \rho_{100}(1+\gamma)$.
From the weighings

$$V_o = W_o/\rho_o \;:\; V_{100} = W_{100}/\rho_{100}.$$

Then
$$V_o(1+G) = W_{100}/\rho_{100}.$$

and
$$\frac{W_o}{\rho_o}(1+G) = \frac{W_{100}}{\rho_{100}}.$$

or
$$1+G = \frac{W_{100}}{W_o} \cdot \frac{\rho_o}{\rho_{100}} = \frac{W_{100}}{W_o}(1+\gamma).$$

Since we know γ we obtain $1+G$.

We can now use the bottle to determine the expansion of any other liquid between 0° and 100°. Let 1 cc. of the liquid at 0° expand to $1+\Delta$ at 100°, and let σ_o, σ_{100} be its densities at 0° and 100°.

Then $$\sigma_o=\sigma_{100}(1+\Delta).$$

Filling the bottle first at 0° and then at 100° let ω_o, ω_{100} be the weights of the liquid.

We have $$\sigma_o=\omega_o/V_o,\text{ and }\sigma_{100}=\omega_{100}/V_{100}.$$

Then $$1+\Delta=\frac{\sigma_o}{\sigma_{100}}=\frac{\omega_o}{\omega_{100}}\frac{V_{100}}{V_o}.$$

$$=\frac{\omega_o}{\omega_{100}}(1+G).$$

$$=\frac{\omega_o}{\omega_{100}}\frac{W_{100}}{W_o}(1+\gamma).$$

Whence we know Δ.

Dilatometer Method.—The dilatometer is now usually made in the form represented in Fig. 26. B is a bulb with a fine graduated stem, *s*, rising from the bulb and open at the top. Below, the bulb is connected with a fine-bore tube with a slight thickening at *t* and open at *e*; *st*, is a screw stopper which can be put over the end of this tube, being sprung over the thickening. On screwing the stopper, a pad effectually closes the opening. This form of apparatus is very readily cleaned and filled.

First we must calibrate the bulb and stem, and measure its expansion. Let W_o be the weight of mercury, density ρ_o, filling the bulb at 0° from *e* up to the zero of the scale. Then the volume $V_o=W_o/\rho_o$.

Let the weight of mercury filling an observed number of divisions of the stem be found. From this the volume of each scale division in terms of the volume of the bulb can be found. Let it be λV_o. λ will be a very small fraction.

Now start with the bulb filled with mercury at 0° up to the zero, and raise the temperature to 100°. Suppose that the mercury rises N divisions.

Let V_o expand to $V_{100}=V_o(1+G)$.

Then the total volume of the mercury is

$$V_{100}+N\lambda V_{100}=V_o(1+G)(1+N\lambda).$$

But if γ is the expansion of mercury we have

$$V_o(1+\gamma)=V_o(1+G)(1+N\lambda).$$

Whence $$1+G=\frac{1+\gamma}{1+N\lambda}.$$

Now repeat this operation with the liquid, of which the expansion is to be determined, and let it rise from the zero division through n divisions when the temperature is raised from 0° to 100°. Let 1 cc. expand to $1+\Delta$ cc. Then the total volume at 100° is $V_{100}(I+n\lambda)$

$$=V_o(1+G)(1+n\lambda).$$

But this is also $$V_o(1+\Delta).$$

Then
$$1+\Delta=(1+G)(1+n\lambda).$$
$$=(1+\gamma)\frac{1+n\lambda}{1+N\lambda}.$$

Whence Δ is determined.

Matthiessen's Hydrostatic Method.—Matthiessen * determined the expansion of water by a hydrostatic method. For this purpose the linear expansion of a glass rod was measured, and its volume expansion was deduced. A piece of the rod was then cut off and weighed in water at different temperatures. The loss of weight gave the weight of a volume of water equal to that of the glass, and since the expansion of the glass was known that of the water could be determined.

Matthiessen also applied the method to find the expansion of mercury when that of water had been determined, weighing a small bucket containing mercury in water at different temperatures. He obtained results very close to those of Regnault. Later he applied the method to other metals.

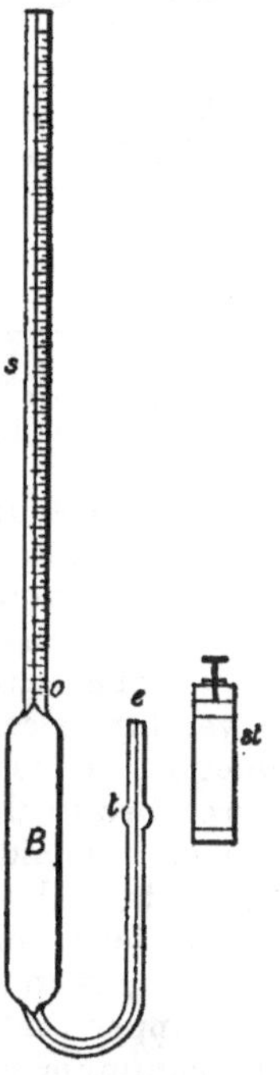

FIG. 26.—Dilatometer.

The Expansion of Water.—Researches on the expansion of water have been made by many experimenters using one or other of the methods already described, that with the dilatometer giving, probably, the best results. By some the dilatometer has been modified so as to have a constant internal capacity. As usually employed, the rise of the water in the dilatometer shows only the so-called apparent expansion—the excess of expansion of water over the containing vessel. But the expansion of mercury is about seven times that of glass, so that if about $\frac{1}{7}$ of the bulb of the dilatometer is filled with mercury, the internal capacity is constant, and the rise of the water shows its true expansion. Though interesting, this modification probably does not give such accurate determinations as the simple instrument used in the ordinary way.

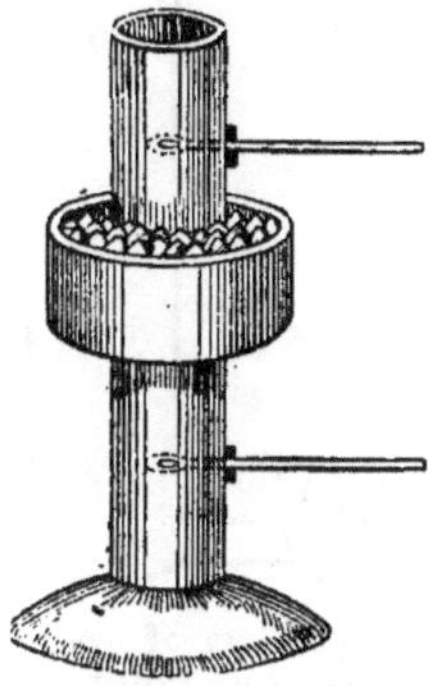
FIG. 27.—Hope's Apparatus.

All the methods concur in showing that water has a point of maximum density at about 4° C., and that the volume is nearly equal for temperatures equidistant on the two sides of this point. But since the rate of change of volume is very small near the maximum density, it is exceedingly difficult to determine the exact position of the point.

Hope's Apparatus.—The existence of a maximum density may be shown by Hope's apparatus, which consists of a cylindrical vessel of tin, surrounded midway by a gallery. The vessel is filled with water, and two thermometers are inserted, one near the top, the other near the bottom, as

* *Phil. Trans.*, clvi., 1866, pp. 231–248, "On the Expansion of Water and Mercury."

indicated in Fig. 27. A freezing mixture is put into the gallery. At first, the water is some few degrees above the point of maximum density. The water in contact with the side near the gallery becomes cooled, and, therefore, more dense, and sinks, while the warmer, lighter water from below rises up and takes its place, and circulation of the liquid is produced. This readjustment of unequally heated matter by its change of density is an instance of "convection." There is, at first, no circulation in the upper half of the vessel, and the higher thermometer is only cooled by conduction, which takes place but slowly. Its fall is, therefore, very gradual. But the denser water falls to the bottom in the lower half, and the lower thermometer falls rapidly. The circulation goes on until all the lower part has become of the maximum density. Then the circulation ceases, for further cooling only makes the water lighter, and it does not descend. The cooling of the lower layers can now only take place by conduction, and is very slow. When uniformity is reached in the lower half, the further cooling of the middle strata leads to circulation upwards, and this will go on to a higher and higher level until the upper thermometer is reached, when its fall will be much more rapid. The upper part may now fall below the point of maximum density, for the lighter colder water goes on rising upwards.

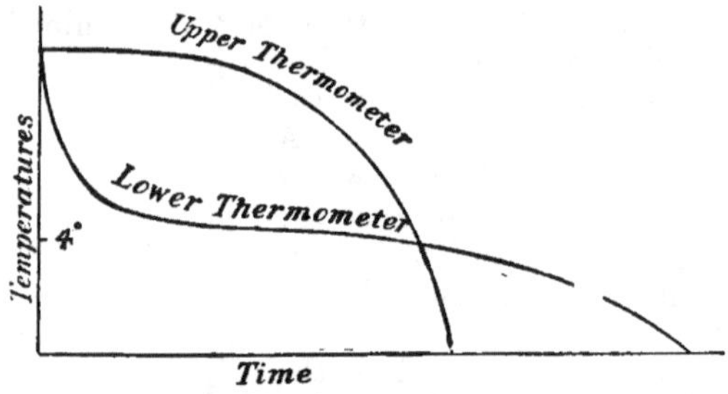

FIG. 28.—Indications of Thermometers in Hope's Apparatus.

Conduction may now begin to affect the lower thermometer, and it may fall somewhat more rapidly. Fig. 28 represents a series of observations on a time-temperature diagram —the long, nearly horizontal portion of the lower thermometer indicating a temperature in the vicinity of that of the maximum density.

Apparatus of Joule and Playfair.—Probably the most accurate mode of determining the point of maximum density is that of Joule and Playfair,* based on a principle similar to that applied in Hope's apparatus. Two tall iron cylinders A and B (Fig. 29) are open above, and connected by a trough T of about one inch cross-section. In this is placed a glass bulb only just floating. Below is a cross pipe P, furnished with a stop-cock.

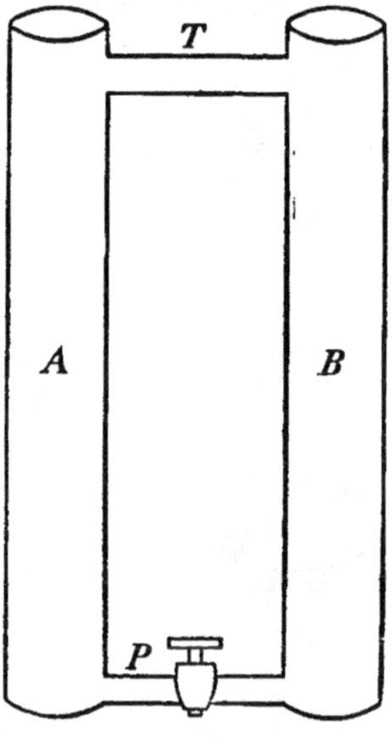

FIG. 29.—Apparatus of Joule and Playfair for Point of Maximum Density of Water.

The cylinders are filled with water, one just above, the other just

* Joule's *Scientific Papers*, vol. ii. p. 173.

below the temperature of maximum density ; the communication is cut off at P, the contents are stirred, and the temperature of each cylinder read. Communication is then made at P, and when the disturbance has subsided,

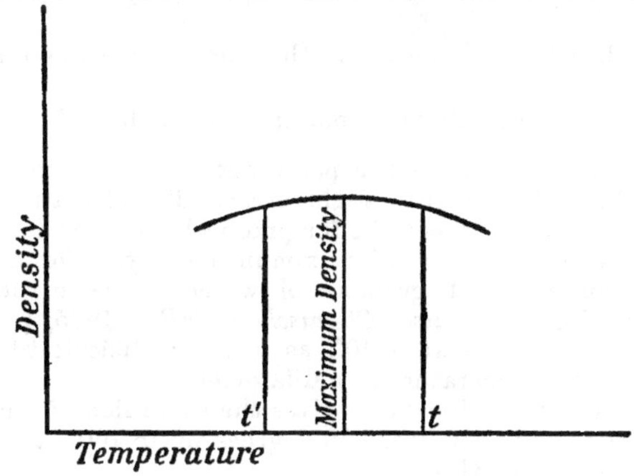

FIG. 30.—Density-Temperature Curve of Water.

the direction of flow is ascertained by the motion of the glass bulb along T.

It will evidently take place, below, from the heavier to the lighter,

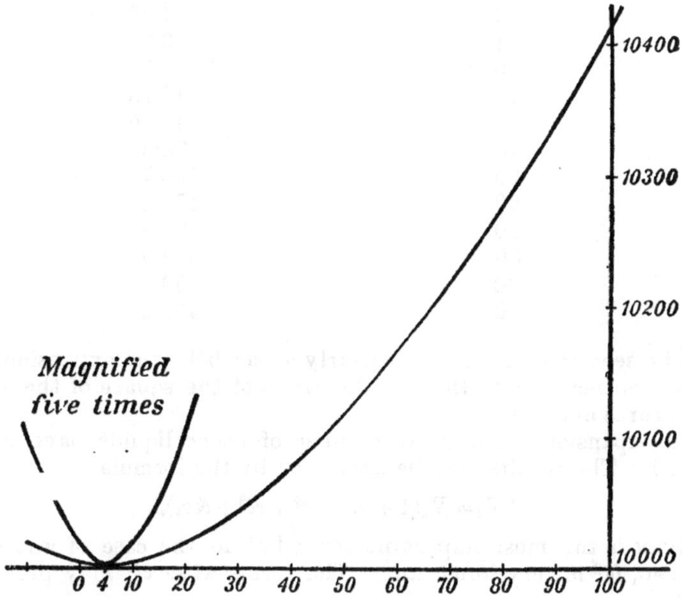

FIG. 31.—Curve of Expansion of Water vol. 10,000 at 4° C.

and above from the lighter to the heavier column. If, for instance, the warmer cylinder has temperature t, the colder temperature t' and the flow through the trough is from colder to warmer, the colder is less dense. Hence, assuming that the density curve (Fig. 30) is symmetrical about the point of maximum density, t' is further below it than t is above it, and $\frac{t+t'}{2}$ is below the maximum density. If the flow is in the opposite direction, $\frac{t+t'}{2}$ is above the point of maximum density. In this manner Joule and Playfair obtained a number of values of $\frac{t+t'}{2}$ respectively above and below the point, and were able to fix the temperature at 3·95° C. with very slight error. The alteration of density from this to 4° is so slight as to be for practical purposes negligible, and we may take 4° as the point of maximum density. For tables of the density and volume of 1 gramme of water we refer the reader to Landolt-Börnstein, *Physikalisch-Chemische Tabellen*, 1905, p. 37. These contain results beginning at − 10°, as it is not difficult to keep water liquid even at that temperature in a dilatometer.

Results.—The following table gives the expansion of a mass having volume 10,000 at 4° C. for every 10 degrees up to 100°, and the results are represented in Fig. 31:—

Temperatures. Degrees.	Expansion of Mass having vol. 10,000 at 4°.
− 10	18·6
0	1·32
4	0·0
10	2·73
20	17·73
30	43·46
40	78·2
50	120·7
60	170·5
70	227·0
80	289·9
90	359·0
100	434·3

It will be seen that the curve is nearly a parabola, the expansions being, however, rather greater than in the ratios of the square of the excess of temperatures above 4°.

The expansions of a great number of other liquids have been determined. The results may be expressed by the formula

$$V_t = V_o(1 + at + bt^2 + ct^3 + \&c.).$$

Usually at is the most important term, but in the case of water, as we have seen, bt^2 is very important. The terms after ct^3 may probably be neglected.

CHAPTER IV.

EXPANSION OF GASES.

Expansion of Gases depends on Pressure Changes—Volume Expansion at Constant Pressure—Gay-Lussac's Method—Regnault's Experiments—Increase of Pressure with Constant Volume—Gas Thermometry—Regnault's Normal Air Thermometer —Hydrogen Thermometer—Bottomley's Air Thermometer—Callendar's Compensated Air Thermometer.

The Expansion of Gases.—Since the volume of gas is very much affected by alteration of the pressure to which it is subjected, the volume expansion with a given rise of temperature may vary widely in different cases through different changes in pressure, and may even be prevented altogether, if the containing vessel is one which preserves a constant capacity. If, for example, a flask containing a small quantity of liquid is closed by a cork through which passes an open tube, dipping under the liquid (Fig. 32), a rise of temperature is accompanied by an expansion of air, which drives the liquid up the tube. But the height to which the liquid rises, and, therefore, the expansion of the air, will differ with different liquids. With mercury it will be much smaller than with water, for the gas inside the flask is exposed to the atmospheric pressure + that due to the column of liquid in the tube. The greater increase of pressure, due to the column of heavier liquid, lessens the expansion of the air.

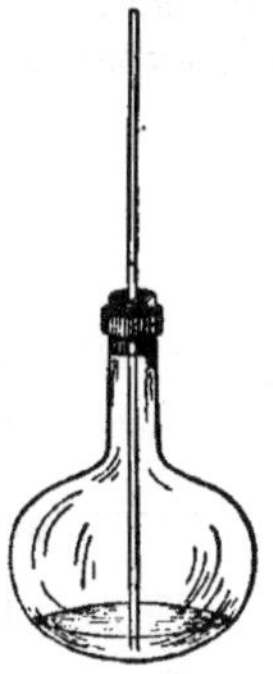

FIG. 32.

We see from this, that the rise of temperature may have two effects:—increase of volume and increase of pressure; and that the effects may differ for different arrangements. It is usual to study the two effects separately, investigating the change of volume when the pressure is kept constant, and the change of pressure when the volume is kept constant.

It may be observed that if Boyle's Law were exactly true, the two effects for a given rise of temperatnre would be equal. For, let a volume V of a gas expand, by a given temperature-rise, to V′, the pressure remaining at its original value, P. Now bring back the volume to V, increasing the pressure to P′. Then the pressure P would increase to P′, with the given temperature-rise, at constant volume V. But by Boyle's Law

$$PV' = P'V$$

and $$\frac{V'}{V} = \frac{P'}{P}$$

Or the volume changes in the same ratio in the one case as the pressure in the other.

But we know that Boyle's Law is not, in fact, quite true, so that it is necessary to investigate the two cases separately.

The Volume Expansion at Constant Pressure—*Gay-Lussac's Method.*—The earliest investigations having any approach to accuracy were made by Dalton and by Gay-Lussac. One arrangement which Gay-Lussac adopted for the purpose is shown in Fig. 33.

Dry air or other gas was admitted to a bulb A through a drying-tube, and a small mercury index *i* was introduced in the tube to indicate the volume, and to cut off communication with the external air. The bulb was then placed, with the stem horizontal, in a vessel of water, as in the figure, the temperature being roughly indicated by the thermometer *t'*, and, more accurately, by the thermometer *t*, which could be drawn out till the column of mercury was just visible. The air tube could also be drawn out till the index *i* was visible. The vessel of water was heated from below. The position of the index was observed at various temperatures, and the bulb and tube having been previously gauged, the air expansion could be found. Gay-Lussac arrived at a result which may

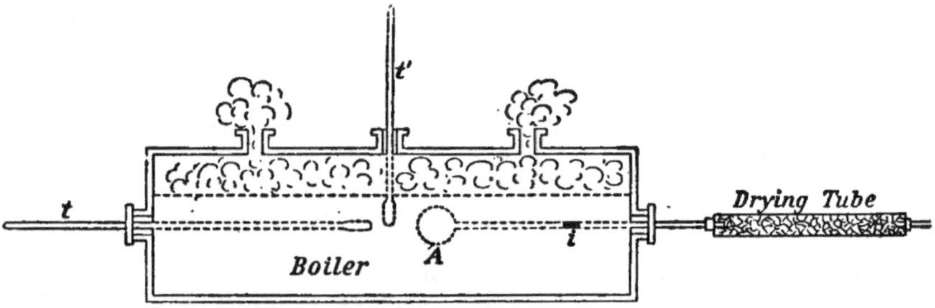

FIG. 33.—Gay-Lussac's Apparatus for Determination of Expansion of Gases at Constant Pressure.

be stated thus: whatever the gas, 267 volumes at 0° increase by 1 volume for each rise of 1° temperature, becoming 367 volumes at 100°.

Dalton had, a short time previously, found that different gases expand equally; he gave the expansion as 1 in 483 volumes at 32° F. for each rise of 1° F. This corresponds to 1 in 269 volumes on the Centigrade scale.

Later investigations have shown that the general conclusions of Dalton and Gay-Lussac are nearly, but only nearly, true, their numerical results giving too large an expansion, while the expansion is also found to vary slightly for different gases. According to Regnault, to whom we owe a very extensive series of researches, 273·1 volumes of hydrogen at 0° expand 100 volumes for a rise of 100°, while the same expansion is given by 272·4 volumes of air and 269·6 volumes of carbon dioxide. Other experimenters have found nearly the same values. It will be sufficient if we here describe

Regnault's Experiments.—Some of Regnault's researches were carried out by a method first used by Rudberg. The two parts of the apparatus are represented in Figs. 34 and 35. The bulb A was placed in a boiler, where it was surrounded with steam and then alternately exhausted by the pump, and filled with dried air many times, till all traces of moisture were removed. The taps *tt* were then turned, so that A was left in communication with the external air, and it was kept at the boiling-point for some time. The junction j was then disconnected and the end of the tube was sealed by a blowpipe flame, the bulb thus being

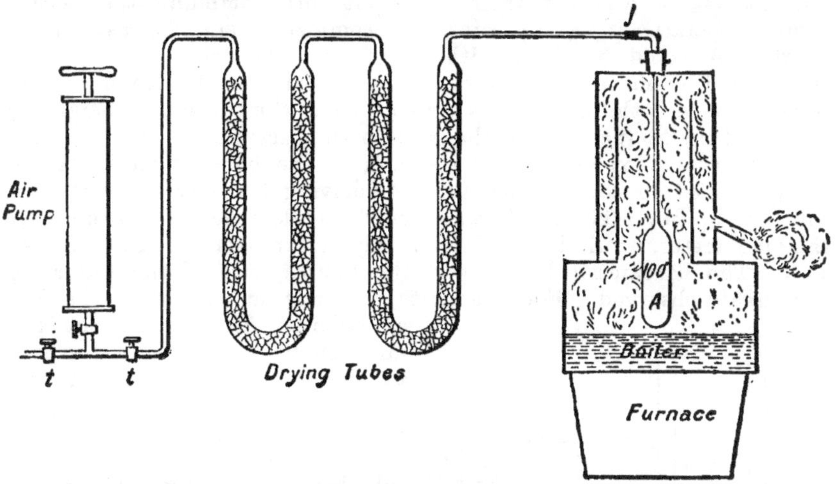

FIG. 34.—Regnault's Apparatus for Expansion of Gas at Constant Pressure (Rudberg's method). Bulb filled with gas at 100° C. and sealed.

filled with dry air at the atmospheric pressure, and at the temperature of the boiling-point. It was then arranged as in Fig. 35, being inverted with the end of the tube under mercury, the bulb being surrounded with melting ice. The end of the tube was broken off, and the mercury rose into the bulb through the contraction of the air. After some time the end of the tube was closed by a small piece of wax, which could be pressed up against it by the arm w. The height of the mercury in the bulb above that in the lower vessel was then determined, and from this could be found the pressure to which the air was subjected. The bulb was taken away and weighed, first when thus partially filled with the mercury which had risen into it, and afterwards when quite filled, and so it was easy to determine what fraction of the volume the air occupied at 0°. By correcting for the reduced pressure, the fraction of the volume which it would have occupied at the atmospheric pressure was calculated. The air occupying this volume at 0° expanded at 100° to fill the whole bulb (itself of larger capacity through the expansion of the glass), and so the expansion could be calculated.

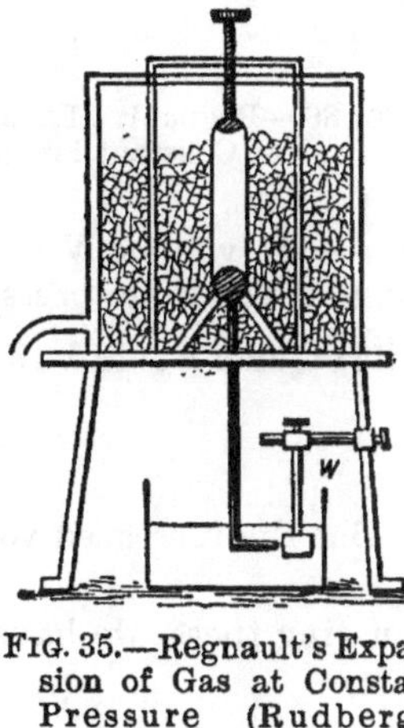

FIG. 35.—Regnault's Expansion of Gas at Constant Pressure (Rudberg's Method). Bulb opened under mercury.

Regnault also devised another method represented in Fig. 36. The bulb A communicated by a capillary tube with the manometer MM′, placed in a constant-temperature water-bath with glass sides. R was a three-way tap, by which M could either be brought into communication with M′ or with the pipe P opening downwards. R′ was a simple tap allowing M′ to be brought into communication with the pipe P′ opening downwards; *p* was a branch tube, put into communication with a drying apparatus and pump before the commencement of measurements, so that A could be filled with air thoroughly dried. A was then surrounded by melting ice, mercury was poured into the manometer till it rose to α in both limbs, *p* was sealed up, and the barometric height was read, and the temperature of the water-bath observed.

A was next exposed to the steam from the boiler, some of the air being pushed in consequence into M, driving the mercury down that tube and up M′. The tap R′ was turned, to allow mercury to run out of the manometer, until the levels were again the same in M and M′, say at β. The barometric height and the temperature of the water-bath were again observed. For simplicity, we may suppose them the same as before. The air in the bulb and the part of the stem exposed to the steam has, in rising from 0° to the boiling-point, filled the increase of volume in these, and also the volume of the manometer tube between α and β. The various volumes being gauged, it is easy to find the expansion of the air. For if V is the volume of the bulb and the part of the stem exposed to change of temperature, κ its mean co-efficient of expansion, v the volume of M between α and β, T the boiling-point, t the temperature of the bath, α the co-efficient of expansion of air, assumed to be constant—in other words, we use the gas scale of temperature to be described hereafter—the volume V of air at 0° has expanded to fill $V(1+\kappa T)$ at T, and v at t, at the same pressure. But if the air contained in v were also at T, its volume would be $v\dfrac{1+\alpha T}{1+\alpha t}$ and the total volume at T would be

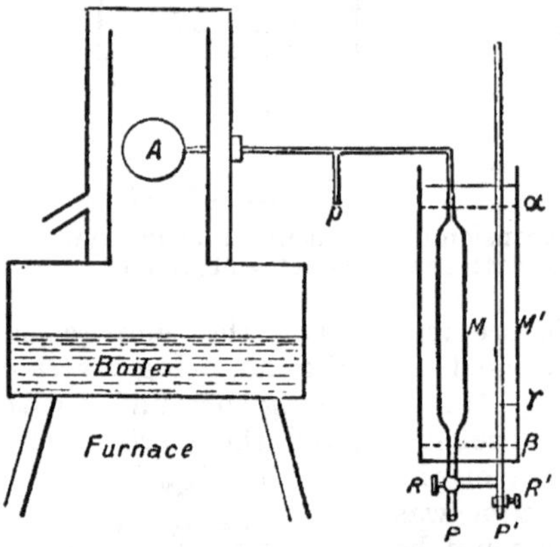

FIG. 36.—Regnault's Expansion of Gas at Constant Pressure.

$$V(1+\kappa T)+v\frac{(1+\alpha T)}{1+\alpha t}$$

But the increased volume of air, all at T, may also be expressed by

$$V(1+\alpha T).$$

Equating these, we have

$$V(1+\kappa T)+v\frac{(1+\alpha T)}{1+\alpha t}=V(1+\alpha T).$$

which determines α.

In practice, the variations of barometer and of temperature of water-bath were allowed for, and the equation was slightly more complicated.

Increase of Pressure with Constant Volume.—Nearly the same form of apparatus was used by Regnault to determine the co-efficient of pressure-increase when the volume is constant. The manometer was, however, somewhat modified in its details, and the water-bath was removed. As before, the bulb was filled with dry air through the tube p, and it was first exposed to the boiling temperature. The mercury being brought to the same level a in both tubes, p was sealed. The temperature was now allowed to fall, and ultimately A was surrounded by melting ice. The air tended to contract and draw the mercury above A, but the pressure was diminished by allowing mercury to run out at P′, so that the level in M was maintained at a, while it fell to γ in M′. The barometric height was read for each of the two temperatures. We will suppose it the same throughout. The temperature of the manometer is that of the surrounding air t, not being disturbed by the introduction of hot air from the bulb. Let the volume of the bulb at 0° be V, and at the boiling-point T let it become $V(1+\kappa T)$. κ may be measured by using the bulb as a mercury thermometer in a preliminary experiment. Let the volume of the connecting-tube and manometer down to the level a be v. Let the barometric height be H and the difference of the levels $a\gamma$ be h. Let β be the co-efficient of pressure-increase at constant-volume—assumed to be constant, that is to say we use the gas scale of temperature. Then, in the first part of the experiment, we have a volume $V(1+\kappa T)$ of air at T° and v at t, all at a pressure H, which would become

$$V(1+\kappa T)+v\frac{1+aT}{1+at}, \text{ if all were at the one temperature T.*}$$

In the second part of the experiment, we have a volume V of air at 0°, and v at t, all at pressure $(H-h)$, which would become

$$V+\frac{v}{1+at}, \text{ if all were at } 0°.$$

Now, these volumes are nearly equal, since κ and v, are both small. Then, without sensible error, we may use Boyle's Law to find what H would become if the former volume were reduced to the latter. It would obviously be

$$H\frac{V(1+\kappa T)+v\dfrac{1+aT}{1+at}}{V+\dfrac{v}{1+at}}=H'$$

and now we have the two pressures $(H-h)$ and H′ with the same volume.

Hence $$H'=(1+\beta T)(H-h)$$

and $$1+\beta T=\frac{H}{H-h}\times\frac{V(1+\kappa T)+v\dfrac{1+aT}{1+at}}{V+\dfrac{v}{1+at}}$$

* We have omitted the correction for increase of volume through increase of internal pressure as this is in general too small to come into account (Callendar, *Phil. Trans.*, A., 1887, p. 170).

We may use either the values of a previously found, or, since a only appears in small terms, we may put it equal to β without sensible error.

Regnault found for air at atmospheric pressure .	$a = \cdot 0036706$
and at 0°	$\beta = \cdot 003665$

From his other results we select those for hydrogen—

For hydrogen at atmospheric pressure .	$a = \cdot 0036613$
" at 0°	$\beta = \cdot 0036678$

The value of a is not quite independent of the pressure, nor is that of β quite the same for different values of the initial pressure at 0° C., but the variations for small changes of pressure are inconsiderable. Thus for hydrogen even when reduced to $\frac{1}{3}$ atmosphere he found $a = \cdot 0036616$; while for air at $\frac{1}{3}$ atmosphere $a = \cdot 0036954$.

Gas Thermometry.—The researches just described showed that the relation between pressure, volume, and temperature on the mercury-glass scale, for the less easily condensed gases such as oxygen, nitrogen, air, and hydrogen, may, without great error, be represented for ordinary temperatures by

$$pv = \kappa(1 + at);$$

where κ is a constant for a given portion of gas and $a = \cdot 00366 = 1/273$ approximately. If we date the temperature from $-273°$ C. as a new zero and write θ for $273 + t$ and R for κa we have

$$pv = \kappa a\left(\frac{1}{a} + t\right) = R\theta.$$

For the same kind of gas R is proportional to the mass dealt with, and if we deal with equal masses of different gases R is inversely as their molecular weights.

If Boyle's Law were exactly true, R would be constant for a given portion of gas at a given temperature, and though it would not be quite constant for different temperatures on the mercury-glass scale, we might arrange a new temperature scale so that R should be constant.

But as Boyle's Law is not a quite correct expression of the relation between pressure and volume for any gas, we cannot give such a simple definition for a gas scale of temperature. We must specify the way in which the pressure or the volume is allowed to vary.

Two methods have been used in practice, corresponding to the two kinds of research described above. In the one the pressure of the gas is kept constant, say at 1 atmosphere, and equal degrees of temperature are defined by equal increments of volume of the gas. In the other the volume of the gas is kept constant and equal degrees of temperature are defined by equal increments of pressure, starting, say, from 1 atmosphere at 0° C. In each there are 100 degrees between 0° C. and 100° C.

In the first case, if a is $\frac{1}{100}$ of the expansion between 0° C. and 100° C. of a volume which is 1 at 0° C., the temperature t, measured from 0° C. is given by

$$v = v_0(1 + at)$$

where v_o is the volume of the gas dealt with at 0° C, v its volume at the temperature to be measured.

Since

$$a=\frac{1}{100}\times\frac{v_{100}-v_o}{v_o}$$

we may put

$$t=\frac{v-v_o}{v_{100}-v_o}\times 100 \qquad (1)$$

If we choose a new zero $\frac{1}{a}$ below 0° C., we may term this the gas zero, and if we put

$$\theta=t+\frac{1}{a}$$

$$=t+\frac{100v_o}{v_{100}-v_o}$$

then

$$\theta=\frac{v}{v_{100}-v_o}\times 100$$

or putting θ_o for $\frac{1}{a}$

$$\frac{\theta}{\theta_o}=\frac{v}{v_o} \qquad (2)$$

In the second case, if β is $\frac{1}{100}$ of the increase between 0° C. and 100° C. of a pressure which is 1 at 0° C., the temperature t, measured from 0° C. is given by

$$p=p_o(1+\beta t)$$

when p_o is the pressure at 0° and p that at the temperature to be measured. Evidently, as with the volume scale, we have

$$t=\frac{p-p_o}{p_{100}-p_o}\times 100 \qquad (3)$$

and dating from a gas zero $\frac{1}{\beta}$ below 0° C., on which 0° C. is $\frac{1}{\beta}=\theta_o$ we have

$$\theta=\frac{p}{p_{100}-p_o}\times 100$$

and

$$\frac{\theta}{p}=\frac{\theta_o}{p_o} \qquad (4)$$

There are therefore two different scales for each gas. Fortunately, however, they are nearly coincident with each other, and nearly coincident, as certain experiments show, with the work scale.

Taking t as the temperature dating from 0° C. on the work scale, t_v the temperature defined by (1) for air, t_p the temperature defined by

(3), also for air, Callendar has calculated that the differences are as in the following table (*Phil. Trans.*, A., 1887, p. 179):—

t	$t_v - t$	$t_p - t$
0	+·0	0
100	0	0
200	0·04	0·084
300	0·09	0·20
500	0·23	0·47
1000	0·62	1·19

A gas thermometer has a very great advantage over a mercury-glass thermometer, in that the expansion of the containing vessel has a comparatively small effect, and an approximate knowledge of its expansion suffices to give the necessary correction unless the temperature be high. The gas, if pure, will always behave in the same way, and therefore the same values should be obtained for a given temperature with different instruments, and direct comparisons of the instruments should not be necessary. The methods used by Regnault for the measurement of expansion at constant pressure, and for pressure increase at constant volume between 0° C. and 100° C., illustrate the two types of instrument. A study of Fig. 36 will show at once that, though the gas scale has advantages, its practical use has serious disadvantages. The apparatus is bulky. It is not "direct reading," *i.e.* the temperature is not at once read off, but manipulation is required, and calculation must be made from the measurements taken. Corrections too must be made for the expansion of the gas reservoir, and for the gas in the tube connecting the reservoir to the manometer. As the temperature of this gas is different at different points, the correction for it is uncertain.

Regnault's Normal Air Thermometer.—Regnault's researches first made exact gas thermometry possible. He employed a thermometer of the constant-volume principle, the bulb containing dry air freed from carbonic acid, and with the pressure at 0° C., equal to 1 atmosphere. The instrument is represented by Fig. 36. The mercury was always brought to the same point on the bulb side of the manometer by adding mercury on the open side, or running it out at the tap below, and the barometric height + or − the difference of level on the two sides of the manometer gave the pressure of the air in the bulb. He termed this instrument "the normal air thermometer."

A simple form of the instrument, devised by Jolly, is represented in Fig. 37.

The two limbs of the manometer are connected by a flexible tube t, and one side, M′, is movable up and down. On the other side, M, is an index mark at a, to which the mercury is always brought before the pressure is measured. At S is a screw by which the bulb and connecting tube can be detached, and after being thoroughly dried they can be connected up again. R is a three-way tap either putting the bulb into connection with the manometer or closing that connection and putting the manometer into connection with the outside air. When the tap is turned on in the latter way, mercury is poured into M′ until it oozes out at R. R is then turned so as to put the bulb in communication with the manometer, and the instrument is ready for use.

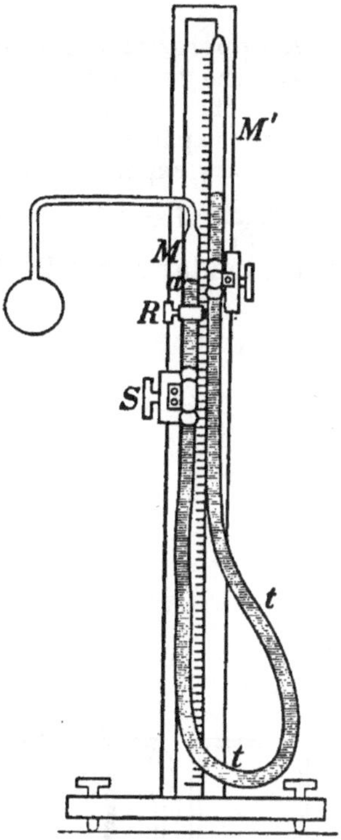

FIG. 37.—Simple Air Thermometer.

Hydrogen Thermometer.—For a long time the normal air thermometer gave the generally accepted scale of temperature. Since 1887, however, it has been superseded by the scale of a hydrogen thermometer at the Bureau International. This instrument is also on the constant-volume principle. The bulb is a cylinder of platinum-iridium 110 cm. long, and with about 1 litre capacity. It is placed in a horizontal position in a vessel with a window, so that mercury thermometers may be laid alongside, and their readings compared with those of the hydrogen thermometer. The open side of the manometer is used as the cistern of a barometer. The difference in level between the mercury at the top of the barometer and that in the manometer where the gas presses on it is the total pressure. Thus the separate reading of the barometer is avoided. The pressure is adjusted to be 1 metre of mercury at 0° C.

The temperature on the gas scale is obtained from the observations with a gas thermometer in the following way. For the small variations of volume allowed, Boyle's Law is sufficiently exact, and, therefore, for a given mass of gas:

Pressure × volume/temperature from gas zero = constant.

The fact that the mass of the gas in the thermometer is constant is therefore expressed by

$$\Sigma\, PV/\theta = \text{constant},$$

where we form the quantity $\frac{PV}{\theta}$ for each part of the volume, and add the results. Let V be the volume of the bulb at θ, V′ its volume at θ'. Let v be the volume of the connecting tube to the manometer at θ, v' its volume while the temperature θ' is being measured. The temperature of v' will be different at different points. Let $\frac{1}{t}$ be the mean of the reciprocals of these temperatures. Let P and P′ be the observed pressures.

Then

$$\frac{P'V'}{\theta'}+\frac{P'v'}{t}=\frac{PV}{\theta}+\frac{Pv}{\theta}$$

whence
$$\frac{\theta'}{\theta} = \frac{P'V'}{P\left(V + v - \frac{P'v'\theta}{Pt}\right)}$$

An approximate value of θ' will suffice to determine V' from V. Unless the temperature θ' be very high, it is not necessary to take into account the increase of volume through increase of internal pressure, and the expansion, with rise of temperature, may be determined by preliminary experiments. The greatest uncertainty is introduced by the term $\frac{P'v'\theta}{Pt}$, for it is difficult to determine t.

Callendar's and Bottomley's Constant-Volume Thermometers. It is evidently advisable to diminish v' as much as possible; at the same time it is advisable to have the mercury manometer as far removed as possible from the high temperature enclosure. To reconcile these two opposing conditions, Callendar* and Bottomley † have both devised air thermometers in which a U tube pressure gauge containing sulphuric acid is interposed between the air bulb and the manometer.

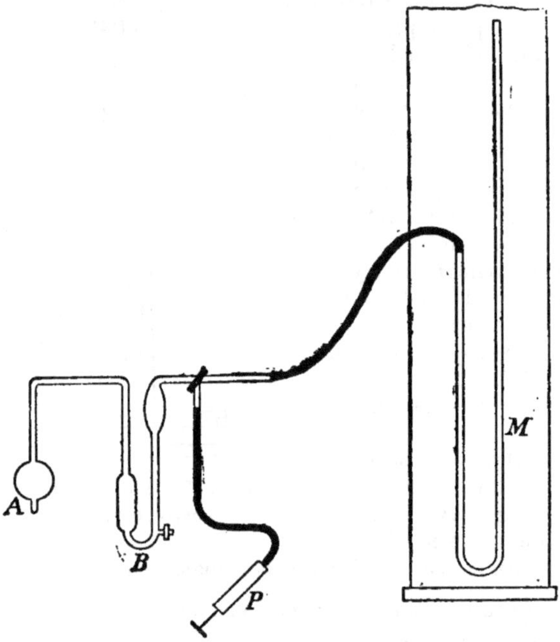

FIG. 37A.—Bottomley's Air Thermometer.—A, Air bulb; B, Sulphuric acid gauge; P, Air pump to make pressure equal on the two sides of B; M, Manometer.

Bottomley's Air Thermometer.—Fig. 37A represents the form used by Bottomley. This will sufficiently illustrate the principle.

The sulphuric acid gauge is at B, as near the bulb A as is convenient. Between this and the manometer M is a force pump, P, by which air can be forced in to make the two sides of the gauge B level. The manometer M then indicates the pressure. The volume v or v' is that of the connecting tube from the bulb to B, and the length of tubing from B to M is immaterial, so that M may be quite protected from any high temperature. The enlargements in B are to allow the air to expand while the temperature is rising without driving the sulphuric acid out of the manometer.

Constant-Pressure Gas Thermometers.—We may take Fig. 36 to represent the ordinary form of this class of thermometer. Since the

* *Phil. Trans.*, A., 1887, p. 166. † *Phil. Mag.*, xxvi., 1888, p. 149.

pressure is only allowed to vary very slightly, we may take Boyle's Law as true, and therefore $\Sigma\ PV/\theta = \text{constant}$, where we add up for each part of the gas, expresses the constancy of the mass of gas in the thermometer. In practice the pressure at the end of the experiment will not be the same as at the beginning, for even if the sides of the manometer are level the barometric pressure is likely to change slightly. And it may be more convenient not to trouble to adjust the levels to equality, but to make them nearly the same, measure the difference and add it to or subtract it from the height of the barometer.

If, then, P be the pressure when the whole apparatus is at θ; P′ the pressure when the bulb is at θ'; if V, V′ be the volumes of the bulb, v, v' the volumes of the connecting tube, $1/t$ the mean reciprocal of the temperature of v' when the bulb is at θ', M the volume of air expelled into the measuring tube when the bulb is at θ', M itself being at θ, we have

$$\frac{P'V'}{\theta'} + \frac{P'v'}{t} + \frac{P'M}{\theta} = \frac{PV}{\theta} + \frac{Pv}{\theta}$$

whence

$$\frac{\theta'}{\theta} = \frac{P'V'}{PV - P'M + Pv - \dfrac{P'v'\theta}{t}}$$

An approximate value of θ' will suffice to calculate V′, and v' and t must be estimated as exactly as possible.

In this class of instrument, then, it is necessary to measure P and P′ and also M, and the corrections for the connecting tube are at least as uncertain as in the other class. The greater number of measurements in this method probably led Regnault to prefer the constant-volume method, though good determinations had already been made of certain high temperatures by Pouillet with the constant-pressure method.

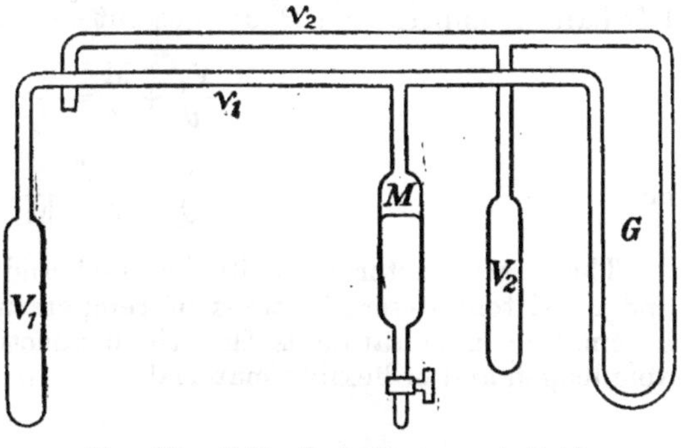

FIG. 38.—Callendar's Compensated Air Thermometer.

Callendar's Compensating Constant-Pressure Thermometer.—Callendar has devised a comparatively simple form of constant-pressure air thermometer, in which measurements of the external pressure are no longer necessary, and in which automatic corrections are made for the connecting tubes.* In this instrument V_1 (Fig. 38) represents the air bulb. M is the measuring tube initially filled with mercury. v_1 is the connecting tube. V_2 is a bulb of volume equal to that of V_1, and with equal "dummy" connecting tube v_2, following as nearly as possible the course of v_1 and closed at the end. The two systems are connected by a U tube pressure gauge G, containing sulphuric acid.

* *Proc. R. S.*, vol. l., 1892, p. 247.

Let us suppose that at first the whole instrument is placed in melting ice at temperature θ_o, M being filled with mercury and the pressure being so adjusted that the levels in G are the same. Now let the temperature of V_1 rise to θ, M and V_2 being still at θ_o. Run out mercury from M till the equality of pressures in G is restored. The volume of air expelled into M is determined by the volume M of the mercury run out. Let the initial pressure be P, the final pressure be P′. The two are nearly equal, so that the constancy of mass of the gas is expressed by

$$\Sigma\frac{PV}{\theta}=\text{constant}.$$

Let V_1 expand to V_1', v_1 and v_2 to v_1' and v_2', and let $\frac{1}{t}$ be the mean of the reciprocals of the temperatures of v_1', v_2'. We have for the bulb and manometer

$$\frac{P'V_1'}{\theta}+\frac{P'v_1'}{t}+\frac{P'M}{\theta_o}=\frac{PV_1}{\theta_o}+\frac{Pv_1}{\theta_o}$$

and for the other system V_2 and v_2.

$$\frac{P'V_2}{\theta_o}+\frac{P'v_2'}{t}=\frac{PV_2}{\theta_o}+\frac{Pv_2}{\theta_o}$$

Now making $V_2=V_1$, $v_2=v_1$, and $v_2'=v_1'$, we may equate the two left-hand members, and P′ divides out so that

$$\frac{V_1'}{\theta}+\frac{M}{\theta_o}=\frac{V_1}{\theta_o}$$

or

$$\frac{\theta}{\theta_o}=\frac{V_1'}{V_1-M}$$

The thermometer is easily handled and gives exceedingly accurate and consistent determinations of temperature. One advantage of the automatic compensation is that the connecting tube may be of considerable length and of flexible material.

CHAPTER V.

CIRCULATION AND CONVECTION IN LIQUIDS AND GASES.

Circulation and Convection of Heat—Hot-Water Heating Systems—Ocean Currents—Convection in Gases—Convection Currents in the Atmosphere—Winds—Land and Sea Breezes—Trade-Winds—Water-Vapour Aids Convection Currents—Weather Forecasting in the Case of Cyclones—Convection in Chimneys and Hot-Air Heating Systems.

Convection of Fluids.—Owing to the ease with which one portion of a fluid can glide past adjacent portions, any local change in density due to expansion by heat, at once results in motion. If, for example, a flask be heated from below, as in Fig. 39, and a little bran be put in the water to show the direction of motion, it is very soon seen that the heated, and therefore lighter, water is rising up from the bottom, its place being taken by a down-current of the colder water from the top. Usually the down-current is along the side; but if the flask be heated at one side only, it is easy to establish the circulation up that side and down the opposite colder side. The circulation may also be seen through the varying refraction of the up and down currents, which give an apparently shimmering motion to any object looked at through the water. This process of circulation through expansion by heat carries the heat from one part of the vessel to another, and this carriage of heat by motion of the heated matter is termed convection. Convection obviously expedites the communication of heat to the liquid as a whole, for not only are fresh portions of liquid being continually brought into contact with the heating surface, but also the heated liquid is continually coming into contact with colder surroundings, with which it shares its heat much more rapidly than with surroundings nearer to it in temperature. As an illustration of this, we may compare the method of boiling water in a test tube by applying heat at the bottom as in Fig. 40 with that represented in Fig. 41. In the latter case the circulation is very local, and the hot water, being the

FIG. 39.—Circulation of Water in a Heated Flask.

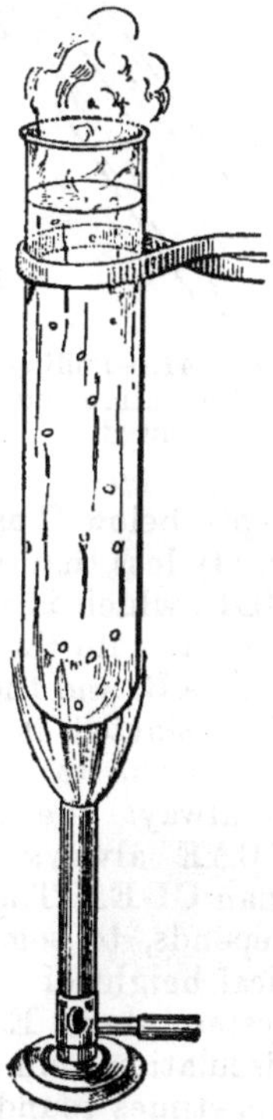

FIG. 40.—Boiling Water by Convection.

lightest, remains near the top, only sending down heat to the lower part of the tube by conduction. Even long after boiling takes place near the surface, no appreciable rise in temperature can be detected by the hand near the bottom of the tube. It is, of course, to make use of convection that boilers or vessels of water are always heated from below.

FIG. 41.—Boiling Water when Circulation and Convection are Prevented.

Hot-Water Heating Systems.—Convection of heat by water is used in warming buildings by hot-water pipes. Fig. 42 shows the principle on which such systems are based. From the heated vessel or "boiler" a pipe leads out at A, near the top, and after circulating round the building along B, C, D, returns into the "boiler" at E, near the bottom. At the highest point of the course, say C, is a cistern by which the whole system is filled with water. The circulation is commenced and maintained thus. The hot water rises in the boiler by convection—not going sideways into the pipe at E—and soon the pressure at E, due to the depth below C as traced through CBA, which is partly warmed, is appreciably less than the pressure due to the depth below C as traced through CDE, which is still all cold. Hence, the cold column presses the water near E into the boiler, and some of the warm water is forced into the pipe AB, and the circulation is started. Once started, it will evidently be kept up, for the water going out of the boiler is always the hottest, and therefore CBAE always gives a less pressure than CDE. The efficacy of the system depends, to some extent, on the vertical height of the highest point of the system above E. If this is small, the circulation will only be slow. It is sometimes found necessary to increase the height artificially by putting in a vertical pipe leading from A to some height above the level to be warmed, and then returning down to that level.

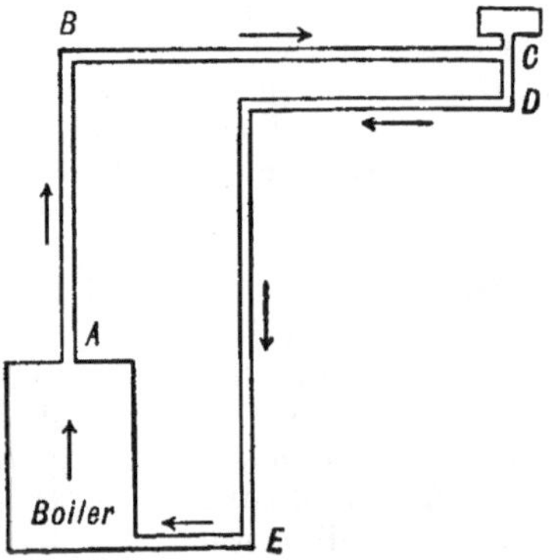

FIG. 42.—Principle of Hot-Water Heating Systems.

Some cases of ocean currents are probably examples of convection due to heat, arising, however, in a slightly different manner, since the heating of the ocean is chiefly from above. To understand the way in which the circulation is maintained, let us imagine a long canal with a horizontal bed stretching from the equator at A (Fig. 43) to the pole at B. If it were all at one temperature to start with, its surface would also be horizontal as AB. But through equatorial heat and polar cold a rise would take place in the surface at A to A′, and a fall at B to B′. The surface, however, thus sloping, could not keep in equilibrium, and at once a surface-current would start from A towards B. The pressure at the bottom at A would, therefore, fall, while that at B would rise through the removal of water from A to B. A reverse current would then be started along the bottom, the excess of pressure urging the cold water from B towards A. In apparent confirmation of this explanation, we know that in the Atlantic the Gulf Stream flows along the surface from the tropics to the polar regions, while it is found, by deep ocean soundings, that the temperature near the bottom, even in the tropics, is not much above the freezing-point, doubtless through a return under-current from the polar regions. But in all probability the surface-current is almost entirely due to winds along the surface. A persistent wind blowing along the surface in one direction will give far more kinetic energy to the water than can be acquired from the potential energy due to heat and expansion.

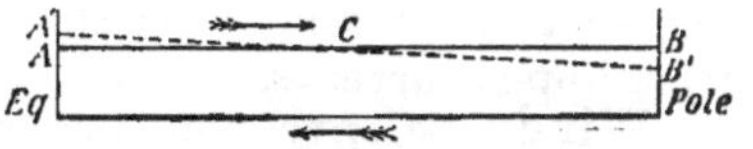

FIG. 43.

Circulation also takes place in lakes and ponds when cold weather sets in. The surface-water cools and falls, its place being taken by the warmer water pushed up from below, and so the water is turned over and over as it were, successive portions of it being cooled. But at 4° the process stops, for at that point the maximum density is attained. When the first mass of water is cooled down to that temperature, it goes to the bottom, and remains there, and the subsequent circulation stops short of this layer of densest water. The non-circulating water gradually increases, the circulating part becoming shallower and shallower, till all is at 4°. Then circulation entirely ceases, and the top layer goes on cooling towards 0°. Hence the existence of a point of maximum density hastens the arrival of the freezing-point by cutting short the process of circulation.

Convection in Gases.—Convection is even more marked in gases than in liquids, partly through their greater expansion with change of temperature, and partly through their smaller frictional resistance to motion.

A very common instance of convection in the atmosphere occurs over the heated surface of the ground on a hot summer day. The air immediately over the surface expands and rises, owing to its diminished density, its place being supplied by downward currents of cooler air from above. The existence of these currents is shown by the tremulous motion of distant objects looked at through the strata of air near the ground. The variations in density produce refraction, and the refraction of the rays is continually altered by the rapid change in position of the up and down currents.

We have similar effects in the tremulous motion of objects looked at through the ascending currents over a gas-flame, and in the flickering shadow of a flame thrown by sunlight.

A special instance of these local convection-currents is afforded by the haze so common on some hot days in summer. If any distant object is looked at through a good telescope on such a hazy day, its outline is found to be continually undulating, owing to the varying refraction. Since variation in refraction is always accompanied by reflection, a considerable quantity of the sun's light is reflected from the surface of these currents —especially at great angles of incidence—and this gives rise to the glare seen, especially, towards the sun. It is very probable that in some cases what is termed "haze" is due to convection-currents started, either by lighter air ascending from the surface, or by heavier cold air descending from upper currents.*

Winds.—Convection also occurs on a much greater scale in the atmosphere, the currents formed being recognised as winds. A well-known example is given by

Land and Sea Breezes.—It is very often noticed at the seaside that there is, during the daytime, a sea-breeze, which changes to a land-breeze at night. In tropical regions these land and sea breezes are even more marked than in higher latitudes. We may explain them as convection-currents. During the day the surface of the land becomes much hotter through the sun's rays than the surface of the sea, the higher specific heat of water, and the mixing up of the surface layers by the waves, both combining to lessen the rise of temperature. The air over the land is, therefore, more heated, and expanding upwards tends to overflow above. The overflow in the upper strata takes place towards the sea, and so the pressure at the sea surface is increased while that at the land surface is diminished. There is therefore a tendency for the surface layer of air to move from sea to land, the motion constituting a sea-breeze. At night, however, the land radiates out its heat more rapidly than the sea, the high specific heat and the agitation of the sea both tending to keep up its temperature. There is, therefore, a contraction of the air over the land, and an overflow in the upper strata from sea to land, accompanied by an opposite flow in the surface strata from land to sea, constituting the land-breeze.

Trade-Winds.—We may also explain in a similar way the well-known trade-winds, which blow in certain latitudes, in our hemisphere, from the north-east towards the equator. As in the land and sea breezes, the equatorial heat expands the air, which overflows in the upper strata towards the polar regions, tending to decrease the pressure at the surface near the equator, and increase the surface pressure in higher latitudes. The surface layers of air are therefore pressed from the north and south towards the equator. The north-easterly direction in the Northern Hemisphere, and the south-easterly direction in the Southern Hemisphere, of these lower currents arise from the rotation of the earth. For, taking the northern trade-winds, the mass of air moving towards the equator continually comes into regions moving faster from west to east than the region just left. There is, therefore, a tendency on the part of the winds to lag behind the earth's surface in its west to east motion; or, the wind has a motion towards the west as well as towards the south, making it a north-east wind.

The corresponding upper current, in its journey towards the pole, ultimately comes down to the surface somewhere about 35° N. latitude, and constitutes a south-west wind. But it is not nearly so constant as the trade wind.

* An explanation of the twinkling of the stars as due to convection-currents in the air has been given by Montigny, Exner, and Rayleigh (*Phil. Mag.*, xxxvi. p. 129, 1893).

Water-Vapour aids Convection-Currents.—Convection arising from heating of the lower layers of the atmosphere is greatly aided by the increased amount of water-vapour which the air takes up, the vapour being much lighter than the air which it displaces. As an example of the joint-effect of water-vapour and expansion through heat to produce convection, we may probably instance the formation of a thundercloud. When a storm is first gathering, an observer a short distance away may see enormous piled-up masses of cloud rising far into the upper regions. These show that a great volume of light, damp air has risen by convection, expanding in the ascent, and therefore cooling through the work done in expansion until the temperature of cloud-deposition is reached. At the same time, it frequently happens that there is a surface indraught towards the storm area from the surrounding region, the upper strata with the clouds, moving in the direction opposite to that of the strata immediately below them.

These examples will prepare us for the general statement that winds are convection-currents in the atmosphere due to local diminutions of air density, either through heat, or increased evaporation, or both. Since the weather depends so largely on the direction of the wind, it is, of course, of the utmost practical importance that the nature and origin of all atmospheric movements should be investigated and explained as thoroughly as possible. But though the above general statements may be made with confidence, meteorologists have not yet succeeded in discovering, except in a few cases, how particular winds arise, *i.e.* what share in their origin is to be assigned to heat, and what share to evaporation. They are still further from foretelling what winds will be formed from a given distribution of the atmosphere with known temperature and amount of water-vapour. The problem is one of enormous difficulty, which will probably only be fully solved in the distant future.

Weather Forecasting in the Case of Cyclones.—At present, the art of weather-forecasting depends largely on the fact that a given distribution of weather travels onwards in a definite course, and that certain definite types of weather have been recognised and their movements studied. As an example, we may give a short account of the simplest type, that of the Cyclone, and the mode in which English weather is forecast when a cyclone is approaching our shores.

There are about fifty meteorological stations scattered over the north-western part of Europe, at which meteorological observations are taken at stated times every day. These include the readings of the barometer, the direction and force of the wind, the kind of weather, and the temperature. The results are at once telegraphed to London. The barometer readings are marked on a map containing all the stations, each reading at its own station. Curves are then drawn on the map, joining all points where the pressure is the same, one curve for every fifth of an inch. There will thus be a curve for 29 inches, another for 29·2, another for 29·4, and so on. It generally happens that these curves do not pass exactly through the stations, but their position may be ascertained from the known readings. If, for example, the reading at London is 29·25, and at Dover 29·15, it is assumed that the 29·2 line passes about halfway between these places. The map with these curves, or isobars, shows the distribution of barometric pressure, and when the direction of the wind is marked on the map, it is found that it blows, in general, from higher to lower pressures, but not straight down the slope. Through the rotation of the earth, the wind has a tendency to go to the right of the area of lowest pressure, as explained already in the case of the trade-winds. One of the commonest arrangements of the isobars is that of a series of oval curves, round the area of lowest pressure. This will be seen from Fig. 44, which gives the map for November 23, 1874, in which the lowest pressure was over the Midlands. The arrow-heads giving the direction of the wind, show us also that it is everywhere blowing towards the right of this low pressure area, and so much so that the whole of the system constitutes a whirlwind or cyclone. The rotation in such a cyclone is, in the northern hemisphere, always counter-clockwise. Through the in-draught of air to the centre, the depression is gradually filled up; but not so rapidly as we should expect, if the air merely moved to the centre, and stopped there. It appears from observations made on high clouds that, in general, it ascends there, and flows outwards again in the upper strata, forming an "anti-cyclone."

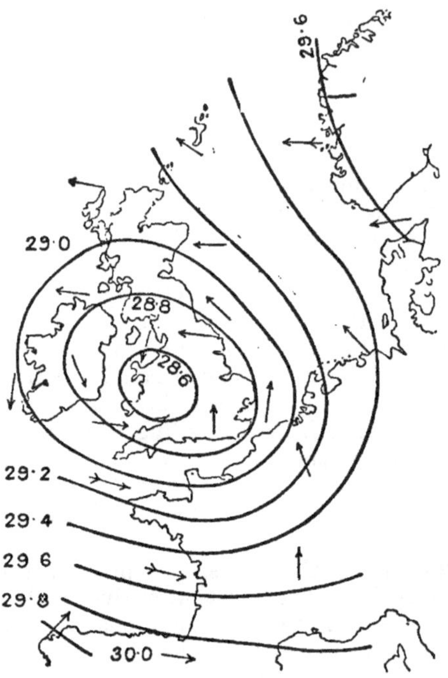

November 23, 1874.

FIG. 44.—The isobars and wind in a Cyclone. (From Abercromby's *Forecasting by Weather Charts.*)

A study of cyclones has shown that in the various parts, there are not only characteristic winds but also characteristic kinds of weather. Thus, to the east of the centre, the wind is usually from the south or south-east, and, far from the centre, the weather is fine. Moving westward towards the centre, the weather gradually gets damper; the sky becomes overcast; and near and at the centre, there is usually rain. Passing the centre, the wind is now from the north or north-west; it is colder and drier; and some distance from the centre the sky usually becomes clear. To the south of the centre, the wind is usually south-west, and north of it north-east.

The distribution of weather in a typical cyclone is represented in Fig. 45.

Cyclones usually move in an easterly direction—most commonly to the north-east in our latitude, the centre travelling at any rate up to 70 miles an hour. But with a given cyclone the rate is often nearly constant for some time.

The figure on following page (Fig. 46) shows the history of a cyclone on four successive days.

It is now easy to see how a forecast may be made. A cyclone is shown by the isobars to be approaching as in Fig. 46, October 26. The track of the centre is either observed or guessed at, and so the subsequent positions of the cyclone are foretold. Each part of the cyclone carries with it its wind and weather, and so the wind and weather may be foretold for the districts passed over by a given part of the cyclone. For instance, if a cyclone-centre passes over Valentia and is moving towards the north of Scotland, south-west winds may be foretold for England. If the isobars are near together, the pressure-slope is steep, and it may be expected that the winds will be strong. If the centre moves more towards the south, say over the Midlands, rain with varying winds may be foretold there, north-easterly winds for Scotland, and south-westerly for the

FIG. 45.—Cyclone Prognostics. (Abercromby.)

southern coasts. It will be seen from this short account that it is exceedingly difficult to forecast English weather with accuracy for any length of time beforehand, as the stations extend so little to the westwards. A cyclone first observed to be approaching in the west of Ireland may already be well on its way towards England, with all its changes of weather, before its course and nature can be accurately observed. When communication with ships in the Atlantic by wireless telegraphy becomes general, no doubt the proportion of successful predictions, already large, will be further increased.

Convection in Chimneys and Hot-Air Heating Systems.—A chimney depends for its successful working upon convection. Taking the case of the ordinary open fireplace, when a fire is first lighted in a room, the column of air in the chimney over the grate is heated and expands, some of it flowing out from the top. The air column in the chimney now weighs less, and the pressure at the grate, as traced down through the chimney, will be less than that at the same level in the rest of the room. The air of the room, therefore, moves towards the grate, and if there is a sufficient communication with the outside, the circulation thus started is maintained, the pressure due to the cold air outside the chimney being always in excess of that due to the warm air and gas in the chimney. In general, the communication with the outside is effected through or under the door, and through the crevices between the window-sashes, the air moving as represented by the arrow-heads in Fig. 47.

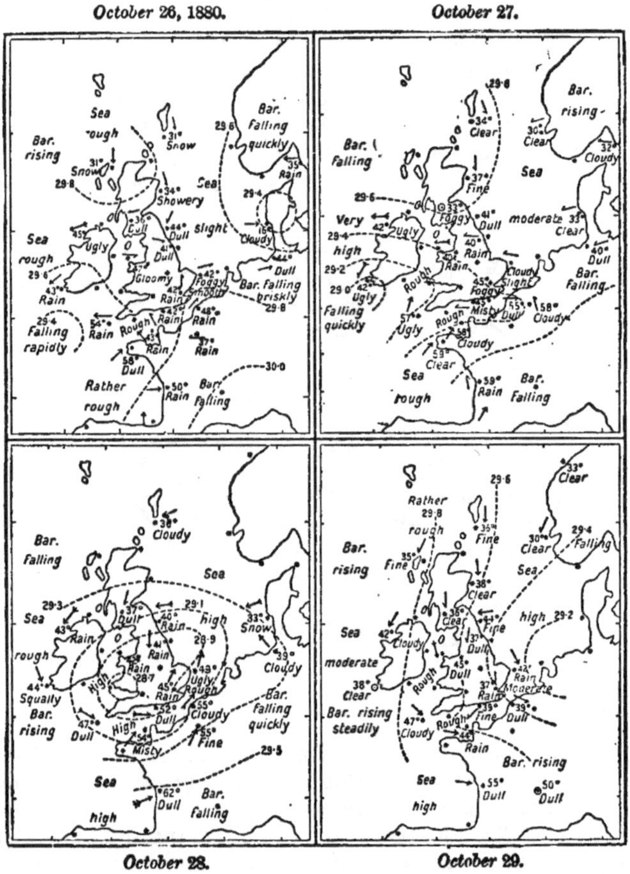

FIG. 46.—Progress of a Cyclone, October 26–29, 1880. (From the *Times*.)

Sometimes this is not sufficient, and unless a passage be made for the admission of cold air, the circulation is not established, and the chimney smokes.

Sometimes the chimney, if a wide one, may even establish within itself a down and up current and so maintain the needed circulation.

We may illustrate these points by lowering a lighted candle into a flask (Fig. 48*a*). No circulation is established, and as soon as all the air is used up, the candle goes out.

If, however, a thin partition be inserted down the neck of the flask

(Fig. 48*b*), the circulation is thoroughly established down one side and up the other. The direction is easily shown by blowing out a lighted taper, and holding it, while still smoking, first on one side, then on the other.

If the candle be put at the bottom of a wide cylinder, it will usually establish its own circulation, but, as the direction of the currents is constantly changing, the flame is very unsteady.

When a chimney with an open fireplace is working properly, a great amount of air in addition to that used to burn the coal is drawn into the chimney through the open space above the grate, and much of the heat of combustion is used to warm this air and send it up the chimney. At first sight, this heat would appear to be wasted, but the additional circulation has one advantage, namely, that it brings into the room a greater supply of fresh air, and so aids ventilation. But in the ordinary arrangement, or rather want of arrangement, by which the cold air is left to find its own way into the room, the greater circulation makes itself evident by the greater draughts along the floor and near the windows. The ventilation in this case is also inefficient, for much of the fresh air rushes straight to the chimney, and the impure air, which is warmer and damper, and therefore lighter, rises to the ceiling, and is only renewed slowly by diffusion. It is much better to provide a special passage for the air into the room by means of some such arrangement as the Tobin ventilator, in which a pipe is brought from the outside somewhat as in Fig. 50.

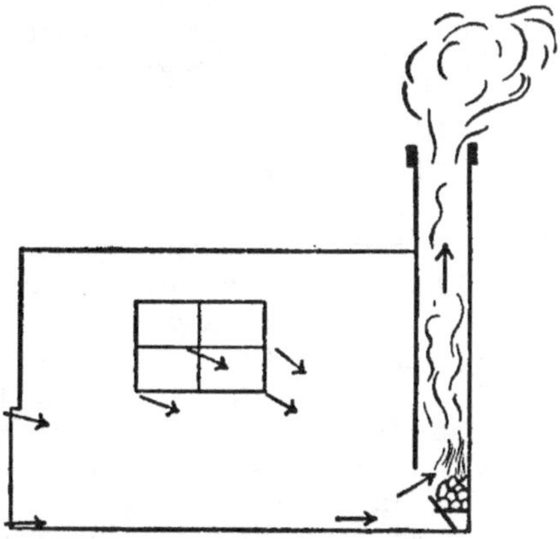

FIG. 47.—Circulation established by an Open Fireplace.

FIG. 48.—Candle in Flask, showing (*a*) no circulation, (*b*) circulation.

The pipe being carried up a few feet, the stream of cold air is directed towards the ceiling, and is warmed there before it descends to the lower part of the room. Since there is thus a more general renewal of the air of the room, the ventilation is much more efficient, while cold draughts are diminished.

Arrangements have been made by which the incoming cold air is warmed by some of the excess of heat which would otherwise be wasted up the chimney. The air is admitted from the outside into boxes placed round the grate and chimney, and then carried into the room—preferably to the furthest corner from the grate, somewhat as in Fig. 51. If the entrance into the room is sufficiently subdivided, no serious draughts will be felt.

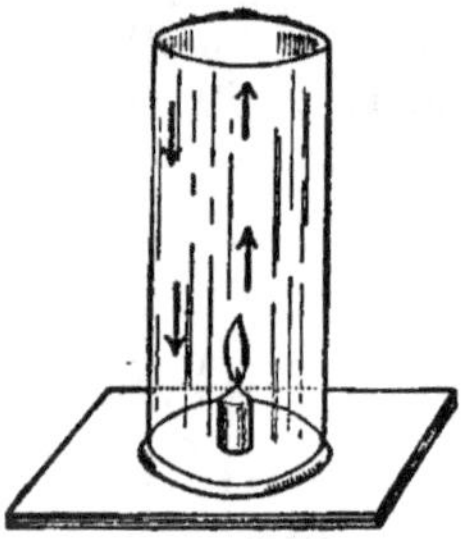

FIG. 49.

In large rooms used for meetings, the problem of warming, and at the same time ventilating, efficiently, is one of great difficulty, and one which architects have only very partially solved. The shape of the building and its surroundings may greatly influence the direction in which convection-currents tend to establish themselves, so that it is almost impossible to foretell the special difficulties to be surmounted. As there is in such halls, generally, nothing corresponding to the open chimney, ventilators are usually provided near or in the ceiling, and the whole room may be in this case regarded as a chimney. The air of the room is warmed by the heating arrangements and by its occupants, so that a circulation is established through door or window, or through other communication, with the external air and out through the ventilators. Frequently, however, especially in cold weather, local circulation is established near the windows, the air in contact with them being cooled, and falling through its increased density, this fall being recognised as a cold down-draught. This may be lessened either by having double windows, or warming the windows by gas-jets or water-pipes placed in the inside sills. Probably, the most common fault in large rooms is that, from motives of economy in construction, the ventilation is left too much to work itself. Some means, such as a fan, should always be provided, by which the impure heated air should be extracted at the top, while fresh air, warmed if necessary, should be introduced at such a level and in such a direction that it will not be felt as an unpleasant draught.

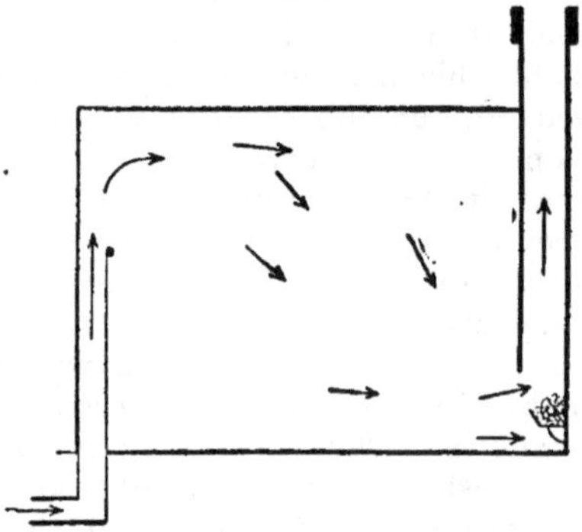

FIG. 50.—Tobin Ventilation.

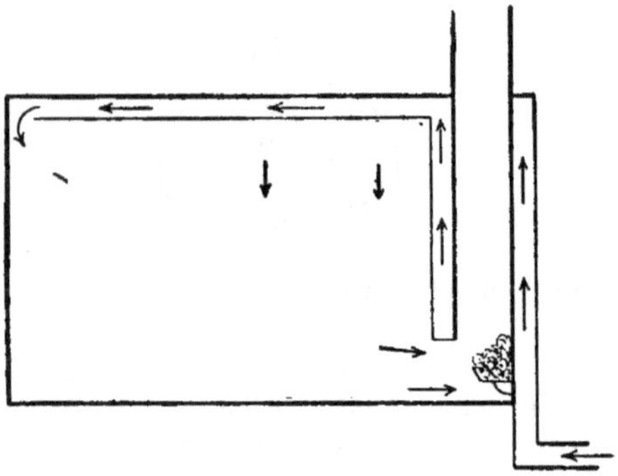

FIG. 51.—Arrangement for Warming Incoming Air.

Inasmuch as our comfort so largely depends on the efficient warming and ventilation of rooms, it is much to be desired that the application of the principles of convection should be more thoroughly studied. The investigation on the scale necessary for its application to large buildings is expensive as well as difficult, and for this reason, probably, it has hitherto been too much neglected. But whatever the expense, it would be worth incurring if we could thereby arrive at some mode which should save us from the too common experience in large buildings, where the means provided for ventilation and warmth are found to be totally inadequate, and where improvement of ventilation means increase of draught, and increase of warmth means absence of ventilation.

CHAPTER VI.

QUANTITY OF HEAT. SPECIFIC HEAT.

Quantity of Heat—Unit Quantity: the Calory—Specific Heat—Water Equivalent and Capacity for Heat—Method of Mixtures—Regnault's Determinations by the Method of Mixtures—Experiments on Solids—On Liquids—On Gases—Specific Heat of a Liquid by Mixture with known Solid—Method of Cooling—Method of Melting Ice—Bunsen's Ice Calorimeter—Method of Condensing Steam—Joly's Steam Calorimeter—Differential Steam Calorimeter—Method of Electric Heating—Specific Heat of Water—General Results—Law of Dulong and Petit.

Quantity of Heat. Specific Heat.—When a cold body is put in contact with a hot body, the colder rises in temperature, while the hotter is cooled. We describe the rise of temperature of the one by saying that it has gained heat, the fall of temperature of the other by saying that it has lost heat. But we go further than this, and regard the heat gained by the one as identical with the heat lost by the other. That is, we regard the heat as something which we can identify at least in thought, and which has been transferred from the one body to the other. This is no doubt metaphysical; for all that we are entitled to assert, from actual experiment, is that one body is hotter, the other cooler; but the conception of the identity of heat enables us to describe the process more shortly, and to think of it much more clearly.

Let us suppose that we have a number of equal vessels, containing equal quantities of water at the same temperature. Let us plunge into these vessels different hot masses, into one iron, into another copper, into a third stone, and so on, and let us further suppose that the masses and their temperatures are so adjusted that in each case the temperature of the water has risen by the same amount in coming to thermal equilibrium. Since the mass of water is in each case the same, the same amount of heat has been gained by each, and, if we take the view that this heat has passed from the body immersed, we see that we can speak of equal amounts of heat lost by quite different substances, though the effects are very different. The iron, for example, may have fallen 10°, the copper 20°, the stone 30°. We need not then speak of iron heat, copper heat, stone heat, but of heat simply, for all these different substances, on parting with their heat to one chosen substance, have the same kind of effect on it.

We are thus led to the idea that we may measure quantity of heat by its effect in raising the temperature of a given mass of some chosen substance, and water is the substance which is usually chosen.

If we make an experiment in which we mix equal quantities of water at different temperatures, we find, after allowing for the heat taken up by the containing vessel, that the final temperature of the mixture is very nearly the mean of the two initial temperatures. If, for example, we mix 50 grammes of water at 15° with 50 grammes at 17°, the resulting temperature is 16°, within the limits of errors of observation. Then the heat given out by a gramme in cooling from 17° to 16° will raise another gramme from 15° to 16°, or the same heat will raise a given gramme from 15° to 16°, and from 16° to 17°. Or, if we take a wider range, and have initial temperatures of 0° and 30°, the mixture is exceedingly near to 15°; or the heat required to raise 1 gramme from 0° to 15° is very nearly the same as the heat required to raise 1 gramme from 15° to 30°. Exact experiments, to be described later on, show that the heat values of successive degrees in the rise of temperature in a quantity of water are not quite the same, and we are therefore obliged to specify the temperature range used.

The most convenient range is 1° at the average laboratory temperature, viz., from 15° to 16°, and we therefore choose the following definition:—

The unit quantity of heat, or the calory, is the quantity which raises 1 gramme of water from 15° C. to 16° C.*

But in rough work, sufficient for many purposes, we may neglect the variation in the heat required for a rise of 1° at different parts of the scale, and take the calory as simply the heat raising 1 gramme of water 1° C. If m grammes of water are raised t°, the heat gained by the water is therefore mt calories.

We are thus enabled to measure and express the heat given up by any other substance in cooling through a definite range of temperature.

Let us suppose, for instance, that 50 grammes of iron are heated to 100° and then plunged into 50 grammes of water at 15°. The iron and water will ultimately come to a common temperature of about 23·5°. Hence, 50 grammes of iron, in cooling 76·5°, have given up heat which raises the temperature of the 50 grammes of water 8·5°; or, the iron has given up $50 \times 8{\cdot}5 = 425$ calories. On the assumption (not quite accurate) that each degree has the same heat value for the iron, each gramme of iron in cooling 1° has given up $\frac{425}{50 \times 76{\cdot}5} = \frac{1}{9}$ calory nearly.

Thus the iron in rising 1° requires $\frac{1}{9}$ of the heat which will raise the same mass of water 1°. This is expressed by saying that the *specific heat* of iron is $\frac{1}{9}$.

Or, let us suppose that 50 grammes of lead at 100° are plunged into 50 grammes of water at 15°. The common temperature will now be about 17·5°; or 50 grammes of lead in cooling 82·5° have raised 50 grammes of water 2·5°, and have given up $50 \times 2{\cdot}5 = 125$ calories. Assuming equality of value of each degree, 1 gramme of lead, in cooling

* The unit of heat used by Regnault was defined as that raising 1 gramme of water from 0° to 1°. Another unit is the $\frac{1}{100}$ part of the heat raising 1 gramme of water from 0° to 100°. We may term the three calories respectively the 15°, the 0°, and the mean calory.

1° gives out, or in rising 1° takes in, $\frac{125}{50 \times 82 \cdot 5} = \frac{1}{32}$ calory, about. This is expressed by saying that the *specific heat* of lead is $\frac{1}{32}$. These illustrations prepare us for the following definitions:—

The Specific Heat of a Substance is the number of calories needed to raise 1 gramme of the substance 1° C.

If the specific heat of a substance over a range $t°$ is s, the quantity of heat required to raise m grammes of the substance $t°$ is mst calories.

Water Equivalent and Capacity for Heat.—Since mst calories would raise ms grammes of water through the same range $t°$, the quantity ms is termed the *Water Equivalent* of the m grammes. When the range is 1° the quantity of heat required is termed the *Capacity for Heat.* The two expressions have the same meaning in practice.

We shall now give an account of the chief methods of determining specific heats. The details of the methods, though of the utmost importance in obtaining exact results, need not be fully described here. These may be best understood from the accounts given by the original workers. Our aim is to point out the general principles.

The method most easily applied is

The Method of Mixtures. — Suppose that we are to find the specific heat s of a certain solid. Then a known mass M of it is raised to a known temperature t', and dropped into a known mass of water W at a known lower temperature t. The experimenter observes the temperature θ at which the mixture stands when the two have come to thermal equilibrium. If all the heat lost by the solid could be assumed to have gone into the water, and to remain there, then expressing the equality

Heat lost by solid = Heat gained by water

we have
$$Ms(t' - \theta) = W(\theta - t)$$
or
$$s = \frac{W}{M} \times \frac{\theta - t}{t' - \theta}$$

But in practice the heat does not all go into the water and remain there. Some of it goes into the containing vessel or *Calorimeter*, into the thermometer, and into the stirrer necessary to mix the water up thoroughly. Some of it passes out through the calorimeter, where it is partly given to the air, and partly radiated out into the surrounding space. Corrections must be determined and applied on both these accounts. We may understand how they are made by considering an example. Let us suppose that we are to find the specific heat of a specimen of brass. It is advisable to have the brass either in a spiral roll, or in a coil of wire, or in pieces, in order that its surface shall be large, and that it shall quickly part with its heat to the water when immersed. The brass may conveniently be heated to the temperature of boiling water in a steam-jacketed chamber—or, for rough work, in a test-tube immersed in boiling water—and the temperature may be taken as that of boiling water at the atmospheric pressure at the time. This will be nearly $100 + \frac{H - 760}{27}$, where H is the barometric height in millimetres.

The vessel or calorimeter containing the water (Fig. 52) should be of thin metal polished on the outside—thin that it should not absorb much heat, polished that it should lose little by radiation. It should have a lid, and should be supported by badly conducting material, with as small surfaces of contact as possible, say pointed wood or ebonite pegs, and should be surrounded by an outer vessel, preferably a water-bath, kept at a constant temperature. An intermediate thin metal vessel, highly polished on both sides, may still further diminish the radiation loss. To allow for the quantity of heat absorbed by the calorimeter, thermometer, and stirrer, we must find their capacity or water equivalent, and regard it as so many grammes of water in addition to that actually contained in the calorimeter. This water equivalent is determined by putting into the calorimeter about the same quantity of water as will be used in the final experiment with the brass, and by adding some hot water so as to produce about the same rise of temperature as is expected in that experiment, and noting the exact rise produced.

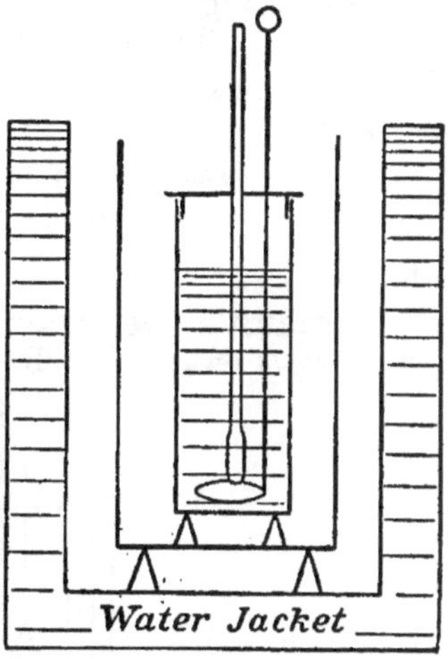

FIG. 52.—Calorimeter.

Thus, suppose that the calorimeter contains 100 grammes of water at 15°,* and that 40 grammes of water at 48° are poured in. The contents are well stirred, and in half-a-minute the temperature has risen to 23·55°. Meanwhile, however, the calorimeter has been losing heat. To estimate this loss the rate of fall is now observed. Suppose that one minute after the last observation the temperature is 23·35°, and one minute later still 23·15°. The rate of loss corresponds to a fall of 0·2° per minute, or of 0·1 per half-minute. But during the rise of temperature the rate of loss may be taken as half this on the average; for supposing the temperature to rise uniformly from the moment of mixture to the maximum observed—only an approximation to the truth, no doubt—the average excess above the surrounding enclosure is only half the final excess, and the loss, which, for small excesses, may be taken as proportional to the excess above the surroundings, is at only half the rate of the final loss observed. Thus, had all the heat been kept in, the temperature would have been $\frac{1}{2} \times 0{\cdot}1° = {\cdot}05$ higher, or the corrected temperature is 23 6°.†

Expressing that the gain of heat by calorimeter and contents = loss by hot water, and putting w for the water equivalent to be found—

$$(w + 100)\,(23{\cdot}6 - 15) = 40(48 - 23{\cdot}6)$$

* In mere demonstration experiments it is easier to work with small quantities, but, if exact results are required, then larger quantities, say not less than 500 grammes of water, should be used. Otherwise the corrections are too large a fraction of the whole effect.

† In accurate work much more care must be taken with the correction for loss of heat. A description of the mode of doing this will be found in Ostwald's *Physico-Chemical Measurements*, p. 126.

whence $$w = 13{\cdot}5 \text{ nearly},$$

or we may regard the calorimeter, thermometer, and stirrer as 13·5 grammes of water extra.

To find the specific heat of the brass, again let us start with the calorimeter containing 100 grammes of water at 15°. The brass, which we may suppose weighs 150 grammes, has been for some time in the steam-jacketed chamber or heated vessel. For simplicity, let us take its temperature as exactly 100°. It is quickly dropped into the water, and the calorimeter is stirred. Let the temperature readings be—

maximum, ½ minute after mixture . . .	23·8
1 minute after maximum	23·6
2 minutes ,, ,, ...	23·4

Final rate of loss of heat corresponds to fall of 0·2 per minute.
,, ,, ,, 0·1 per ½ minute.
Loss during rise . . . . . . 0·05 ,,
The corrected maximum is therefore 23·85°.

If s is the specific heat of brass on the average of the range from 100 to 23·85, then

Heat lost by brass = Heat gained by calorimeter

gives $$150s(100 - 23{\cdot}85) = (100 + 13{\cdot}5)(23{\cdot}85 - 15),$$

whence $$s = 0{\cdot}0879.$$

Regnault's Determinations by the Method of Mixtures.—*Experiments on Solids.* Regnault used the method of mixtures for the

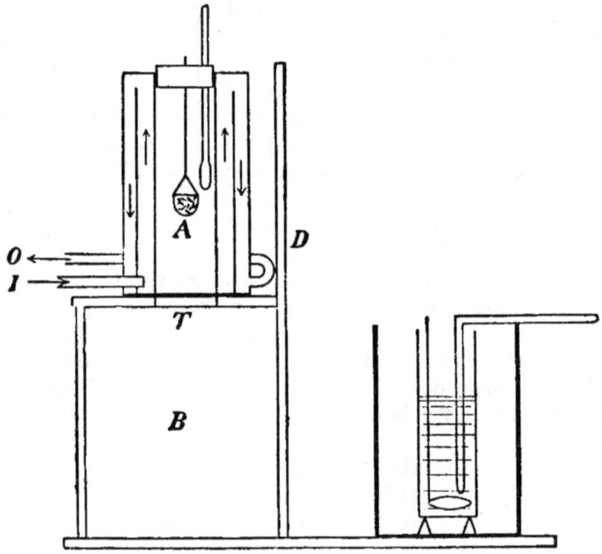

FIG. 53.—Regnault's Apparatus for Determination of Specific Heat of Solids by Method of Mixtures.

determination of the specific heats of a large number of solids, liquids, and gases. When working with solids he used an apparatus, the principle of which is shown by Fig. 53.

The substance to be experimented on was broken in pieces, and placed in a little metal basket hung in the steam-jacketed chamber A, placed on a box B. A was closed above by a cork, through which passed the thread supporting the basket, and a thermometer with its bulb close to the basket, so as to indicate the temperature of the contents. The steam was admitted at I and taken out at O. Below, the chamber was closed by a trap-door T, in the top of the supporting box. Steam was passed through the jacket surrounding A for one or two hours, so that the substance was at the temperature of the steam. The side D of the box was prolonged upwards to screen the steam-chamber from the calorimeter. A trap door moving vertically in this side was drawn up when the substance was sufficiently heated, and the calorimeter C was pushed in so as come immediately under A; the trap-door beneath A was opened, the thread cut, and the basket dropped into the calorimeter. The calorimeter was then withdrawn, and the door in D at once shut down. The rise in temperature of the calorimeter could then be noted, and by subsequent observations the loss of heat to the surroundings could be determined and allowed for. The capacity of the basket was, of course, found by subsidiary experiments, and the heat which it gave to the calorimeter was subtracted.

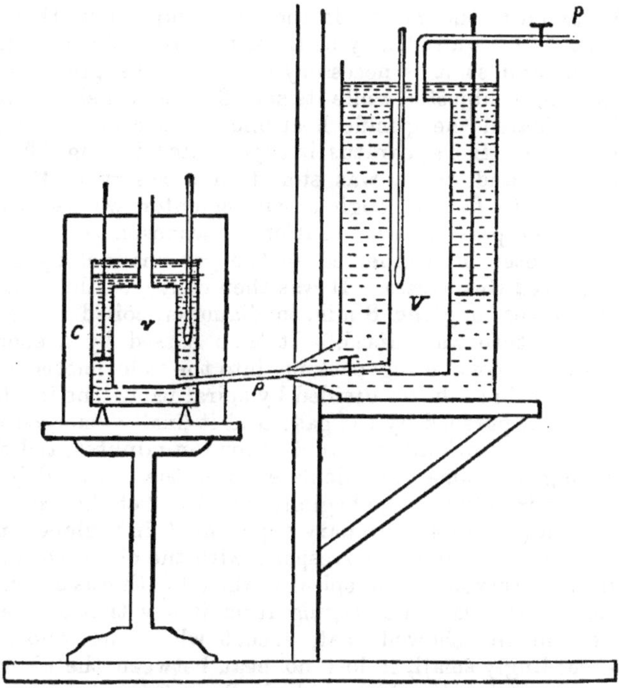

FIG. 54.—Regnault's Apparatus for Determination of Specific Heat of Liquids by Method of Mixtures.

Experiments on Liquids.—One form of apparatus which Regnault used to determine the specific heat of liquids is represented in Fig. 54. The actual mixture of the liquid with water was not permissible in many cases, and therefore a thin metal vessel was fixed within the calorimeter. Into this the liquid was poured, and ultimately it came to temperature equilibrium with the water in the calorimeter. Virtually, then, the method may be described as one of "mixture." The liquid was initially brought to some desired temperature, above or below that of the calorimeter, in a vessel V contained in a constant-temperature bath, placed close to the calorimeter, but screened from it by a badly conducting partition. A pipe p, with a stopcock in it, led from this vessel into the vessel v, within the calorimeter C, and when "mixture" was to take place the cock was turned on and pressure applied through the pipe P to the surface of the liquid, which was then forced into the calorimeter. Besides the heat brought into the calorimeter by the liquid some would be conducted by the connecting pipe, but this could be determined and allowed for.

Experiments on Gases.—As we have seen, the expansion of a gas with rise of temperature depends on the pressure to which it is subjected. In the expansion, the surrounding material is pressed out, and heat has to be given to the gas to do the work implied in this pressing out. The heat thus required may be a very appreciable fraction of the whole heat given, and so it is necessary to specify the pressure condition to which the gas is subjected while its specific heat is being found. Regnault only investigated the specific heat under one condition, viz., that of constant pressure. His apparatus is represented in Fig. 55. The gas, carefully purified and dried, was stored in a reservoir R, from which it was allowed to flow through a gas-regulator worked by hand, so that its excess of pressure over that of the atmosphere was constant. A water-manometer M, connected to the gas channel by a very narrow tube, indicated this excess. It was then conveyed through a spiral metal tube, 10 metres long and 8 mm. in diameter, coiled in an oil-bath, where its temperature was raised. It then passed by a short tube surrounded with non-conducting packing into the calorimeter, which consisted of a series of brass boxes divided by spiral partitions inside, so as to lengthen the path pursued by the gas; and it finally emerged into the air.

The gas was allowed to flow for ten minutes, and the quantity flowing during that time was calculated from the observed fall of pressure in the reservoir between the beginning and end of the experiment. By collecting the gas in a subsidiary experiment in a globe, and weighing it, the weight was found to correspond with the observed difference of pressure in the reservoir. The spiral in the oil-bath was so long that the temperature of the gas on emerging from it was that of the oil, and subsidiary experiments showed that, except when the velocity of the gas was exceedingly small, it lost no heat between the oil-bath and the calorimeter, and entered the calorimeter at the temperature of the oil. It left it at the temperature of the calorimeter. Its pressure at entry and emergence was shown by subsidiary experiments to differ by not more than 1 mm. of water. Hence, the pressure was practically constant.* We see then that a known weight of gas at constant pressure was cooled in the calorimeter by an observed mean amount. This was again virtually the method of mixtures. Knowing the capacity of the calorimeter, the experiment enables us to determine the specific heat of the gas.

* Searle (*Proc. Camb. Phil. Soc.*, xiii. Pt. V. p. 244) has shown that the heat given up by unit mass of gas would be equal to specific heat at constant pressure × temperature fall, even if there were a considerable difference of pressure between entry and exit.

Corrections had to be made, however, for gain or loss of heat in other ways during the experiment. On the one hand, the calorimeter gains heat by conduction and radiation from the heating part of the apparatus. This Regnault * assumed to be the same per minute throughout the experiment. On the other hand, as the calorimeter rises in temperature, it parts with heat by radiation and by conduction to the surrounding air, the quantity lost being proportional to the excess of its temperature. By observing the change of temperature for ten minutes before the gas flows, and for ten minutes afterwards, the quantity of heat conducted and radiated from the heating part of the

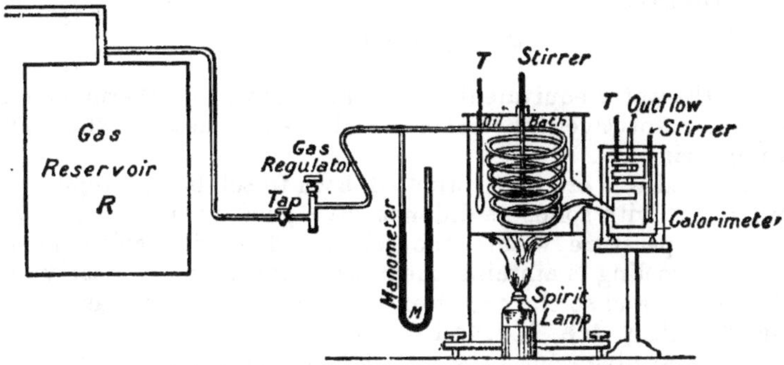

FIG. 55.—Regnault's Apparatus for the Determination of Specific Heat of Gases at Constant Pressure.

apparatus, and the quantity lost to the surroundings per 1° excess of temperature, is calculated, and so the result is corrected.

Neglecting corrections, if W = the weight of gas flowing, T = temperature of the oil-bath, t_1, t_2 = initial and final temperatures of calorimeter, p = water-equivalent of calorimeter and contents, s = Specific Heat of the gas,

$$Ws\left(T - \frac{t_1 + t_2}{2}\right) = p(t_2 - t_1).$$

Determination of the Specific Heat of a Liquid by Mixture with a known Solid.—An obvious mode of determining the specific heat of a liquid consists in heating a known weight of a solid of known specific heat, and immersing it in the liquid contained in a calorimeter. An equation like that of p. 66 (modified, of course, by the necessary corrections) then serves to give us W, the water equivalent of the liquid. Knowing its weight, we have at once its specific heat.

Method of Cooling.—In this method the liquid to be experimented on is placed with a thermometer in a highly polished metal vessel, serving as a calorimeter, and this is suspended by silk threads in a colder enclosure, the walls of which are kept at some constant temperature. This may

* Swann (*Phil. Trans.*, A. 210, p. 231) has pointed out that Regnault's assumption that the conduction from the heating apparatus to the calorimeter is the same during the flow as before and after is an overestimate. The flow of hot gas through the connecting tube will tend to raise the temperature of the part of the tube more distant from the heater, and will so reduce the gradient and lessen the flow of heat. Regnault therefore put down to conduction heat which was really carried by the gas, and so he underestimated the specific heat (see below, p. 86).

conveniently be made 0° C. by surrounding the outside of the enclosure with melting ice. In some experiments the space between the calorimeter and the enclosure has been exhausted. The quantity of heat given out by the surface of the containing vessel in a given time depends not on the nature of its contents, but on the temperature of its surface alone. If, for example, in two different cases, the temperature is observed to fall at twice the rate in one case that it falls in the other when the mean temperature is the same, the heat given out in the same time is the same in the two cases, therefore the capacity of the calorimeter and its contents in the first case is only half as great as in the second case.

Hence we have

$$w + m_1 s_1 = \tfrac{1}{2}(w + m_2 s_2)$$

where w is the water equivalent of the calorimeter and thermometer, and $m_1 m_2 s_1 s_2$ are respectively the masses and specific heats of the liquids in the two experiments.

The method was originally applied both to solids and liquids, but it was found that with solids it did not give good results, owing to variations of temperature within the solids. The circulation in liquids during their cooling maintains their temperature more nearly uniform at each instant, and so the objection is much less in their case.

The Melting Ice Method—*Bunsen's Ice Calorimeter.*—In this method the heat given out by a body in cooling down from some higher temperature to 0° C. is measured by the quantity of ice which it will melt. It was first used by Black and afterwards by Lavoisier and Laplace. They collected and weighed the water resulting from the melting. But as it was practically impossible to collect the whole of the water, the method failed to give very good results. It was, however, modified by Bunsen in such a way as to make it of very great service. In his calorimeter, instead of collecting the water resulting from the melting, the contraction which takes place in the change from solid to liquid is observed. The amount of this change was measured by a separate experiment in which a known weight of ice at 0° was contained in a bulb, the rest of the space being filled with mercury. The ice was then melted to water at 0°, mercury being drawn into the bulb to occupy the space left by the ice in melting; the additional weight of mercury gave the contraction. He found that a gramme of ice at 0° contracted from 1·09082 cc. to 1·00012 cc. of water at 0°.*

The construction of the calorimeter is illustrated by Fig. 56.

A is a test-tube fused into the glass vessel B, which is continued into the narrow tube C. B is nearly filled with water which has been previously deprived of air by boiling, and the remaining space is occupied by mercury, which also fills the tube C round the bend and some way along the horizontal part which lies against a graduated scale. To prepare the calorimeter for use, a stream of alcohol, cooled by a

* Barnes (Physical Constants of Ice, *Trans. Roy. Soc. Canada,* Ser. III., vol. iii., Sect. III., 1909–10) gives a summary of the work on Ice, and concludes that the best value for the density of natural ice is 0·91704, while that of artificial ice is 0·91676. He takes as the best value of the latent heat of fusion a determination by Professor A. W. Smith in 1903 corrected to 79·818.

freezing mixture, is passed through the test-tube A until a cap of ice is formed round it in the vessel B. The calorimeter is now placed in melting ice or snow, and left, preferably for some days, till the ice cap is all at 0°. A layer of wool is placed at the bottom of A to prevent breakage, and some distilled water is added. When all is at 0°, the calorimeter—still surrounded by melting ice—is ready for use. It is better, as pointed out by Professor Boys, *Phil. Mag.*, xxiv., 1887, p. 214, to surround the calorimeter by an air-jacket by enclosing it in an outer tube, which is immersed in the melting ice. A small piece of the substance to be experimented on is raised to any desired temperature, and then dropped into A, giving its heat to the water. But as the water rises in temperature it becomes more dense, and therefore remains at the bottom of the tube round the substance. Gradually heat is conducted out into B, where some of the ice is melted by it, and the melting continues till all is again at 0°. The tube C having been calibrated, the recession of the mercury along it gives the weight of ice melted, a contraction of $1{\cdot}09082 - 1{\cdot}00012 = {\cdot}0907$ cc., corresponding to the melting of 1 gramme. For comparative measurements it is only necessary to observe the contraction in different cases, but for the determination of the amount of heat actually yielded, it is necessary to know the latent heat of water or the number of calories taken up by 1 gramme of ice at 0° in melting to water at 0°. By using a known weight of water at a known boiling temperature, that is, by adding a known number of calories to A, Bunsen determined the latent heat of water as 80·025.* The calory, in terms of which this is expressed, is $\frac{1}{100}$ of the heat required to raise 1 gramme of water from 0° to 100°. It will be observed that in this method the radiation correction disappears, for the substance taking in the heat always remains at the same temperature, and is always in equilibrium with its surroundings.

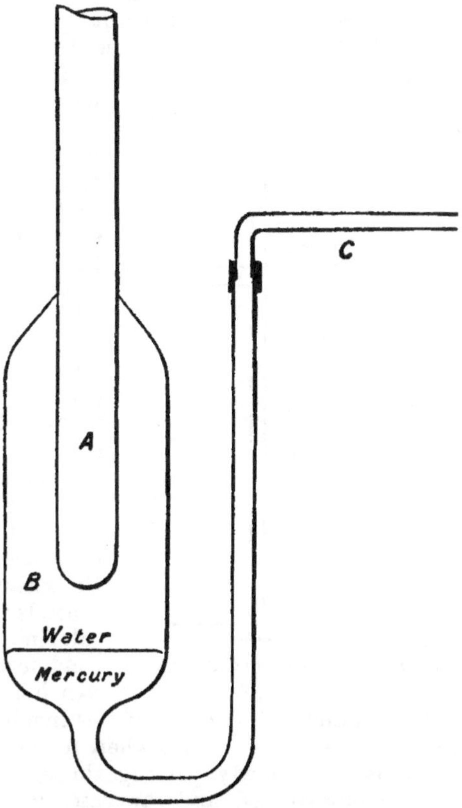

FIG. 56.—Bunsen's Ice Calorimeter.

Joly's Steam Calorimeter.—Professor Joly has introduced and

* In Ostwald's *Physico-Chemical Measurements*, p. 136, will be found a description of modifications of the calorimeter to make it more convenient in use.

perfected a calorimeter in which the latent heat of steam is used somewhat as the latent heat of water is used in Bunsen's ice calorimeter.* The instrument is represented in Fig. 57. The calorimeter C is a light metal vessel, double walled and covered with cloth, placed beneath a delicate balance which is not represented in the figure. A wire hangs from one arm of the balance, and passing through a very small hole at the top of the calorimeter sustains a light wire platform *wp* within it. Beneath and attached to the platform is a catchwater *cw* of platinum foil. The substance of which the specific heat is to be found is placed on the wire platform and counterpoised. There is a wide pipe S entering at the top of the calorimeter through which steam can be introduced. During the counterpoising, the steam pipe is not connected to the calorimeter and the entrance hole S is plugged up. There is a large exit hole at the bottom of the calorimeter, which is also stopped at this stage by a cap *c*, shown in the figure out of position. Before an experiment a thermometer is inserted in the calorimeter, and left till all has come to a steady temperature t_1. Meanwhile steam is got up in the boiler, and during the experiment it must be coming off very freely. It is passed through the coupling tube for some time before connection to the calorimeter, to drive out all air. Then the thermometer in the calorimeter is read and withdrawn, and the hole for its insertion is plugged up. The exit at the bottom is unstopped, the entrance for the steam is unplugged, and the steam pipe connected. For 30 or 40 seconds the steam rushes freely down the calorimeter, driving out all the air, and condensing on all the surfaces. Then the exit hole is nearly closed, and in from one to four minutes the whole of the inside of the calorimeter rises to the temperature of the steam, when no more should condense. While the steam is still slowly passing through the calorimeter, the balance is again counterpoised, and the gain in weight gives the weight of steam condensed by the platform, catchwater, and substance in rising from t_1 to the temperature t_2 of the steam. A previous experiment has given the weight condensed by the platform and catchwater, and the excess above this is due to the substance alone.

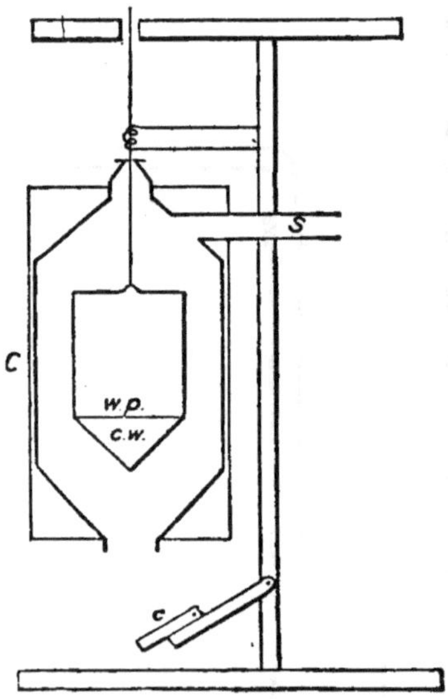

FIG. 57.—Joly's Steam Calorimeter.

If W is the weight of the substance, s its specific heat, ω the weight of steam condensed by it in rising from t_1 to t_2, λ the latent heat of steam, then s is found from

$$Ws(t_2 - t_1) = \omega\lambda.$$

Loss or gain of heat by radiation during the rise in temperature is almost entirely eliminated by the sudden inrush and condensation of

* The perfected form is fully described in *Proc. R. S.*, xlvii., 1890 p. 218

steam on all the surfaces, which very rapidly attain the temperature t_2. There is, however, a slight continuous radiation from the suspended part, and a corresponding continuous condensation and gain of weight which must be determined and allowed for. In estimating ω, correction must be made for the difference in buoyancy of air at t_1 and steam at t_2.

To keep the wire free from the sides of the hole through which it enters the calorimeter, and yet to keep that hole sufficiently small to prevent serious leakage of steam, the following construction is adopted. The top of the calorimeter is conical, and is ground flat so as to make a circular hole. On this rests a little copper disc weighing about 22 mgm., and drilled centrally with a hole about $\frac{2}{3}$ mm. in diameter. The suspending wire of platinum, which may be about 0·1 mm. in diameter, passes down through the central hole in the disc, and as the wire swings from side to side it pushes the disc about with it until the swings have diminished to less than the diameter of the hole, when the disc is left and the wire finally hangs centrally. Its weight suffices to prevent lifting by the steam. A platinum spiral in an electric circuit surrounds the wire, and during the experiment this is made to glow just visibly. The heat it gives to the suspending wire suffices to prevent the condensation of steam on the wire.

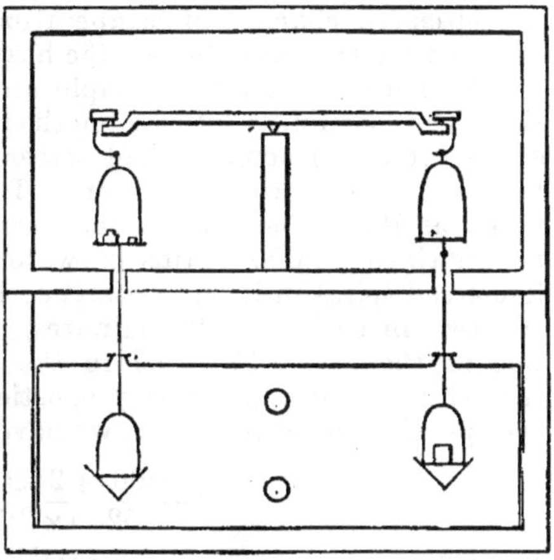

FIG. 58.—Joly's Differential Steam Calorimeter.

The Differential Steam Calorimeter.—Dr. Joly has devised a form of calorimeter represented diagrammatically in Fig. 58. In this a platform or holder and catchwater depends from each arm of the balance into a common steam chamber. The two holders are made to have equal thermal capacities, so that if the substance to be tested is placed on one, the excess of condensation on that side is due entirely to the substance. The corrections for the holder and catchwater and also for the slow radiation from them are eliminated.

This apparatus is the only one which has as yet been used to measure the specific heat of gases at constant volume quite directly.* For this purpose Joly used two equal copper globes, each 6·7 cm. in diameter and about 160 cc. capacity. In the experiments on air, one of them was filled with air which, at the lower temperature, was at normal pressure, while the other was filled with air at a much higher pressure. In some cases the mean higher pressure during an experiment was 26 atmospheres. The excess of condensation on the one side was due to the excess of weight of air on that side, the volume of which was constant, except for the expansion of the copper with rise of temperature and its

* *Phil. Trans.*, A., 1891, p. 73.

extension with increase of internal pressure. These could be determined and allowed for, and the heat needed to raise a known weight of air at constant volume from about 15° to the temperature of steam was determined. We shall give some of Joly's results later.

The Method of Electrical Heating.—This method was first suggested and used by Joule * after he had discovered the law of heat development in a wire carrying an electric current, viz., that the heat is proportional to C^2Rt where C is the current, R the resistance, and t the time for which the current runs.

Joule passed the same current through two similar platinum spirals, one immersed in a standard calorimeter containing water, and the other in the calorimeter containing the substance to be experimented on. The heats developed were proportional to the resistances of the spirals. These were measured beforehand. The heats could also be expressed in terms of the heat capacities of the calorimeters and their contents and the observed changes of temperature, and the equation of the two sets of measurement gave the specific heat.

We may take as an example an experiment described by Joule to show the possibility of the method. The standard calorimeter, with stirrer and thermometer, had water equivalent 280 grains and it contained 35,000 grains of water. The immersed spiral had resistance taken as 100. The other calorimeter had water equivalent 260 grains; it contained 28,000 grains of water and 80,500 grains of lead. The immersed spiral in it had resistance 106. A current was passed through the two in series for 20 minutes, when the increase of temperature was in the first 3·575° and in the second 4·35°. Expressing the fact that the heat developed is proportional to the resistance, and denoting the specific heat of lead by s, we have

$$\frac{(80500s+28260)4\cdot35}{35280\times3\cdot575}=\frac{106}{100}$$

whence

$$s=0\cdot03073$$

This method has since been used by many experimenters, and has been brought to great perfection by Griffiths in his research on the mechanical equivalent of heat, to be described hereafter. It is not necessary to employ two calorimeters. For, under proper conditions, both the current and the resistance can be measured with very great accuracy, and the value of C^2Rt is thus known in different experiments, which may be carried out successively with the same calorimeter. Equating the ratio of the heats developed in two experiments as expressed in terms of heat capacity and rise of temperature to the ratio of C^2Rt in the two experiments, we have the data for the determination of specific heat.

The Specific Heat of Water.—Nearly all determinations of specific heat have been made in terms of a water unit, such as the heat required to raise 1 gramme of water through 1° C. at some named part of the scale. It is, therefore, of the utmost importance to determine whether the choice of the particular degree affects the value of the unit, that is, to determine whether the specific heat of water itself varies as the temperature changes.

* *Scientific Papers*, vol. i. p. 192.

It has long been recognised that, at any rate, there is no evidence for its constancy, but determinations of the nature and magnitude of the variation have been so conflicting till recently that experimenters have been too often content to leave it out of account. But measurements of temperature and quantity of heat have so greatly advanced in accuracy, that now the results of different workers begin to show agreement, and there is no longer any doubt as to the existence of a variation, and even its magnitude at ordinary temperatures is probably fairly determined.

Before giving an account of more trustworthy work, we may illustrate the difficulties of exact calorimetry by briefly describing two earlier researches.

In 1847 Regnault published an account of experiments in which he sought to determine the specific heat of water at different temperatures by the method of mixtures. A large boiler was so arranged that the water in it could be boiled under different pressures, and therefore at different temperatures, these ranging from 107° to 190°. When the water was at the boiling temperature a quantity of about 10 kgm. was rapidly forced into a large gauged calorimeter containing a known quantity—about 100 kgm.—of cold water, at an observed temperature ranging from 8° to 14°. From the temperature of the mixture the ratio of the mean specific heat over the rise of temperature in the one case to that over the fall in the other could be determined. A series of experiments led Regnault to express the specific heat at $t°$, taking the specific heat between 0° and 1° as 1, by

$$\text{Specific heat at } t° = 1 + \cdot 00004t + \cdot 0000009t^2.$$

From this the specific heat at 15° is 1·0008, while the mean specific heat from 0° to 100° is 1·005.

In 1870 Jamin and Amaury described an electrical method. The calorimeter was of thin copper and contained 350 grammes of water. It was surrounded by a spiral of insulated German-silver wire of known resistance, through which a known current could be passed, to supply a known quantity of heat. Outside the spiral was a layer of swansdown of such low conductivity that nearly all the heat generated was conducted inwards to the water. Outside the swansdown packing was a thin polished copper vessel. This was placed in the middle of a double-walled enclosure, containing water between the walls. In this water was another spiral through which a current could be passed, so as to make the temperature of the enclosure rise at the same rate as that of the outside of the calorimeter, and thus eliminate any radiation correction. The results obtained were expressed by

$$\text{Specific heat at } t° = 1 + \cdot 0011t + \cdot 0000012t^2.$$

We shall realise how widely this differs from Regnault's value by noting that it gives the specific heat at 15° as 1·0168, and the mean specific heat from 0° to 100° as 1·06.

The first results to which any value can now be attached were published by Rowland in 1879 * in an account of experiments on the mechanical equivalent of heat, to be described later. Here it is sufficient

* *Physical Papers*, p. 343. A recalculation of the results in terms of the scale of the Paris hydrogen thermometer is given by Dr. W. S. Day, *Phil. Mag.*, July 1898.

to say that the heat was supplied to water in a calorimeter by violent stirring, the energy put in by the stirring being transformed to heat proportional to the work done. Rowland found that the same amount of work produced a different temperature rise in the same quantity of water at different parts of the scale. The range was from 5° to 36° C., and there was distinct evidence for the existence of a minimum specific heat about 30°. The recalculated results are represented in Fig. 60. Rowland verified the variation by experiments by the method of mixtures, from which he found the coefficient of decrease between 0° to 30° to be ·000236.

In 1893 a most careful determination of the variation between 15° and 25° was published by Griffiths,* who used the method of electric heating. The research was really on the mechanical equivalent of heat, but we may regard it here as one on the specific heat of water.

Taking the specific heat at 15° as 1, Griffiths found that if t lies between 15° and 25° then

$$\text{Specific heat at } t^\circ = 1 - \cdot 000266(t - 15)$$

where t is the temperature on the hydrogen scale. This is represented in Fig. 60.

In 1894 † some experiments on the latent heat of steam led Griffiths to suppose that the mean specific heat of water between 0° and 100° is exceedingly near to that at 15°. Since the specific heat is decreasing at 15°, this implies, of course, that it must reach a minimum before 100°, and then increase again. Griffiths' supposition was confirmed by Dr. Joly by an experiment with his steam calorimeter, in which a known weight of water was raised from about 12° to 100°, so that its mean specific heat between 12° and 100° was determined. Griffiths found from Joly's work that

$$\frac{\text{Mean specific heat}}{\text{Specific heat at } 15^\circ} = \cdot 9975$$

instead of $\frac{1 \cdot 005}{1 \cdot 0008} = 1 \cdot 004$ as found by Regnault.

Meanwhile Bartoli and Stracciati ‡ had been carrying out a very extensive series of researches on the specific heat of water by the method of mixtures. They made three sets of experiments. In all the calorimeter contained a known weight of water initially at the temperature of the surroundings. In the first set, different metals of specific heat known through the range used were heated to 100° and dropped into the calorimeter. In the second set, a known weight of water at 0° was dropped in; and in the third set, a known weight of water at a temperature above that of the calorimeter was mixed. The three sets showed results very much of the same nature. The mean is shown in Fig. 60. It should be noted that the metal set gave a

* *Phil. Trans.*, 1893, A., p. 361.

† *Phil. Trans.*, 1895, A., p. 320. Dr. Joly's experiment referred to in the text is described in the paper.

‡ *Nuovo Cimento*, 32. A brief account is given in *Beiblätter*, xvii., 1893, pp. 542, 638, 1038.

gentler slope down from 0° to 20°, and a steeper slope up after that than the water sets. Thus, at 0° the metal mixtures gave 1·00551, and the water mixtures 1·00777; while at 31° the metals gave 1·00337 and the water 1·00145. The mean of these results gives a slope from 0° to 15°, very nearly the same as those obtained by Rowland and Griffiths (see Fig. 60), but Bartoli and Stracciati found a minimum at 20°.

Lüdin in 1895 (*Beiblätter*, 1897) also used the method of mixtures, and his results are represented on Fig. 60.

Dr. Barnes has made the most complete series of experiments up to the present (*Phil. Trans.*, A. 199, 1902, p. 149). The method, that of electric heating, was suggested by Prof. Callendar, and the apparatus was devised by him (*Phil. Trans.*, *loc. cit.*, p. 55). But as Prof. Callendar, with whom Dr. Barnes was associated at first, was unable to continue the experiments, they were carried out by Dr. Barnes. In this method a stream of water is led through a narrow tube t (Fig. 59), through which passes a fine platinum wire. This wire carries an electric current introduced and taken away by the thick wires *cc*. The temperatures of the water on entering and leaving t are taken by the platinum thermometers

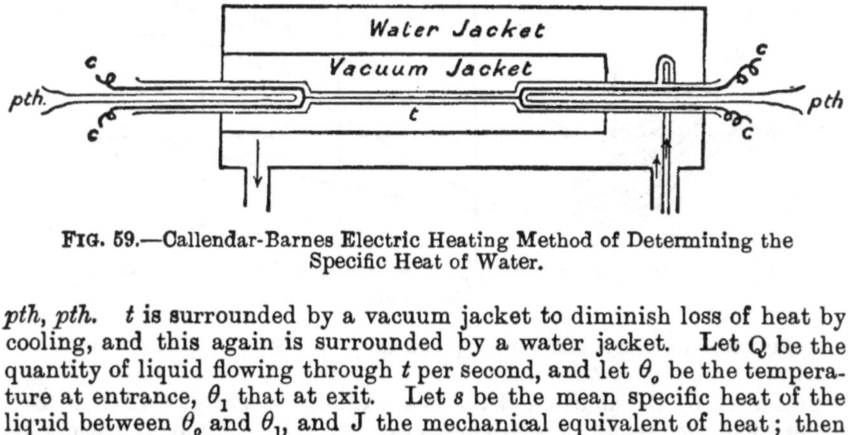

Fig. 59.—Callendar-Barnes Electric Heating Method of Determining the Specific Heat of Water.

pth, *pth*. t is surrounded by a vacuum jacket to diminish loss of heat by cooling, and this again is surrounded by a water jacket. Let Q be the quantity of liquid flowing through t per second, and let θ_0 be the temperature at entrance, θ_1 that at exit. Let s be the mean specific heat of the liquid between θ_0 and θ_1, and J the mechanical equivalent of heat; then the work value of the heat gained by the water is $JQs(\theta_1 - \theta_0)$. But if E is the potential difference between the ends of the fine wire, and C is the current in it, and if, for simplicity, we suppose all the heat given by the current to remain in the water, the heat is EC in work measure. Hence we have

$$JQs(\theta_1 - \theta_0) = EC,$$

and measuring Q, θ_1, θ_0, E and C, we can determine s. We must refer the reader to the original papers for the account of the various corrections and their determinations. There is a minimum value at about 40° C. Callendar (*Phil. Trans.*, A. 199, p. 142) gives the following formula for the specific heat: From 0° to 20° C., $s = {\cdot}9982 + {\cdot}0000045(t-40)^2 + {\cdot}00000005(20-t)^3$. From 20° to 60° C. the last term is omitted and $s = {\cdot}9982 + {\cdot}0000045(t-40)^2$. From 60° to 200° C., $s = {\cdot}9944 + {\cdot}00004t + {\cdot}0000009t^2$ (Regnault's formula corrected).

Fig. 60, representing all the results, is taken from his paper.

General Results. — Probably in all substances the specific heat changes with the temperature. In general it increases as the temperature rises, so long as the substance does not change its state. A specific heat increasing uniformly with rise of temperature, would be represented by

$$s = \alpha + \beta t$$

but it is only over small ranges that such a simple formula represents the results of experiments. Over larger ranges they may be better represented by

$$s = \alpha + \beta t + \gamma t^2$$

and no doubt, if the results were accurate, they would be still better

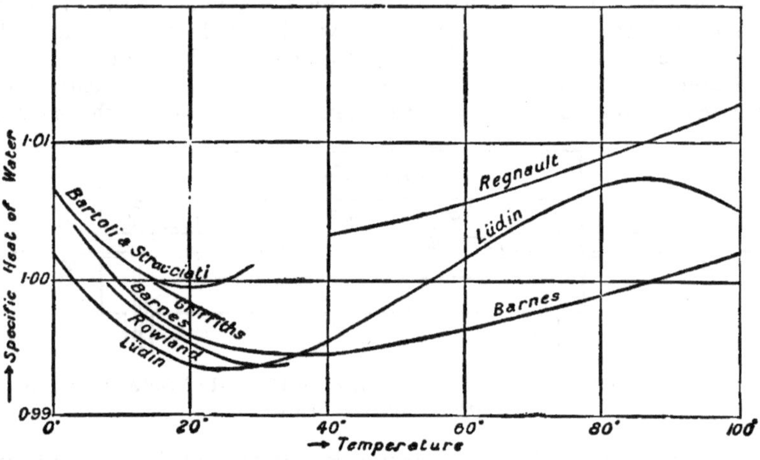

FIG. 60.—Results of Different Experiments on the Specific Heat of Water.

represented by the addition of terms containing higher powers of t. But at present the errors of experiment are so great that it is useless to trouble about these higher powers.

We probably have the most accurate knowledge of the variation of specific heat in the case of aniline over the range from 15° to 52°. This was determined by Griffiths (*Phys. Soc.*, 13, 1894) by the method of electrical heating. He selected aniline in place of water, on account of its great suitability as a calorimetric liquid, for it is easily obtainable in a fairly pure state, it has a low vapour pressure at ordinary temperatures, and with its specific gravity, 1·02, and specific heat, ·52, its capacity for heat per unit volume is only half that of water. The specific heat at $t°$ is given by Griffiths in terms of the calory at 15° as

$$s = \cdot 5156 + (t - 20) \cdot 0004 + (t - 20)^2 \cdot 000002.$$

Regnault has obtained the specific heat of a number of liquids. From these we may select alcohol, for which he gives, in terms of the calory at 0°,

$$s = \cdot 54755 + \cdot 002242t + \cdot 000006618t^2$$

Alcohol may be used in the inner tube of the Bunsen calorimeter in place of water, for the determination of specific heats below 0°.

The change in specific heat of a number of metals with change of temperature has also been found. Among the most important of these is platinum, on account of its use to determine high temperatures. Violle,* using the method of mixtures, has obtained for this metal over the range 0° to 1200°

$$s = \cdot 0317 + \cdot 000012t.$$

Naccari † has found the value for various metals by the method of mixtures, between ordinary temperatures and 320°, in the form

$$s = a(1 + bt)$$

The following results are given by him. The unit is the calory at 0°:—

	a	$10^6 b$
Copper	·09205	230·8
Silver	·05449	392·9
Aluminium	·21116	449·3
Lead	·02973	456·9
Zinc	·09070	489·5
Nickel	·10427	907·0
Iron	·10442	1029·1

Tilden ‡ has investigated the specific heat of iron, nickel, cobalt, aluminium, silver, gold, and platinum, over the range from − 182° C. to + 100° C., extended in some cases to 630° C., and his results show that the specific heat decreases as the temperature of determination decreases. In the case of platinum the decrease is regular, or the relation between specific heat and temperature is linear. But in other cases the decrease is more and more rapid as the temperature falls.

The most remarkable changes of specific heat are those which occur with carbon, boron, and silicon. These were investigated by H. F. Weber.§ He used the Bunsen Ice Calorimeter from − 50° to + 250°, employing alcohol in the inner tube below 0°, and for carbon he extended the experiments to the range between 600° and 1000°, when he used the method of mixtures. His plan at this higher range consisted in heating a known weight of platinum and the carbon to the same temperature and dropping the two simultaneously, one into each of two water calorimeters. The platinum gave the temperature, while the rise in the other calorimeter, when the temperature was thus known, gave the specific heat sought. His results for diamond between − 50° and + 250° are nearly represented by

$$s = \cdot 104 + \cdot 0008t,$$

though the rate of increase is appreciably diminishing as the temperature

* *Phil. Mag.*, vol. iv., 1877, p. 318.

† *Atti R.A. de Torino*, 23, 1889; *Beiblätter*, xii., 1888, p. 326.

‡ *Phil. Trans.*, A. 194, p. 233, 1900, and A. 201, p. 37, 1903. References to other work will be found in these papers.

§ *Pogg. Ann.*, cliv., pp. 367 and 553.

rises. In the range from 250° to 600°, in which he did not experiment, the diminution must have become much more rapid, for he found that the specific heat in the range from 600° to 1000° only very slowly increased towards ·46.

With graphite, the specific heat about 0° may be represented nearly by

$$s = \cdot 152 + \cdot 0007t.$$

The rate of increase falls off somewhat more rapidly than with diamond, and the values for the two are not very different over the higher range, as the following table shows:—

SPECIFIC HEATS OF DIAMOND AND GRAPHITE.

	606·7°	806·5°	985·0°
Diamond . .	·4408	·4489	·4589
Graphite . .	·4431	·4529	·4674

With crystallised boron the specific heat rises in a similar way, and between 0° and 250° it is fairly, but not exactly, represented by

$$s = \cdot 22 + \cdot 00071t.$$

Here again the rate of increase falls off as t rises, and from the exact results over the range Weber deduced a limiting value at high temperatures of 0·50.

With crystallised silicon the limiting value was nearly reached in the lower range of experiments. We may take the following values:—

Specific heat of crystallised silicon at		−50° .	·13
,,	,,	0 .	·16
,,	,,	50 .	·18
,,	,,	100 .	·195
,,	,,	200 .	·202

and the value to which the results tend as 0·205. We shall see the bearing of these values directly.

Dewar,* working at the same time as Weber, also found that the specific heat of gas-carbon increased very considerably with rise of temperature. Between 20° C. and 1040° C. he obtained a mean value of 0·32, and between 20° C. and 2000° a mean value of 0·42, concluding that at 2000° it must be at least 0·5.

Dewar † has also found the specific heat of diamond, graphite, and a number of other substances at low temperatures by means of a liquid air or a liquid hydrogen calorimeter. The calorimeter consists of a vacuum vessel of 25 to 50 c.c. capacity, containing liquid air or liquid hydrogen, and it is immersed in a large vacuum vessel containing the same liquid. From the calorimeter a long narrow tube rises up, and to its end a small test tube containing the substance to be experimented on is attached by a short rubber tube. The rubber tube acts as a valve cutting off the test tube from the calorimeter, except when the experiment is to be made.

* *Phil. Mag.*, xliv., 1872, p. 461.

† *Proc. R.S.*, A. 76, 1905, p. 325.

When the test tube is held vertical, the substance, which has been brought to any desired temperature, drops into the calorimeter, and some of the liquid boils off. A branch tube leads from near the upper end of the vertical tube, and the evolved gas, passing through it, is collected in a receiver and measured. The arrangement is similar in principle to V. Meyer's Vapour-Density Apparatus (Fig. 102, p. 177). A gramme calory was found to evolve 13·2 c.c. of oxygen, 15·9 c.c. of nitrogen, and 88·9 c.c. of hydrogen, measured in each case at 0° C. and 760 mm. The observations were reduced by comparison with lead, of which the specific heat had previously been found to increase very nearly uniformly from 0·0280 at − 220·5° C. to 0·0295 at − 85° C. After an experiment had been made with the substance under investigation, a similar experiment was made with a quantity of lead, so chosen that about the same quantity of gas was evolved.

The specific heats of a number of substances at various ranges down to − 188° C. were determined, and the following results were found for diamond, graphite, and ice:—

SUBSTANCE.	18° to − 78°.	− 78° to − 188°.	− 188° to − 252·5°.
Diamond . . .	0·0794	0·0190	0·0043
Graphite . . .	0·1341	0·0599	0·0133
Ice	0·463	0·285	0·146

Referring to the original paper for details, we may note here that Dewar was able to determine with the calorimeter the latent heats of oxygen, nitrogen, and hydrogen. He also found that the specific heat of liquid hydrogen is 3·4, the value which it has in the gaseous condition.

Influence of Change of State on Specific Heat.—The difference between the specific heats of diamond and graphite is an illustration of the fact that the particular condition of a substance, crystalline or amorphous, softened or hardened, affects its specific heat, though, as a rule, the variation with such condition is not great. But as a substance changes from the solid to the liquid or from the liquid to the gaseous state, the specific heat may change very considerably. As typical, we may take water and lead. We have:—

Ice .	·5	Water	1·0	Steam (constant pressure)	·34
Solid lead	·0314	Molten lead	·0402		

As a rule, the change is in the same direction for other substances, but there is no known relation between the specific heats of the same substance in the different conditions.

Atomic and Molecular Heats.—There is undoubtedly some relation between the specific heat of a substance and its atomic or molecular weight. Thus for the less condensable simple gases the product specific heat × atomic weight is very nearly constant for different gases and independent of the temperature, when the specific heat is taken in all cases at constant pressure, or in all cases at constant volume. Again, for a very large number of solid elements the product is not far from constant, though the constant is not the same as for gases.

These relations imply that if we take quantities of different elementary substances proportional to their atomic weights, and therefore presumably containing the same number of atoms, the heat capacities of these quantities will be nearly equal. Or the heat capacity per atom is the same for different elements, the gas atom, however, having a different capacity from the solid atom. The product atomic weight × specific heat is termed the "atomic heat."

In many classes of compounds, too, the product specific heat × molecular weight, the "molecular heat," is not far from constant, each class having as a rule its own constant, though in some cases each constituent atom may be regarded as having its atomic heat—as a gas if the compound is gaseous, as a solid if it is solid.

Without entering into minute detail, we shall give examples of the evidence on which these statements are founded.

Taking first the case of gaseous bodies. Regnault found by the method already described that at constant pressure:—

1. The specific heat is nearly independent of the temperature.
2. It is nearly independent of the pressure so long as this is constant during an experiment.
3. The capacities for heat of equal volumes of different gases at equal pressures are nearly equal—whence on the molecular hypothesis the capacity for heat of different molecules is the same.

But the further a gas departs from the behaviour indicated by $pv = R\theta$ the less nearly does it fall in with these laws.

The following are some of Regnault's results. The value for nitrogen was calculated from those for air and oxygen:—

Simple Gases.

Gas.	Atomic Weight.	Specific Heat.	Atomic Heat or Product of Atomic Weight × Specific Heat.
Hydrogen . . .	1	3·409	3·409
Air	—	0·2374	—
Oxygen	16	0·2175	3·4800
Nitrogen . . .	14	0·2438	3·4132
Chlorine . . .	35·5	0·1210	4·2955
Bromine . . .	80	0·0555	4·4400

Compound Gases.

Gas.	Molecular Weight.	Specific Heat.	Molecular Heat or Product of Molecular Weight × Specific Heat.
Type AB—			
CO	28	0·2450	6·8600
NO	30	0·2317	6·9510
HCl	36·5	0·1852	6·7598
Type AB_2—			
CO_2	44	0·2169	9·5436
N_2O	44	0·2262	9·9456
SO_2	64	0·1554	9·9528
More easily condensable—			
H_2O . . .	18	0·4803	8·6454
H_2S	34	0·2342	7·9628

The last column confirms the result that for less condensable gases the capacity for heat of equal volumes at equal pressures is constant, for the weights of equal volumes of these gases are then very nearly proportional to their atomic weights if elements, to their molecular weights if compounds. The heat capacities of these equal volumes are therefore proportional to the numbers in the last column. Dividing the "Molecular Heat" for gases of the type AB by 2, and that for gases of the type AB_2 by 3, we get atomic heats 3·3 or 3·4, nearly the same as those found for hydrogen and oxygen, a result indicating that the atoms have the same capacity for heat even in combination. It will be noticed that the atomic and molecular heats for the more easily condensable gases are somewhat widely different from those of the more permanent gases.

E. Wiedemann * has simplified Regnault's method of experiment on gases, and has obtained results in close agreement. His value for air at 0° was 0·2389, and he found no change between 0° C. and 200° C. for air, hydrogen, and carbon monoxide. The capacities for heat of equal volumes of these gases at the same pressure were nearly the same, but the more condensable gases gave very different values from the others, and as their temperature rose the specific heat increased.

No doubt the specific heat at constant volume is the more appropriate quantity for comparison, inasmuch as at constant pressure a certain amount of the heat goes to push out the containing surface against the external pressure. As we shall see later, the ratio of the two specific heats is nearly constant for air, hydrogen, oxygen, and nitrogen, so that we may regard the results obtained in the one case as

* *Phil. Mag.* (5), 2, p. 81, 1876. The remarks in the footnote on p. 71 apply also to Wiedemann's experiments.

proportional to those which may be expected in the other. But only in very few cases have the specific heats at constant volume been worked out, and as yet only by Joly with the steam calorimeter.* He has found that the specific heat alters with the density. Denoting the density by ρ, he found that for air about a mean temperature of 50° specific heat at constant volume $= 0\cdot17151 + 0\cdot02788\rho$, giving 0·17154 as the value at 0° and 760 mm. For carbon dioxide about a mean temperature of 55° and a range of pressure up to 80 atmospheres, he found $0\cdot1650 + 0\cdot2125\rho + 0\cdot340\rho^2$. For hydrogen he found 2·40, with decided indications of a decrease with increasing density. By varying the initial temperature he was able to determine that there was no change in the specific heat of carbon dioxide at constant volume between 10° C. and 100° C. if the density did not exceed ·08, but that there was a rapid increase at higher densities as the temperature decreased below 30°.

Swann† has employed the method of electrical heating to determine the specific heats of air and carbon dioxide at atmospheric pressure. A measured quantity of the gas was driven past a coil heated by an electric current, and the energy given by the coil to the gas was measured electrically. The rise in temperature of the gas was determined by platinum resistance thermometers placed in the stream before it came to and after it left the heating coil. Swann found that in terms of the calory at 20° C. the specific heats were

Air at 20° C., 0·24173; at 100° C., 0·24301.
CO_2 at 20° C., 0·20202; at 100° C., 0·22141.

Extrapolating, we obtain for air at 0° C. the value 0·24141, a result nearly 2 per cent. higher than that of Regnault.

Dulong and Petit's Law.—From their researches on specific heat Dulong and Petit were led to conclude that for many solid elements the product

$$\text{Atomic weight} \times \text{specific heat} = \text{constant},$$

a law now known by their name. As pointed out already, it implies that if we take a weight of each element equal in grammes to the number expressing the atomic weight, and therefore, on the atomic theory, containing the same number of atoms, the capacity for heat is the same. Or the heat capacity per atom is constant.

Regnault made a very extensive series of researches to test this law, and found that for most solid elements the product is nearly, but only nearly, constant. This might be expected. For, as we have seen, the specific heat changes with condition and temperature, and we could only expect to find any exact relation for the different elements when we had them in corresponding condition and at corresponding temperature, and we do not yet know in what correspondence consists. On page 87 we give a selection from Regnault's values of the specific heat, adding carbon, boron, and silicon, with the limiting specific heats as given by Weber. The product, Atomic weight × specific heat, the "Atomic Heat," is given in the last column.

* *Phil. Trans.*, A., 1891, p. 73, and A., 1894, p. 943.
† *Phil. Trans.*, A. 210, p. 199, 1910.

Element.	Atomic Weight.	Specific Heat.	Atomic Heat.
Sulphur . . .	32	·1776	5·6832
Phosphorus . .	31	·1887	5·8497
Zinc	65	·0955	6·2075
Aluminium . .	27·5	·2143	5·8932
Iron	56	·1138	6·3728
Nickel . . .	58·5	·1091	6·3823
Tin	118	·0562	6·6316
Copper . . .	63·5	·0951	6·0389
Lead	207	·0314	6·4998
Mercury (solid) . .	200	·0319	6·3800
Platinum . . .	197	·0324	6·3828
Iodine . . .	127	·0541	6·8707
Bromine (solid) . .	80	·0843	6·7440
Sodium . . .	23	·2934	6·7482
Silver . . .	108	·0570	6·1566
Gold	196	·0324	6·3504
Carbon . . .	12	·46	5·52
Boron . . .	11	·50	5·50
Silicon . . .	28	·205	5·74

Neuman first pointed out that in certain compounds of similar constitution the product, Molecular weight × specific heat, or "molecular heat," is nearly constant. Regnault investigated the molecular heats of a very large number of constants, and found that each class had its own molecular heat. The results for the members of a class differ, however, several per cent. from the mean.

Later, Kopp took up the investigation, and found that by assigning to each atom its own atomic heat—not the same for all elements—a very great number of quite different compounds in the solid state come under the following rule (known as Kopp's Law): "The molecular heat of a solid compound is the sum of the atomic heats of the constituents."

The following are the atomic heats assigned: *—

C . . .	1·8	Be . . .	3·7	P . . .	5·4
H . . .	2·3	Si . . .	3·8	S . . .	5·4
B . . .	2·7	O . . .	4·0	Ge . . .	5·5

and for other elements, 6·4.

Thus for ice the molecular heat is, by Kopp's Law,

$$2 \times 2{\cdot}3 + 4 = 8{\cdot}6$$

whereas experiment gives $0{\cdot}474 \times 18 = 8{\cdot}5$.

For calcium carbonate, $CaCO_3$, the calculated molecular heat is—

$$6{\cdot}4 + 1{\cdot}8 + 3 \times 4 = 20{\cdot}2$$

while experiment gives $0{\cdot}203 \times 99{\cdot}9 = 20{\cdot}3$.

For sodium chloride, NaCl, the calculated molecular heat is—

$$6{\cdot}4 + 6{\cdot}4 = 12{\cdot}8$$

while experiment gives $\cdot 214 \times 58{\cdot}4 = 12{\cdot}5$

* Nernst, *Theoretical Chemistry*, p. 154.

CHAPTER VII.

CONDUCTIVITY.

The Passage of Heat from one Body to Another—Conductivity—Differs enormously in different Substances—General Remarks on Conductivity in the Three States—Definition of Conductivity—Diffusivity—Emissivity—Measurements of Conductivity—Péclet's Method—Bar Methods of Despretz, Forbes, Neumann, and Ångström—Gray's Method—Berget's Experiment on Mercury—Experiments of Wiedemann and Franz—Kundt's Experiments—Senarmont's Experiments on Crystals—Lees's Experiments—Lundquist—Weber—Conductivity of Gases—Experiments of Stefan, Kundt, and Warburg.

Transference of Heat by Conduction.

The Passage of Heat from one Body to Another.—There are two modes in which heat is transferred from one portion of matter to another—conduction and radiation. In **conduction**, the matter receiving the heat is in contact with the matter from which it receives it, and the temperature falls continuously along the course by which the heat is flowing. If, for example, I put one end of a poker in the fire and hold the other end, the heat is conducted along the poker from the warmer to the colder portions, the heat passing down the slope of temperature, warming the iron as it travels, so that all the intervening portions of the poker are intermediate in temperature between that of the fire and that of my hand.

In **radiation**, the matter receiving the heat is *not* in contact with the matter from which it receives it. If I warm my hands before a fire, I do so by radiation, the heat received by my hands passing through the intervening air without warming it. In fact, in the case of radiation, any matter through which the radiation passes may be colder or hotter than either or both of the bodies between which it is passing. We cannot, therefore, suppose that the energy passes from one to the other as heat, but that, on leaving the sender, it is converted from heat-energy into another form which we term *radiant* energy, to be reconverted into heat on reaching the receiver.

Conductivity.—We have already noted as the chief characteristic of heat conduction, that the heat always travels from hotter to colder matter. The greater the slope—that is, the difference of temperature between neighbouring points a given distance apart—the greater the amount of heat conducted. If a thermometer at the temperature of the room be placed in a vessel of hot water, it rises much more rapidly at first, when the temperature slope between the water and the mercury is great. The rate of rise gradually slackens till, ultimately, when the mercury and the water are at the same temperature, there is no further passage of heat. Without at present giving a precise significance to the term, we may call this power of conducting heat "Conductivity."

The conductivity varies enormously in different substances. We may, *e.g.*, light a wooden match, and let it burn down nearly to the fingers, without receiving any appreciable quantity of heat through the wood, while an iron wire held by the side of the match rapidly becomes uncomfortably hot, and a copper wire can only be held for a few seconds. One end of a glass-rod may be melted in a flame while the rod is held in the fingers two or three inches away from the melted part, while a copper rod of the same diameter, with one end in the flame, will be too hot to handle at a point many inches from the flame. The difference in conductivity may be illustrated by smearing the two rods with wax, and noting the difference in the times taken by the wax to melt along the two.

The apparent coldness to the touch of metals, as compared with other solids, is explained at once by their greater conductivity. The skin is generally at a higher temperature than the metal, rapid conduction ensues on contact, and the hand loses much heat. If we touch a piece of wood at the same temperature, the amount conducted for the same slope is very much less, and the hand loses heat much more slowly. If the hand, however, is colder than the metal and wood, then it receives more heat from the metal, which feels much hotter than the wood.

If we paste thin pieces of paper on to two blocks, one of wood and one of iron, and hold them with the paper exposed to the flame of a Bunsen burner, the piece on the wood rapidly chars, while that on the iron remains unburnt. The flame may be regarded as supplying nearly the same quantity of heat in each case. Though the paper is itself a bad conductor, it is thin, and a comparatively small difference of temperature between the two sides will establish a sufficient slope to carry away all the heat supplied. If then, on the farther side, there is a good conductor, such as iron, it will rapidly convey away all the heat supplied to it, and will not allow the temperature of the side of the paper in contact with it to rise very high. The other side of the paper, not being very much higher in temperature, does not rise to the charring-point. If, however, on the farther side of the paper, we have a block of wood, it does not rapidly convey away the heat supplied, and so the temperature of the side of the paper in contact with the wood rises, that of the other side rising still higher, and soon the charring-point is reached.

We may illustrate this in another way. If a thin paper cup or tray is constructed so as to hold water, the water in the cup may be boiled over a flame without burning the paper. The water prevents the inner surface of the paper from rising above 100°, while a sufficient slope will be established in the paper by a not very much higher temperature for the outer face, to carry through all the heat supplied by the flame, and the paper, therefore, remains unburnt. If we replace the paper by a thin copper vessel, the copper probably does not rise more than a fraction of a degree above the temperature of the water. As an effect of this, it may easily be observed that the flame does not come in contact with the heated surface. The temperature of the surface remains too low to permit of combustion, and there is a layer of unburnt gas, of greater or less thickness, according to the temperature of the surface with which it is in contact.

In steam boilers we have a somewhat similar effect. A temperature slope is established from the flame to the water within the boiler, but the outside temperature of the metal may be not greatly above that of the water, if the thickness is not very great. If, however, there is a badly conducting incrustation within the boiler, the conditions are altered. In order that the water may receive the same quantity of heat, the outer side of the incrustation must be much hotter than the inner; the metal must therefore be much hotter than the water. It may therefore rise to the temperature at which it is easily burned by the flame, or it may even be softened through heat so much as to give way to the internal pressure.

Another effect of the incrustation may be noted. In order to get the same quantity of heat into the water through the badly conducting wall, a higher temperature may be necessary within the furnace, and then there will be a greater quantity of heat passing through every outlet at which there is waste.

On the other hand, if we surround the outside of a boiler with a badly conducting layer of boiler "clothing" to keep its heat in, a very large difference of temperature between the two sides of the layer will correspond to only a small flow of heat from the boiler, so that though the inside may be at the temperature of the water, and the outside at that of the air, no very great quantity of heat escapes.

Of course, the same principle is used in our own clothing. We put on non-conducting substances of such thickness and quality that the temperature slope from our bodies to the outer air shall not carry off more heat than we are willing to part with.

General Remarks on Conductivity in the Three States.—Common observation tells us that solids differ enormously in their conductivity, the metals being the best conductors. They range from copper and silver, which are the best, downwards through minerals and wood to furs and feathers, which are probably the worst.

Liquids are Bad Conductors.—Though we may readily heat liquids they are in general bad conductors. When we boil water in a vessel, we always supply heat from below, so that the heated water may rise and give place to colder water from above, which is heated in its turn. The hot water rising, comes in contact with much colder water, so that the temperature-slope is very steep, and the heat is soon shared between the hotter and colder portions, even though the liquid is a bad conductor. This process—termed "**convection**"—is therefore really only conduction aided by a transportation of the hotter matter—as it were, an artificial deepening of the temperature-slope. If, instead of heating the water from below, we heat it from above, then it takes a very much greater time to get heated throughout. It is quite easy to boil water on the top of a test-tube, holding the bottom part in the hand, and receiving no sensible amount of heat. The experiment is still more striking if a small quantity of ice is kept at the bottom of the tube by a little wire gauze (Fig. 61).

If convection is prevented, we find that liquids are generally to be ranked with the worse conducting solids.

Gases are also bad conductors, worse even than liquids. Here again the badness of conduction is masked by convection. The air gets hot on a summer day, not solely by conduction from the warm ground, but through the joint effect of convection and conduction.

The tremulous motion of objects at a distance, seen through hot air, is due to this convection, as already explained in Chapter V. Unequally heated masses of air are moving up and down irregularly, and refracting the light, making it appear to the eye to come now from one point now from another.

When convection is prevented, it is found that air is a bad conductor, and probably, on this bad conduction of air, depends the non-conducting power of woollen clothes, blankets, and other loosely woven textures. The wool is in itself a very bad conductor, while, through its being matted together, it entangles the air, which is a still worse conductor, so as to prevent convection currents, and thus the whole layer of wool and air conducts badly. Even were the wool itself a good conductor, the same effect might be produced, for the path by which the heat must get from one side to the other, travelling through the wool, is an indirect one, and so the wool slope of temperature is very gradual.* This probably explains the efficacy as a non-conductor of slag wool, a material consisting of blast-furnace slag, blown out into fine fibres by steam. The slag is itself not so bad a conductor, but, when loosely packed, the path from one side to the other of a layer of the wool is very much longer than the thickness of the layer. For example, the path from A to B (Fig. 62) through the wool is much greater than the direct path. Then, again, the cross-section of the material is in reality only a fraction of the whole cross-section of the layer, both conditions combining to diminish the conduction through the wool, while the

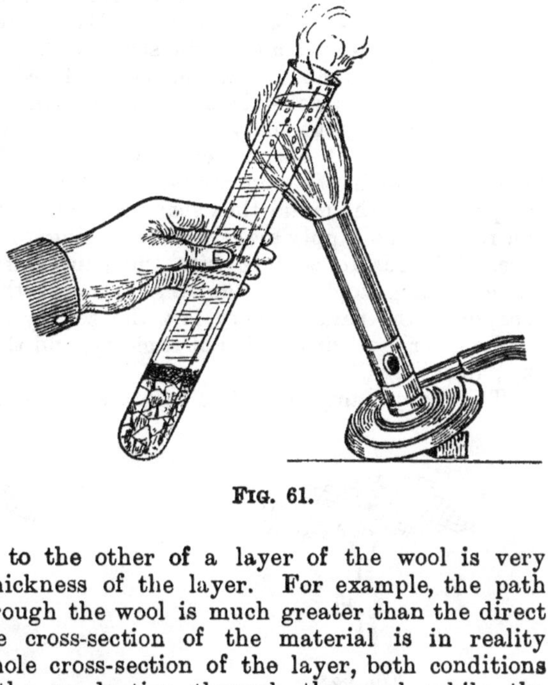

FIG. 61.

* From the table given later, it will be seen that wool and cotton are nearly equal in their conductivities, yet undoubtedly woollen clothes are the warmer. This is in great measure due to the fact that woollen cloth is more open in texture, and therefore holds a thicker layer of air for the same weight. Probably, also, the more open texture of the wool allows a rather freer exchange between the air within the meshes and the air outside, so that the vapour-laden air can pass from the skin to the outside. The more closely woven cotton hinders the passage of the air, and it is more likely, in cooling down, to deposit its moisture in the cloth, this deposition being aided by the hygroscopic property of cotton. If now evaporation takes place freely from the outer surface of the moistened cloth, the temperature is lowered there, and heat is more rapidly conducted from the skin, which may thus be chilled.

air imprisoned in the fibres cannot form convection currents to aid the transport of heat.

The low-conducting power of air thus entangled by a loosely packed solid may be illustrated by filling several test-tubes respectively with wool, slag wool, loose sand, and coarse copper filings, and inserting a thermometer in each. When all are plunged suddenly in boiling water, there is no very great difference in the rate of rise of the thermometers, for the heat comes chiefly through the air in each case.

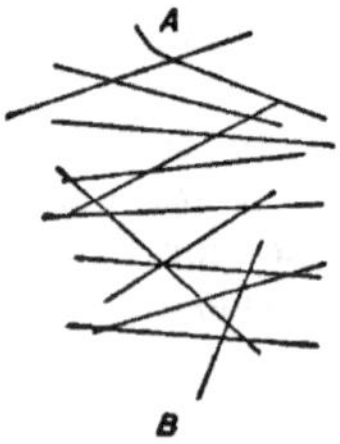

FIG. 62.

Safety Lamp.—We have already mentioned that a flame does not actually come in contact with a thin vessel containing water, or, rather, that the metal conducts the heat away so readily that the gas in contact with it is lowered below combustion point. This is made use of in the Davy safety lamp. A very simple experiment will serve to illustrate the principle of the lamp. If a burner is placed a short distance below a piece of iron gauze, and the flame is lighted, it usually burns under the gauze only (as in Fig. 63A). Or, if the flame is lighted above the gauze, it burns over it only (as in Fig. 63B). The gauze is a fairly good conductor, and carries the heat away rapidly to the surrounding parts, and the metal thus sharing the heat has a large surface, which can radiate the supply away without rising to the combustion temperature. The gas in contact with the gauze on the other side from the flame is thus kept from igniting. But if the gauze has not a large area, or if the flame is very hot, the supply of heat to the gauze may not be got rid of with sufficient rapidity, and the gas on the other side of the gauze is ignited.

The Davy Lamp consists of a brass base containing the oil-reservoir,

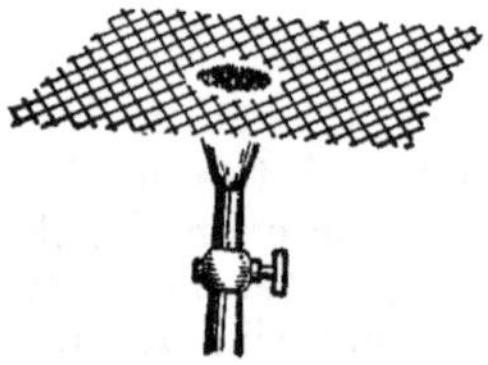
FIG. 63A.

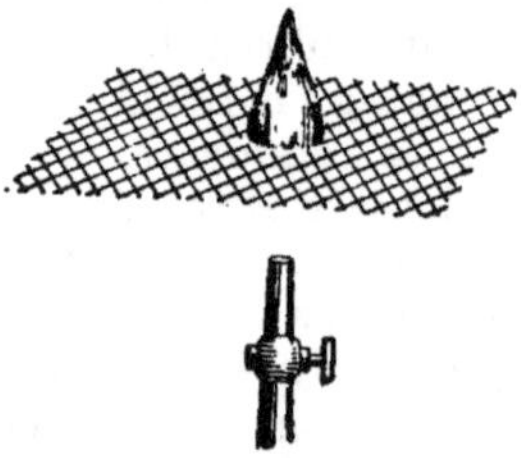
FIG. 63B.

Showing Principle of Safety Lamp.

and an iron gauze chamber, in which the flame burns. If fire-damp is present in the air in any quantity, it burns inside the lamp over the oil flame with a bluish light, but the gauze is at first able to radiate off the heat, or carry it away to the body of the lamp sufficiently rapidly to prevent the gas outside from rising to ignition point. If, however, the flame inside the lamp gets very large, there is great danger of the lamp becoming overheated. The lamp is, therefore, not by any means absolutely safe, but its indications are sufficient to give warning of danger, and on such warning it should be extinguished.

Definition of Conductivity.—We may now give a precise signification to the term as follows:—The conductivity of a substance is the quantity of heat conducted per second through a square centimetre in the substance, when the temperature changes in a direction perpendicular to the area at the rate of 1° C. per centimetre. If then AB (Fig. 64) is a plane over which the temperature is $\theta°$; CD, EF two parallel planes at $\frac{1}{2}$ cm. distance, one on each side, over which the temperatures are $\theta+\frac{1}{2}$ and $\theta-\frac{1}{2}$ respectively, the quantity of heat passing in one second through 1 square centimetre of the plane AB is the conductivity of the substance at the temperature θ. We shall denote this quantity by k. If the area be A square centimetres and the time t seconds, the quantity passing through must be $k\mathrm{A}t$, for, the circumstances are the same for each square centimetre and for each successive second.

It is not easy to make direct exact experiments on the flow of heat with different slopes of temperature, but general experience might lead us to expect it to be proportional to the steepness of slope, or the fall of temperature per centimetre. We may roughly verify this by immersing a thermometer in warm water, and noting the rate of rise. If we keep the thermometer moving in the liquid all the time, the outside layer of the glass has probably nearly the same temperature as the liquid, and the rise per second is nearly proportional to the distance of the top of the column from the final point reached—that is, the quantity of heat received per second is nearly proportional to the difference of temperature between the mercury and the water outside.

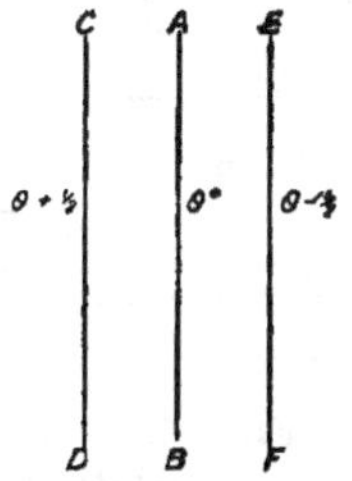

FIG. 64.

This is evidently complicated by the expansion of the glass, a thermometer giving true indications only when both glass and mercury are at the same temperature. A still better verification consists in immersing a small calorimeter containing water in an outer vessel also containing water, but at a different temperature, keeping the contents of both vessels well stirred, and it will be found that the quantity of heat passing from one vessel to the other is roughly proportional to the difference of temperature. We shall, therefore, assume that the quantity of heat conducted is proportional to the slope of temperature. Such experiments as these bear to the verification of the law the same relation that experiments with Atwood's machine bear to the verification of the laws of motion. We must regard them as suggesting, rather than verifying. The more exact verification is in the agreement of experiment with calculations based on the assumption of the truth of the law.

If the temperature-difference between two neighbouring points, d apart at equal distances, one on each side of the area A through which heat is flowing, is equal to τ, then $\frac{\tau}{d}$ is the rate of variation per centimetre or the slope of temperature, and assuming that the flow of heat is proportional to this, the total quantity flowing through A in time t is given by

$$\mathrm{Q}=k\mathrm{A}t\frac{\tau}{d},$$

where k is the conductivity at θ, the temperature of A.

Diffusivity.—A well-known piece of apparatus used to illustrate conductivity, was devised by Ingenhousz. In this a number of equal metal bars are placed in a row, with their lower ends in a vessel into which hot water can be put (Fig. 65). The rods may be smeared with beeswax, and the rate of melting along the different rods compared. But it is clear that the propagation of the high temperature along the rods depends not only on their conductivity, but also on their specific heats. If, for example, a cubic centimetre of one rod has twice the heat capacity of the same volume of another, it requires twice the quantity of heat to raise its temperature to the same extent, and if the melting extends equally quickly along the two rods, the conductivity must be twice as great in the one case as it is in the other: that is, we must consider, not only the conductivity, but the ratio—

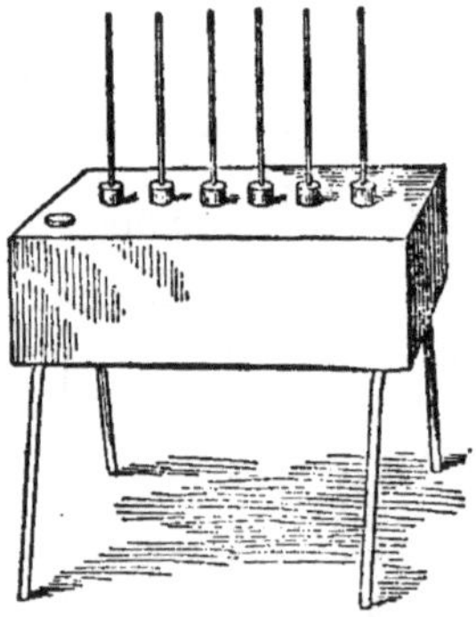

FIG. 65.—Ingenhousz's Apparatus.

$$\frac{\text{Conductivity}}{\text{Heat capacity per cubic centimetre}} = \frac{\kappa}{\rho s},$$

where κ is the conductivity, ρ the density, and s the specific heat. This quantity is termed the *diffusivity* of the substance. We may regard it as the conductivity for temperature as distinguished from the conductivity for heat.

Another quantity which plays an important part in researches on conductivity is the *emissivity* of a surface. We may define this as the heat lost by the surface per square centimetre per second, per degree of excess of temperature of that surface above the surroundings.

Measurements of Conductivity.—From the definition of conductivity we have

$$k = \frac{\text{quantity of heat passing through a sq. cm. per sec.}}{\text{slope of temperature perpendicular to the area}}$$

and if we can separately determine numerator and denominator, we obtain k.

It would appear then that the simplest way to measure conductivity would consist in catching the heat passing through some known area and measuring at different instants its rate of flow, and at the same instants observing the temperatures at two points near the area to give the temperature-slope. But while it is easy to measure the total quantity of heat which a body has gained in a given time, *i.e.* the difference between its gain and loss, it is quite another matter to measure the rate at which it is gaining at any instant. Indeed, the difficulty is not unlike that which a merchant might experience in trying to estimate his actual rate of profit at any instant. He may succeed in finding his exact profit or loss during a year, but he can hardly trace all the transactions and estimate the profit or the loss which is accruing in any one minute. And, again, while various methods of measuring the temperature at a point may be successful, it is not always easy to carry out the measurement without affecting and even diverting the flow of heat. Through these difficulties there is no measurement in which more widely different results have been obtained by different observers; the value of the conductivity for a substance as given by one kind of experiment being perhaps several times the conductivity as given by another kind of experiment on the same substance. The chief difficulties have only comparatively recently been so far surmounted that different methods give fairly accordant results. We shall not attempt to give here a complete account of the subject,* but rather seek to illustrate it by describing a few typical experiments which have given good results.

Solids—*Péclet's Method.*—Pioneer work was done in the measurement of conductivities by Péclet. He used a calorimeter, the bottom of which consisted of a plate of the substance to be tested, while the sides had very low conductivity. The calorimeter contained a known quantity of water at a known temperature, and when it was plunged into a vessel of warm water at a known temperature, the rate of rise of temperature of the water in the calorimeter was observed. If A was the area of the plate, κ its conductivity, d its thickness, Q the quantity of heat coming through per second, θ the temperature difference of its two faces, then if the heat-flow was normal to the surface

$$Q=\frac{\kappa A\theta}{d}$$

whence

$$\kappa=\frac{Qd}{A\theta}$$

The chief difficulty consists in measuring θ. Let us suppose that the water is at rest and that the heat flows steadily, *i.e.* that the same quantity flows across each cross-section of plate and water alike. Let the temperature in the water on the two sides at distances d_1, d_2 from the nearest faces differ from the temperature of the faces by θ_1, θ_2, and let the conductivity of water be κ^1. We have

$$Q=\frac{\kappa A\theta}{d}=\frac{\kappa^1 A\theta_1}{d_1}=\frac{\kappa^1 A\theta_2}{d_2}$$

If we observe the water temperatures at the distances d_1, d_2 from the plate, and assume that these are the temperatures of the two faces of the plate, we get for the conductivity

$$\kappa=\frac{Qd}{A(\theta+\theta_1+\theta_2)}=\frac{Qd}{A\theta}\times\frac{\theta}{\theta+\theta_1+\theta_2}$$

or the value we find ought to be multiplied by

$$\frac{\theta+\theta_1+\theta_2}{\theta}=1+\frac{\kappa}{\kappa^1}\left(\frac{d_1+d_2}{d}\right)$$

to give the true value.

Péclet was well aware of this source of error. With the cooler vessel above, convection would come into play, and so reduce the difference of temperature between the plate surfaces and the water round the thermometers and virtually reduce d_1+d_2, but not sufficiently. Péclet aided convection and virtually reduced d_1+d_2 still farther by brushing the surfaces all the time. He supposed that thus $\frac{d_1+d_2}{d}$ was rendered negligible, *i.e.* that the water against the plate, being continually and

* An account will be found in Winkelmann's *Handbuch der Physik*, vol. ii.

violently renewed, had the same temperature as that further off round the thermometer. But though, no doubt, $\frac{d_1 + d_2}{d}$ was small, it is to be remembered that with metals $\frac{\kappa}{\kappa^1}$ might be great. Thus for copper and water, we now know that it is about 700; so that a layer of water which remained against the plate on each side of it, and $\frac{1}{1400}$ of its thickness would make the result about double the true value. That Péclet failed to remove such a layer is shown by the fact that for copper he obtained a value about five times that now accepted. In fact the layer of water of sloping temperature was practically 2/700 of the thickness of the plate on each side. If $\frac{\kappa}{\kappa^1}$ were not large the layer might be still so thick, and yet the results would not be far wrong. Accordingly, Péclet's work for low conductors is fairly confirmed by later work.

Bar Methods of Finding the Conductivity of Metals.—There are

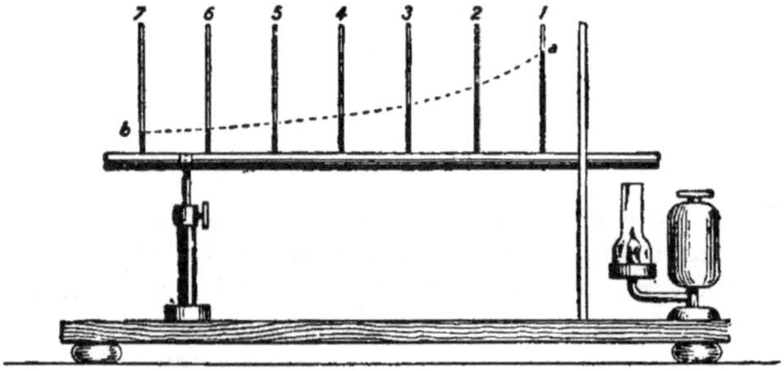

FIG. 66.—Despretz's Bar Experiment.

two types of experiment in which a long metal bar is heated at one end, and the conductivity is deduced from temperature observations along the bar. In one—the steady-flow method—one end is kept at a constant high temperature until the temperature at each point has come to a steady value. As no part of the bar is now gaining or losing heat, the heat conducted per second through any cross-section must equal that emitted per second from the surface beyond that section, and if the rate of emission can be determined, an observation of the slope of temperature at the cross-section will give the conductivity necessary to supply heat at the rate at which it is emitted beyond. In the other type, one end of the bar is subjected to periodic variations of temperature, and consequently waves of varying temperature travel down the bar. The conductivity is calculated from the march of these waves.

To the first method belong the researches of Despretz, Forbes, Wiedemann and Franz, and, more recently, Tait and others.

Despretz's Experiments.—Despretz used bars of different metals of the same dimensions, and with the surfaces varnished in the same way, so that the loss of heat to the surroundings should be the same for the same excess of temperature. Thermometers 1, 2, 3, 4, 5, 6, 7 (Fig. 66) were placed in holes in the bar filled with mercury.

Fourier had calculated the flow of heat along such a bar, assuming that the law of conductivity is true, and that the loss to the surroundings from any part of the bar is proportional to the excess of temperature above the surroundings. The curve of temperature *ab* enabled Despretz to verify Fourier's calculation, and at the same time to obtain comparative results for the conductivity in the different bars. The following special case may serve to show that it is possible to compare conductivities by means of the temperature-curves.

Suppose that two bars have the same cross-section, and that their surfaces are similarly treated, so as to lose the same amount of heat per sq. cm. for the same excess of temperature above the surroundings. Let the hot end of each be at the same temperature, but let the temperature-curve in the first case slope n times as quickly as in the second, so that if T, T′ (Fig. 67) represent equal temperatures on the two bars, and the first is at a distance OM from the end of the bar, then the other is at a distance ON $= n$.OM. Hence, we may divide the bars into corresponding elements at the same temperature, those of the first bar having $\frac{1}{n}$ the length of those of the second. Then the heat lost beyond any point of the first bar is $\frac{1}{n}$ of the heat lost beyond the point of the second bar having the same temperature.

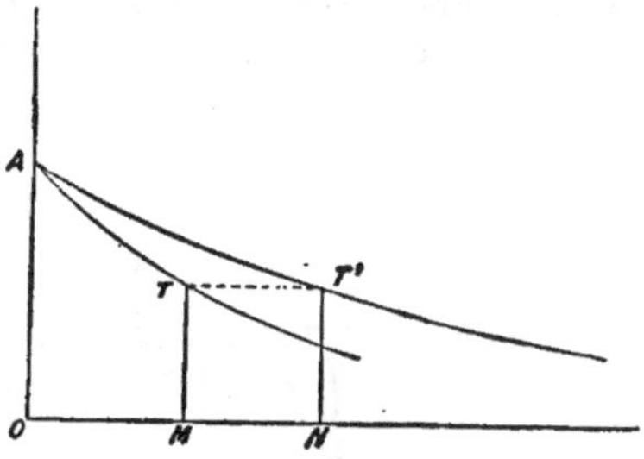

FIG. 67.

If, now, the conductivities are k_1 and k_2, the heat conducted across M on the first bar is to the heat conducted across N on the second as $k_1 \times$ slope at M is to $k_2 \times$ slope at N. But the total fall of temperature at M in length 1 is equal to the total fall at N in length n, or the slope at M is n times that at N. Then the quantities of heat conducted across M and N are as $k_1 n$ to k_2.

The heat conducted across a section is equal to the heat lost beyond that section, and we have just shown that the ratio of the heats lost beyond corresponding points is as 1 to n.

Therefore, $$1 : n = k_1 n : k_2$$

or, $$k_2 = k_1 n^2.$$

Forbes's Experiment.—Forbes modified the method so as to make it give absolute results for the conductivity of an iron bar. His apparatus is shown in Fig. 68.

The end of the bar is immersed in a constant high temperature bath, and after some hours' heating, the steady state is reached, the temperature at various points of the bar being shown by thermometers, as in Despretz' experiment, inserted in small cavities in the bar containing mercury.

If the temperature falls through τ in a small distance d at any point of the bar, then, as we have seen, the quantity of heat conducted across the section of area A in t seconds is

$$Q = kA\frac{\tau}{d}.$$

The thermometers give τ; A is the cross-section of the bar, and if we can find Q, this equation gives k.

In order to find Q, Forbes made a subsidiary experiment with a second bar of the same material, but of much smaller dimensions, also furnished with a thermometer. The bar was raised to a high temperature and then placed near the other, which was now cool, so that the surface of the small bar was exposed as nearly as possible to the same conditions, as to loss of heat for the same temperature, as the large bar, and the rate of cooling was observed. Knowing the capacity for heat of the bar and the area of its surface, the rate of cooling at any temperature gave the quantity of heat lost by each unit of area in a given time for that temperature, and thus the quantity of heat lost per second by every part of the larger bar in the first part of the experiment could be determined, and so Q could be calculated.

Determining the conductivity at different sections of the bar at different temperatures, Forbes was further able to show that the conductivity of iron decreases with rise of temperature.

Tait repeated Forbes's work on the same bar, and used also bars of copper and other metals in the same manner.* Stewart † has also carried out experiments on the conductivities of iron and copper by the

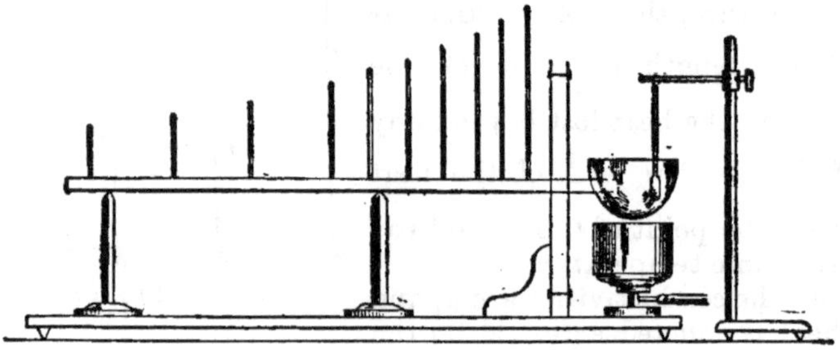

FIG. 68.—Forbes's Bar Experiment.

method of Forbes, using in place of thermometers a thermo-electric junction inserted in small holes in the bar under observation.

Lees, in the course of some experiments to be described below, has also used the same method for brass.

Neumann and Ångström's Method.—The second or temperature-wave method of measuring conductivities was used by Neumann, who heated one end of a bar of the substance experimented on, and then cooled it. Observation of the march of the cooling along the bar enabled him to determine the conductivity.

Ångström alternately heated and cooled one end of the bar, and after a time waves of temperature-disturbance travelled regularly along the bar. The rate of progress at different points and the rate of diminution in the amplitude of the disturbance enabled him to calculate the conductivity.‡

Gray's Experiments.—A simple mode of determining conductivity of metals has been used by J. H. Gray.§ A wire a few centimetres long, and about 0·2 cm. diameter, was hung vertically with its upper end soldered into the bottom of a copper vessel containing boiling water and its lower end into a copper ball, 5·5 cm. in diameter, drilled with a hole in which a thermometer was inserted. The wire was surrounded by cotton wool, and the loss from the sides was negligible.

* Tait's *Heat*, chap. xiv.

† *Phil. Trans.*, 1893, A., p. 569.

‡ "Heat," *Ency. Brit.*, 9th ed., or Tait's *Heat*, chap. xiv.

§ *Phil. Trans.*, A., Part I., 1895, p. 165.

A thermometer in the boiler gave the temperature of the upper end of the wire, and that in the ball gave practically the temperature of the lower end. The mean slope of temperature along the wire was thus known. The quantity of heat conducted per second was determined by the rate of rise in temperature of the copper ball. Assuming that the conductivity varies uniformly with the temperature, the experiments gave the conductivity at the mean temperature of the wire, about 53°. The value of the method lies in its applicability to metals which cannot easily be obtained in large masses.

Berget's Experiments on Mercury.—Berget * determined the conductivity of mercury, using a column of the liquid in a vertical glass tube surrounded by a much wider column of the same liquid, both being heated at the top by steam or mercury vapour, and conducting the heat downwards to a base kept at 0°. The central column terminated in an ice calorimeter, the central column and the wider surrounding column being at the same temperature at the same level. The heat flow was vertically downwards, a conclusion confirmed by the temperature-slope, which was uniform down the column. The quantity of heat arriving at the lower end was determined by the amount of ice melted. The temperature at various points along the column was determined by thermo-electric junctions, and thus the temperature slope was known. He found that the conductivity was constant between 0° and 100°, and that it then diminished.

The following table gives some of the results obtained for metals, the different values for the same metal by the same observer being obtained with different specimens.

Metals.	Conductivity.	Observer.
Iron	·209(1 − ·00147t)	Forbes.
,, ...	·197(1 − ·00002t)	Tait.
,, ...	·175(1 − ·0015t)	R. W. Stewart.
,, ...	·164	Neumann.
,, ...	·199(1 − ·00287t)	Ångström.
Copper (1) . . .	1·08(1 + ·0013t)	Tait.
,, (2) . . .	·71(1 + ·0014t)	,,
,, ...	1·108	Neumann.
,, (1) . . .	1·027(1 − ·00214t)	Ångström.
,, (2) . . .	·983(1 − ·00152t)	,,
,, ...	1·12(1 − ·001t)	R. W. Stewart.
,, (1) . . .	·9594 } between 10° and 97° say at 53°	J. H. Gray.
,, (2) . . .	·8884	,,
,, (3) . . .	·8612	
Very impure copper (4)	·3497	
,, ,, (5)	·3198	,,
Silver	·9628	
Gold	·7464	,,
Platinum . . .	·1861	,,
Mercury . . .	·02015 between 0° and 100°	Berget.
,, ...	·02(1 − ·00045t) between 0° and 300°	,,
,, ...	·01516 at 4·5°	H. F. Weber.
,, ...	·01620 at 17°	,,
Brass	·27 between 25° and 35°	Lees.

* *Journal de Physique*, viii., 1888, p. 503.

Experiments of Wiedemann and Franz.—*Relation between Heat Conductivity and Electric Conductivity.** Wiedemann and Franz experimented on the relative conductivities of metal bars working by the same general method as Despretz, but using a thermopile instead of thermometers to give the temperatures along the bar, the one junction of the pile being brought by a suitable arrangement in contact with any desired point of the bar. They found afterwards that the same results were given by using thermometers. The bars were silver-plated, and polished so as to have the same emissivity, and the chamber in which they were placed could be exhausted. The results obtained were comparative. They appeared to show that there is some connection between conducting-power for heat and for electricity. For the metals were found not only to follow the same order for the two conductivities, but in many cases the numbers bore nearly the same ratio to each other.

More recent work has confirmed this supposition. The following are some of the values for metals and alloys of the ratio thermal conductivity/electrical conductivity or k/c at 18° C. as determined by Jaeger and Diesselhorst (*Phys. Tech. Reichsanstalt Wiss.*, Abh. 3, 1900), together with the temperature co-efficient of the ratio.

	$10^{-10}k/c$.	Temperature. Co-efficient.		$10^{-10}k/c$.	Temperature. Co-efficient.
Copper, pure (1) .	6·65	0·0039	Platinum, pure .	7·53	0·0046
„ „ (2) .	6·71	0·0039	Iron (1) . .	8·02	0·0043
Silver „ .	6·86	0·0037	„ (2) . .	8·38	0·0044
Gold „ .	7·09	0·0037	Steel . . .	9·03	0·0035
Zinc „ .	6·72	0·0038	Bismuth . .	9·64	0·0015
Cadmium „ .	7·06	0·0037	Constantan . .	11·06	0·0023
Lead „ .	7·15	0·0040	Manganin . .	9·14	0·0027
Tin „ .	7·35	0·0034			

The electron theory of conduction for heat and for electricity (J. J. Thomson, *Corpuscular Theory of Matter*) gives an explanation of the connection between the two quantities. According to that theory the ratio should be proportional to the absolute temperature, *i.e.* should have a temperature co-efficient 0·00367, and at 0° C. its value should be $6 \cdot 3 \times 10^{10}$. The table shows that for many metals the values are not very different from those given by the theory. With alloys considerations of thermo-electric effects probably come in to add to the effective resistance, and so to diminish the conductivity and increase the ratio.

Kundt's Experiments *on the Relation between the Velocity of Light in Metals and their Electric and Heat Conductivities.*—Kundt † determined the refraction of light by exceedingly thin prisms of various metals and taking the velocity in each metal as being inversely as the refractive index, he found for red light—

Silver	100	Platinum	15·3	Nickel	12·4
Gold	71	Iron	14·9	Bismuth	10·3
Copper	60				

values not very different from the relative values for heat or electrical conductivity. The actual refractive index from air into silver was 0·27, and there was in this case but little dispersion. In other

* *Pogg. Ann.*, lxxxix., 1853, p. 497.

† *Phil. Mag.*, xxvi., 1888, p. 1.

metals there was considerable dispersion. Probably the refractive index for very long waves should be compared to bring out any true physical relation, and red light is only the best approximation which could be made to such long waves. At present we can only say that Kundt's results point in the direction of some connection between the three sets of phenomena, light velocity, heat conductivity, and electric conductivity.

Solids of Low Conducting Power and Crystals.—Many experiments have been made to determine the conductivity of non-metallic solids. With amorphous solids usually some method similar to that of Péclet has been used, and the range of conductivity has always been found to be far below that of metals. Taking first the conductivity of crystals, we might expect that this would be different along the different axes, an expectation verified by experiment. The subject was first studied by Senarmont, who used a very simple method, preparing a plate of the crystal with a small hole through it. The plate was covered with a film of beeswax, and a silver wire passing through the hole was heated. The heat was conducted through the crystal, and the beeswax

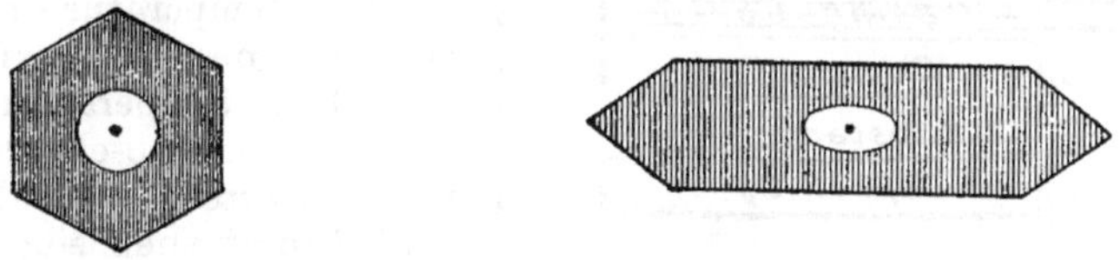

FIG. 69 (*a* and *b*).—Showing Conductivity in Crystals.

was melted. When the conduction was the same in all directions in the plane of the section, as in a plate of quartz cut perpendicular to the axis, the figure of the melted wax was circular, as in Fig. 69*a*. When it differed in different directions the figure was elliptical as in Fig. 69*b*. The conductivities along the two axes of the ellipse may be shown to be proportional to the squares of the axes.

A method of experiment developed by Lees * is especially suitable for the determination of the conductivity of crystals. A long brass bar with diameter 1·93 cm. was used as in Forbes's experiment, and its conductivity was determined by his method as $\cdot 268\{1 + \cdot 002(t - 17°)\}$. It was then cut in the middle and a plate of the crystal, of the same area as the cross-section of the bar, was inserted between the cut faces, these being amalgamated to give good contact with the crystal. Temperature observations along the bar, made by means of thermo-electric junctions, gave the temperature of each face of the plate and therefore the temperature slope through it, while the known conductivity of the bar and the temperature slope in it adjacent to the plate gave the rate of passage of heat into and out of it. From these data the conductivity could be determined.

Lees found the following conductivities between 25° and 35°:—

Quartz, along the axis	·0299
,, perpendicular to the axis . .	·0158
Iceland spar, along ,, ,, . .	·0100
,, perpendicular to ,, ,, . .	·0084
Mica, ,, cleavage planes .	·0018

* *Phil. Trans.*, A., 1892, p. 481.

The method was used for various other solids and the results obtained were in close agreement with those obtained by another and more exact method which he devised later, now to be shortly described.

*Lees's Disc Experiments.**—The substance to be tested was formed into a disc X (Fig. 70), say 4 cm. in diameter and 2 or 3 mm. thick. This was placed between two copper discs C_1 C_2 of the same order of thickness, continuity being ensured by a negligible layer of glycerine. Against one copper disc was laid a flat coil through which a current could be passed so as to supply heat at a determinate rate, and on the coil was a third copper disc C_3. Fig. 70 represents diagrammatically the pile of discs thus made, suspended in a constant-temperature enclosure. The surfaces of the pile were varnished to give them the same emissivity, h say, so that the rate of heat emission from a square centimetre v degrees above the enclosure would be hv. The copper discs would each be of uniform temperature throughout to a near approximation, and their temperatures were taken by thermo-electric junctions inserted in small holes drilled in at their edges. Let their temperatures, measured as excesses over the temperature of the enclosure, be as indicated on the right hand of the figure, and let their emitting surfaces have areas as indicated on the left hand. Let the rate of heat supply by the coil be H. Then taking the mean temperature of X as the mean of the temperatures of C_1 and C_2, we have

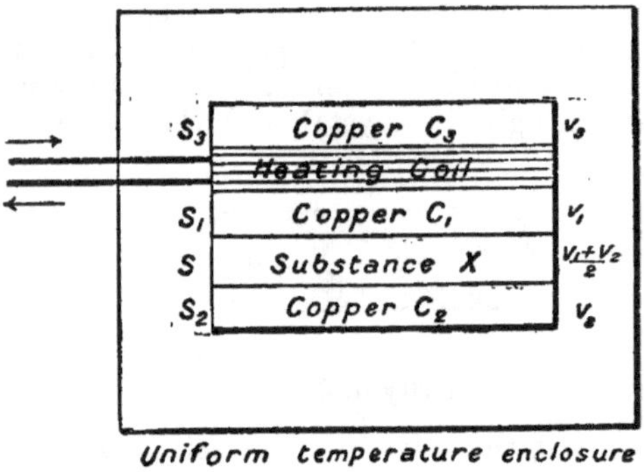

FIG. 70.—Lees's Disc Experiments on Conductivity.—$C_1C_2C_3$, copper discs; X, disc to be tested; $v_1v_2v_3$, temperatures above the enclosure; s, $s_1s_2s_3$, emitting surfaces.

$$H = hs_1v_1 + hs\frac{v_1 + v_2}{2} + hs_2v_2 + hs_3v_3,$$

whence h could be determined, since H was known electrically.

The heat flowing through X may be put as

$$Q = \kappa A\frac{v_1 - v_2}{d}$$

and it may be taken as equal to the mean of the heat flowing into it from C_1 and that flowing out of it to C_2. But the heat flowing from C_1 into X is equal to that emitted from X and C_2, and that flowing from X into C_2 is equal to that emitted from C_2. Whence we have

$$\kappa A\frac{v_1 - v_2}{d} = \frac{1}{2}\left\{ hs\frac{v_1 + v_2}{2} + hs_2v_2 + hs_2v_2 \right\}$$

$$= hs\frac{v_1 + v_2}{4} + hs_2v_2$$

* *Phil. Trans.*, A., 1898, p. 399.

Knowing h from the equation for the heat supply H, we can hence find κ. We have omitted all account of corrections. For these the original paper may be consulted.

The method gave the conductivity at different temperatures, and assuming that it might be put in the form

$$k = k_{35^\circ}\{1 + \alpha(t - 35)\}$$

the values which Lees found for the conductivity at 35°, and for the co-efficient α are

	k_{35°	α
Window glass	·00245	+ ·0025
Sulphur	·00067	− ·0036
Ebonite	·00042	− ·0019
Shellac	·00058	− ·0055

We add some conductivities which he found by the divided bar method for a mean temperature of 30° :—

Marble	·00709
Slate	·00475
Paraffin	·00061
Paper	·000315
Silk	·00022
Mahogany	·000465
Cork	·000129

The following conductivities are taken from Lord Kelvin's article on Heat in the *Encyclopædia Britannica* (9th ed.). They probably need revision :—

Substance.		Authority.
Sandstone of Craigleith Quarry	·01068	Forbes and Lord Kelvin obtained by observations on underground thermometers, noting the delay of the heat of summer and the cold of winter.
Underground strata	·005	,,
Trap rock of Calton Hill	·00415	,,
Porphyritic trachyte	·0059	Ayrton and Perry.
Oak across fibres	·00059	Péclet.
Fir along fibres	·00047	,,
Fir across fibres	·00026	,,
Cork	·000029	,,
Writing paper	·000119	,,
Carded wool	·000122	,,
Finely carded cotton wool	·000111	,,
Eider-down	·000108	,,

Conductivity of Liquids. — Usually conduction in a liquid is greatly assisted by convection. If, however, convection be prevented by heating the liquid from the top, it is found that except in the case of mercury the conducting power is low. The first quantitative experiments appear to have been made by Despretz, who heated a column of water from above, and observed the temperature at various depths when the steady state was reached. He found that the slope of temperature was in accordance with Fourier's calculation founded upon the law of conductivity, and thus proved that the law was the same as for solids.

Other experimenters have since made determinations of the conductivity of liquids. Thus Lundquist, employing Ångström's method, obtained for the conductivity of water at 40·8° $\kappa = \cdot 00156$. H. F. Weber * used a disc method, the liquid layer, ·231 cm. thick, lying between two copper discs 16 cm. diameter, and being kept in position by capillarity, or by a glass rim to the lower plate. The upper plate was about 1 cm. thick, and was assumed to be at one temperature throughout, the temperature being given by a thermo-electric junction attached to it. The lower plate was half the thickness of the upper. When the whole apparatus was at one temperature it was suddenly placed on a block of ice at 0°, in an enclosure at 0°, and the rate of cooling of the upper plate was observed. The mathematical deduction of the conductivity

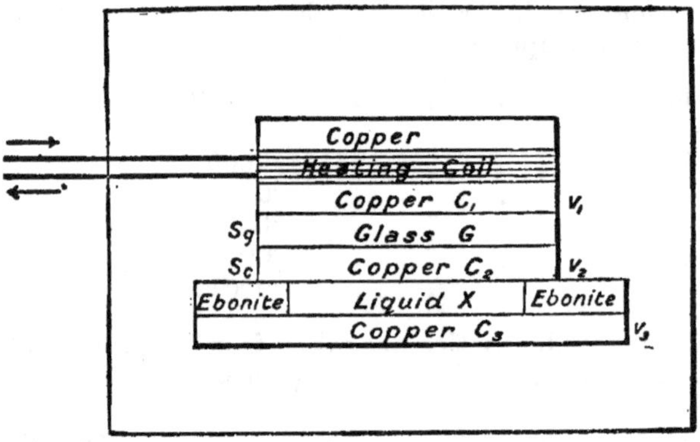

FIG. 71.—Lees's Disc Experiments on Liquids.—$C_1C_3C_2$, copper discs; G, glass disc; X, liquid surrounded by ebonite ring.

from the rate of cooling is not simple, and Weber's calculation has been criticised and amended by Lorberg, whose recalculation gives a value for water nearly 10 per cent. greater than that of Weber. But, where so much uncertainty exists, it is sufficient to state Weber's results for water in the form

$$\kappa_t = \cdot 0012(1 + \cdot 008t)$$

where the assumption is made that the conductivity changes uniformly with the temperature.

Other experimenters have obtained values of the same order of magnitude. We shall describe only the method used by Lees,† a modification of his disc method already described.

The liquid tested filled the cavity made by an ebonite ring between two copper discs. The principle of the method consisted in sending a known quantity of heat down through the upper of these discs, and from the temperature of the lower disc determining the conductivity of the liquid layer. The heat thus sent down was determined by putting a

* Chree, *Phil. Mag.*, xxiv., 1887, p. 1, gives an account of this and other work.
† *Phil. Trans.*, A., 1898, p. 399.

glass disc of known conductivity (as in Fig. 71) on the upper disc, and then a copper plate on the glass, a heating coil on this, and a final copper disc on the coil. The pile thus built up was varnished to give it uniform and known emissivity, and placed in horizontal position in a uniform temperature enclosure. The temperature excesses of the three discs $v_1 v_2 v_3$ were determined by thermo-electric junctions inserted in small holes, and were taken as uniform in each disc. Let the conducting area of each face of the glass disc be a_g, that of its emitting surface s_g; let its conductivity be κ_g, and let its thickness be g. We may take the heat passing through its middle surface as

$$\kappa_g a_g \frac{v_1 - v_2}{g}$$

The heat emitted by its surface is

$$h s_g \frac{v_1 + v_2}{2}$$

and that emitted by the lower half is approximately

$$h s_g \frac{v_1 + v_2}{4}$$

Hence the heat passing through the lower face is

$$Q_1 = \kappa_g a_g \frac{v_1 - v_2}{g} - h_g \frac{v_1 + v_2}{4}$$

which is determinate.

Now, passing to the copper disc with temperature excess v_2, let its emitting surface be s_c, the heat passing from this with the liquid and ebonite is equal to the heat passing in minus the heat emitted by the disc, or is determinate as

$$Q_2 = Q_1 - h s_c v_2$$

Assuming that the flow is vertically down through the liquid, that its conductivity is κ, its area a, and its thickness l, the heat passing through it is

$$\kappa a \frac{v_2 - v_3}{l}$$

The heat certainly does not flow vertically through the ebonite, but we may put it as

$$A(v_2 - v_3)$$

where A is some constant, and we may then put

$$Q_2 = \kappa a \frac{v_2 - v_3}{l} + A(v_2 \quad v_3)$$

If, then, we can determine A, we can find κ, since Q_2 is known.

To find A a separate experiment was made, in which the liquid was replaced by air, of which the conductivity is approximately known.

The results obtained, between 25° and 45°, agreed with the formula

$$\kappa = \kappa_{25}\{1 + \alpha(t - 25)\}.$$

The following table is taken from Lees's paper :—

Liquid.	κ_{25}	a	Weber's values of κ between 9° and 15°.
Water . . .	·00136	–·0055	·00136
Glycerine . . .	·00068	–·0044	·00067
Methyl alcohol . .	·00048	–·0031	·000495
Ethyl alcohol . .	·00043	–·0058	·000423

At first sight the results obtained by Lees appear in fair agreement with those given in the last column as obtained by Weber. But while Lees's results for water, for example, may be put in the form $\kappa_t = \cdot00155(1 - \cdot0048t)$ where, it must be remembered, the actual range of observation lies between $t = 25°$ and $t = 45°$, Weber's results give $\kappa_t = \cdot0012(1 + \cdot008t)$ where the range of observation lies between $t = 4°$ and $t = 24°$. They agree nearly at $t = 25°$, but have opposite signs for the temperature co-efficient. Further experiment is urgently needed to find whether there is any reality in this change of sign. Indeed, until different methods give closer agreement, all the results must be regarded as uncertain.

Conductivity of Gases.—The investigation of the conductivity of gases is complicated, not only by the ease with which convection occurs but also by their transparency to radiation. It is necessary, then, to disentangle the three effects of convection, radiation, and conduction, and to find how much is due to conduction alone. Passing over earlier work, which was merely qualitative, we shall describe briefly the experiments of Stefan,* Winkelmann, Kundt and Warburg, and Todd.

Maxwell calculated the conductivity of gases from the kinetic theory of gases (chap. ix.), and showed that over a wide range of pressure the conductivity should be nearly independent of the pressure. The experiments to be described were made with the view of testing Maxwell's results.

Stefan used two coaxal cylinders of thin copper. The inner one served as an air thermometer, its tube passing through the end of the outer and dipping into a vessel of water. The outer vessel was everywhere separated by the same distance from the inner one, and the space between them was occupied by the gas to be experimented on. The outer cylinder was then surrounded by a mixture of snow and water, and the rate of fall of temperature of the air in the inner cylinder was observed. Hence, the conductivity could be calculated. For Air Stefan found $k = \cdot0000558$. Maxwell had already calculated the value $k = \cdot000054$. The result for Hydrogen was seven times as great, in accordance with Maxwell's result. The conductivity was also found to be independent of the pressure, in accordance with Maxwell's prediction. Stefan, however, took no account of the heat radiated.

Winkelmann † experimented both with cylinders and spheres in a method closely resembling that of Stefan, and obtained with different apparatus a value of the conductivity of air very close to that of Stefan.

Kundt and Warburg ‡ sought to eliminate convection and radiation. They noted the cooling of a thermometer in an enclosure containing the gas, the heat capacity of the thermometer being known. As the pressure of a gas is reduced, there is a limit beyond which the conduc-

* *Journal de Physique*, ii. p. 147, 1873. † *Wied. Ann.*, xlviii., p. 181.
‡ *Pogg. Ann.*, clv. and clvi.

tivity is no longer independent of the pressure but falls very rapidly with it. In order, then, to determine the effect of radiation they reduced the pressure far beyond this limit, so that conduction was negligible, and at such low pressure convection also vanished. This was borne out by the fact that the loss of heat was the same wherever the thermometer was placed in the enclosure. Radiation would be the same, while conduction, if it existed, would depend on the distance of the thermometer from the walls. Gas was then admitted, but only to such a pressure, that while conduction was active, convection had not yet come into play. The radiation effect could now be allowed for, and it was found that between an upper limit of 150 mm., and a lower limit of 1 mm. for air and 9 mm. for hydrogen, the rate of cooling was quite independent of the pressure, and they assumed that, within this range, convection did not exist and that conduction was constant. Their result for hydrogen was seven times that for air, as Maxwell had expected, and they estimated the conductivities as

Air	·000048
Hydrogen . .	·000341

Todd* experimented on the heat conducted through a layer of gas between two horizontal metal plates. The upper one was the base of a steam chamber, and was so maintained at 100° C. The lower was maintained at a constant temperature, about 10° C., by a stream of water flowing against its under surface. The amount of water flowing in any time and the difference of temperature at inlet and outlet gave the heat absorbed by the lower plate. This heat was partly conducted, partly radiated. Convection was eliminated by having the hot plate uppermost. To understand the principle of the method let us suppose that the plates are of indefinitely large area to eliminate edge effects. Let Q be the heat received below per sq. cm. per second. Let κ be the conductivity of the gas, θ the temperature difference of its two surfaces x apart, and let R be the heat received by radiation per sq. cm. per second. Then $Q = R + \kappa\theta/x$. But if x be varied R is constant, so that we have $(Q - R)x = \kappa\theta$. Plotting Q against x, we have a rectangular hyperbola with $Q = R$ as asymptote, and the curve will give this, and therefore $\kappa\theta$ and hence κ.

Todd found for air $\kappa = 0{\cdot}0000571$, for carbon dioxide $\kappa = 0{\cdot}0000411$, and for nitric oxide $\kappa = 0{\cdot}0000888$, all at 55° C.

The subject of conductivity is one of which the mathematical development—based on certain assumptions—has outstripped experimental verifications. Fourier, the founder of the mathematical theory, in his *Théorie Analytique de la Chaleur*, discussed many problems, such as that of the motion of heat in bars, which have been made use of by Despretz and succeeding experimenters, and of the motion of heat in spheres, of which we have a special case in the earth.

The reader will find an account of the problem presented by the penetration of the sun's heat into the earth in Tait's *Heat*, p. 218, and a sketch of Fourier's theory in Maxwell's *Theory of Heat*, 5th ed., p. 288.

* *Proc. Roy. Soc.*, A., vol. lxxiii. p. 19, 1909.

CHAPTER VIII.

THE FORMS OF ENERGY. CONSERVATION OF ENERGY. MECHANICAL EQUIVALENT OF HEAT. FIRST LAW OF THERMODYNAMICS.

Introductory Remarks—The Various Forms of Energy—The Identity of Energy—The Conservation or Constancy of Energy—Statement of the Principle—Mayer's Calculation of the Mechanical Equivalent—Joule's Researches—Later Repetition—Experiments of Rowland—of Miculescu—of Reynolds and Morby—of Griffiths—of Schuster and Gannon—The First Law of Thermodynamics.

Introductory Remarks.—The investigation of the conditions under which heat appears in a system, or disappears from it, leads us to regard heat as one among various forms of energy, and when it appears we find that some other form disappears, and *vice versa*. The investigation of the relation of heat to other forms of energy constitutes the subject of Thermodynamics. In this chapter we shall set forth the evidence which leads us to recognise various forms of energy, and to adopt the principle known as the Conservation of Energy. This principle, as applied to heat, is known as the First Law of Thermodynamics.

The study of mechanics leads to the recognition of two great principles of conservation or constancy—the conservation of mass and the constancy of momentum in a given line. These principles have been recognised from the time when Newton placed the science of dynamics on a firm foundation. We have now to add a third great principle of conservation, the Conservation of Energy. This principle was, naturally, only recognised at a much later date, for its recognition depended on a much wider knowledge of physical phenomena and their mutual relations than was possible when the other principles were first enunciated.

The fundamental idea that there is some identity underlying the apparently different phenomena of mechanics, heat, light, and electricity only assumed prominence at the beginning of the nineteenth century, though it was sometimes vaguely perceived in earlier times. At first it was expressed merely qualitatively, by saying that there was some mutual relation between the various "forces of nature," or, as we should now say, between the various forms of energy, so that one "force" was convertible into other "forces." But as modes of measurement improved, and numerical relations accumulated, it was gradually perceived that quantitative relations held between the various correlated "forces," and in 1845 we find Faraday saying: "I have long held an opinion, almost amounting to conviction, in common, I believe, with many other lovers of natural knowledge, that the various forms under which the forces of matter are made manifest have one common origin; or, in other words, are so directly related and materially dependent that they are convertible, as it were, one into another, and possess equivalents of power in their action. In modern times the proofs of their convertibility have been accumulated to a very considerable extent, and a commencement made of the determination of their equivalent forces" (*Exp. Res.*, iii. p. 1). *

This is a full statement in the language of the time of the principle of the conservation of energy, made just when the principle was struggling into general recognition, and before it was placed on a firm experimental foundation by the work of Joule and others. Faraday's statement is divisible into two parts, the first asserting the existence and mutual convertibility of "forces," the second asserting this convertibility in definite ratios. Our account will naturally divide into two corresponding parts. We shall give

(1) An account of each of the forms of energy hitherto recognised, and a statement of the evidence which leads to the belief that they are all forms of one "something" which we term *energy.* †

(2) An account of the modes of measuring the amount of each form in a system, and an examination of the evidence which leads to the belief that one form changes into another in a definite ratio or at a fixed "rate of exchange." We shall then see in what sense we can hold that the total quantity of energy is constant.

The Various Forms of Energy.—We say that a man possesses "energy" when he can do work in overcoming obstacles, either mental or physical. By analogy, the same word is used in physics, and we say that a body possesses energy, when by virtue of its motion or condition it can do work in moving either itself or other bodies against resistance.

When the body can do work by virtue of its motion, it is said to possess

Kinetic Energy.—If a body of mass m starts with velocity v, and moves through a distance s against a uniform force which would produce in it acceleration a, and therefore be measured by ma, we know that its velocity v' at the end of s is given by

$$\frac{v^2 - v'^2}{2} = as.$$

If we multiply each side of this equation by m we get—

$$\frac{mv^2 - mv'^2}{2} = mas$$

$$= \text{force} \times \text{distance travelled against it},$$
$$= \text{work done against the force}.$$

If $v' = 0$ the whole motion is exhausted, and

$$\frac{mv^2}{2} = \text{work which the body does against the force in coming to rest.}$$

* An excellent history of the growth of the doctrine of energy is given in Helm's *Die Energetik.*

† The footnote on p. 116 may be read here.

The quantity $\frac{mv^2}{2}$ is defined to be the Kinetic Energy or Energy of Motion of the body. It measures the work which the body can do in exhausting its motion. Since it is equal to work it can be measured in foot-pounds, foot-poundals, kilogramme-metres, ergs, &c., according to the units chosen.

Potential Energy.—Let us consider the special case in which a body of mass m is projected straight upwards against its own weight with velocity v. For simplicity let us suppose that there is no air-resistance, so that the weight alone acts. As the body rises its kinetic energy gradually disappears, and at the highest point reached the body is for an instant at rest and without kinetic energy. But we do not suppose that this energy is gone out of existence. For, as the body falls it regains energy, and when it has come back to the starting point its velocity is again v and its kinetic energy again $\frac{mv^2}{2}$.

We suppose that the energy did not cease to exist, but that it took a new form no longer manifested in motion but in change of position or change of configuration with respect to the earth. This new form we term Potential Energy or Energy of Position.

We recognise the existence of kinetic energy by our sense of sight—we see the body moving, but we think of potential energy in terms of the muscular sense as well as in terms of the sense of sight. For constraint is always needed to preserve at rest a configuration involving potential energy, and we think of ourselves as upholding by pull or push a body possessing such energy.

It is convenient to measure the gain in potential energy of a body when rising against its weight, by the work done in moving it from its original to its new position. The gain in potential energy is then equal to the loss in kinetic energy, and the sum of the two energies, potential and kinetic, remains constant.

It might seem at first sight that this constancy is merely a result of definition and does not involve any observation or experiment. But it is to be remembered that the force acting is the weight of the body, which is the same whatever the velocity or the direction of its motion, and whenever the motion takes place. Hence the velocity and the kinetic energy are the same at any given point, both in the rise and fall, and the possibility of regaining in the fall all the kinetic energy lost in the rise depends entirely on the nature of the force acting. Were the force dependent on the velocity or its direction, or did it change with time, then the kinetic energy at a given point would be no longer the same in the rise and fall. Indeed, in reality, the air-resistance is always opposed to the motion and the kinetic energy lost against this is not regained in the fall, so that at re-arrival at the starting-point there is a diminution. The loss of kinetic energy is still, of course, equal to the work done against the forces, but this work can no longer be regarded as measuring potential energy stored.

This simple case will serve as an illustration of the general principle that the kinetic energy of a system is only wholly regained on return to the original configuration, when the forces depend solely on the configuration and not on the motion of the parts, or on the time at which the motions occur. When and only when this condition is fulfilled, the work done against the forces in moving from one configuration to another will be independent of the mode in which the change is effected, and for each configuration there will be a definite amount of kinetic energy. The loss of kinetic energy in any change is thus recoverable on changing back again, so that we can assume that potential energy is gained equal to the kinetic energy lost. A system of this kind is termed a "conservative system," and the forces are termed "conservative forces." The experimental basis, then, of the assertion that the sum of the potential and kinetic energies of a system is constant is the observation that the forces depend solely on the configuration.

Probably no system is exactly conservative. But the planets and the sun form a system in which we have not as yet been able to detect any departure from conservation of kinetic + potential energies, in the comparatively short time over which astronomical observations have extended, and it gives us the best illustration of such a system. Even in the solar system, however, there are tidal effects which depend on the velocity of the bodies producing them, and we know that they must decrease the sum total of the kinetic and potential energies though we have not actually observed the decrease. If we include comets in the system, there appear to be some cases of observed decrease and diminution of orbits.

We continually use stores of potential energy in practical life to obtain work. Thus we use a head of water to turn a turbine or a water-wheel. We wind up a clock-weight to keep a clock going, and so on. But we can never use the potential energy directly. We must always allow the body or system to move in the direction of the forces acting, and convert the potential in the first place into kinetic energy, and thus get work from the kinetic energy. It would be absurd to attempt to get energy from a reservoir of water without allowing the water to run down hill, or to attempt to keep a clock going from the potential energy of the weights without allowing them to fall.

Heat Energy.—In almost all cases of motion with which we are concerned on the surface of the earth, the forces are not conservative. We generally find friction of some kind coming into play, some force opposed to the motion. If a body is projected along a horizontal table its kinetic energy gradually disappears, work being done against the friction. When the body comes to rest, the whole of the kinetic energy is lost without any gain of potential energy, for there is no tendency on the part of the body to return to its original position. But a new phenomenon is observed. The body and the table are both slightly heated. This appearance of heat is illustrated by the familiar experiment of rubbing a button on a table and then applying it to the skin, when the heating is found to be quite appreciable. In Sir Humphry Davy's celebrated experiment, ice was melted by rubbing together two pieces in a vacuum. Kinetic energy was continually supplied to the ice. It disappeared in doing work against friction and *heat* appeared in the melting of the ice. Many other examples of the appearance of heat on the loss of kinetic energy by friction, will occur to the reader. In all these cases the kinetic energy has disappeared, and there is no means of regaining it by allowing the body to retrace its path. It has, therefore, not been converted into potential energy. But we suppose the heat which has appeared to be itself a form of energy, and that the energy which was formerly kinetic has taken this new form. We are strengthened in this conclusion by the fact that the appearance of kinetic energy is frequently accompanied by loss of heat. For example, in the steam-engine, the steam cools in moving the piston and in setting the machinery in motion.

Strain Energy.—Often, when kinetic energy disappears, we observe the appearance of strain in matter.

If an indiarubber cord, of unstretched length OA (Fig. 72), is fixed at O, and has a mass attached to the end A, this mass has potential energy. If it is now allowed to fall it acquires kinetic energy, and will move down until it comes to rest for a moment at B, a point below the final position of equilibrium. If it is detained in this position it has lost both potential and kinetic energy, but the indiarubber is stretched, and in this stretched condition we must recognise the cord as still possessing energy. For, when the mass is released, it begins to move upwards, and, as the cord contracts, the mass acquires both kinetic and potential energy. We may term this new form of energy, recognised in the stretched or strained condition of the cord, *Strain Energy.* On allowing the mass to oscillate up and down, we have continual interchanges between the three forms of energy—kinetic, potential, and strain. At B the energy is all strain. As the body moves upwards, the strain energy is converted into kinetic and then into potential, so that at an intermediate point C all three co-exist. When the body returns to rest at A, the strain and kinetic energies have disappeared, and we have only potential energy left, and so on.

FIG. 72.

We have another case of strain energy when one end of a wire is fixed and the other end is twisted. Thus, if the upper end of a vertical wire is fixed, and a mass is attached to the lower end, when the mass is twisted round and released it oscillates to and fro round the wire as axis. We again recognise the wire as possessing strain energy when in this twisted condition. On the molecular theory of matter we may probably describe this strain energy as, at least in part, potential energy of molecular grouping—*i.e.* as energy depending on the relative positions of the molecules to each other. But we cannot give a complete account of it on this theory until we can say how the molecules are arranged in the strain, for the energy does not depend merely on the strain as a whole, but on the mode in which it was effected and the time which has elapsed since it was made. In potential energy due to gravitation, the forces acting are the same for the same grouping at all times, and the potential energy cannot be obtained in any other form, unless the members of the system are allowed to move. But if a wire is strained by twisting, and kept strained, the energy of the strain in general slowly decreases, though the twist of the wire remains the same. If, then, the energy is potential energy of molecular grouping, we must suppose that the molecules do not remain in the same distorted position, but gradually undergo rearrangement of some kind.

Sound Energy.—We frequently find that strain energy disappears, giving rise to sound. If a wire is stretched between two points and pulled transversely, we have strain energy; but, on letting the wire go, it gives a musical note. Or, if the prongs of a tuning fork are pinched together, and then freed, the fork sounds its note. The strained body is observed to oscillate, and an interchange of strain and kinetic energies takes place in it. The oscillations gradually die away, and their energy is transferred to the surrounding air or other bodies, where it produces the peculiar condition which is capable of affecting our sense of hearing. We know that this condition consists of a combination of motion and strain travelling out in wave form from the strained source through the surrounding medium. There is, therefore, in the waves a mixture of kinetic and strain energies, forms already recognised. But since a special sense—that of hearing—is affected by the combination, we give it a special name, that of *Sound Energy.*

Light Energy, a Particular Case of Radiant Energy.—When a body has been raised to a sufficiently high temperature, it becomes incandescent, *i.e.* it is sending out light through the surrounding medium. Meanwhile there is a loss of heat in the body, for if the supply of heat is cut off, the body is observed to cool. Conversely, if light is allowed to fall on an opaque surface, the surface is found to be heated. We are familiar with this heating effect in the warmth derived from sunlight, or from the light of a glowing fire. We know that light is some modification of the medium through which it passes, propagated out in waves from the source with a definite velocity, and this modification on reaching our eyes affects our sense of sight. We also know that the waves must be of lengths within certain narrow limits, in order that they may affect us as light, somewhere between 300 and 800 millionths of a millimetre. But there are other waves sent out from bodies, similar in kind to light waves, of lengths both above and below these limits to which the retina is insensitive. We include the whole series of waves under the general term of "radiation." Since their propagation from a source accompanies loss of heat, and their absorption by a body accompanies gain of heat, we must regard them as possessing a form of energy which we term *Radiant Energy.* Formerly this energy was described as radiant heat, but it is much more convenient to reserve the term heat for the heat energy in the radiating body, or in the heated receiving body, and to use "radiant energy" for the form into which the heat energy changes when associated with the wave motion which travels out into the surrounding medium.

Electrical Energy.—Many bodies on being rubbed are observed to possess new mechanical properties. If a piece of sealing-wax is rubbed with fur, or if a piece of very dry paper is stroked with the finger-nails, it is found that small bodies, such as small pieces of paper, are urged towards the rubbed surface. In other words, we have a development of kinetic energy not accounted for by the immediate disappearance of potential, strain, or heat energy. We suppose, therefore, that we have here a new form of energy, which we term *Electrical*, and by making this supposition we may connect the kinetic energy of the moving small bodies with the kinetic energy of the arm when rubbing the electrified body, the electrical energy being the intermediate form. It is slightly more difficult to rub the body which is being electrified than it would be were there no electrification of its surface. The resistance to rubbing absorbs kinetic energy, which, we suppose, takes the form of electrical energy, and finally reappears as kinetic energy in the particles moving to or from the electrified surface. The increased resistance when electrification is taking place, *i.e.* when kinetic is being turned into electrical energy, is well illustrated in friction and induction electrical machines. It is a common experience that they are more difficult to turn when they are in good order and are generating electricity. The existence of electrical energy is further supported by the occurrence of light and heat in the electrical discharge, *i.e.* by the appearance of light-energy and heat-energy, which would otherwise have to be thought of as arising *de novo.*

In certain cases we have also the appearance of electrical phenomena as the result of heating or straining bodies, and we must then suppose that we have conversions of heat or strain energy into electrical energy.

Magnetic Energy.—If a bar of steel is stroked with one pole of a magnet in one direction, the motion of the magnet is resisted and kinetic energy is lost; it is then found that small pieces of iron are attracted by the ends of the bar, and that attractions and repulsions are experienced between the ends of the steel bar and those of the magnet. We have here another case of the development of kinetic energy not accounted for by the immediate disappearance of potential, strain, or heat energy. We have, then, to suppose another form of energy—*Magnetic Energy* — differing from electrical energy in its mode of production, and in the phenomena accompanying its existence, but similar to it in that we have to suppose its existence in order to connect the kinetic energy lost in magnetisation with the kinetic energy appearing in the motion which we term magnetic action.

The last two forms of energy, electric and magnetic, are closely associated with each other, so that, in general, when we have transformation of the one, we find that the other is present. In the electric current, for example, we have every reason to suppose that there is a transformation of electrical energy into heat taking place within the wire; and, accompanying this transformation, we have magnetic actions round the wire, revealing the presence of magnetic energy there. Or, in the dynamo-machine, we have the transformation of kinetic energy into electric and magnetic energies, both being associated to form the energy of the current. Both electric and magnetic energies are forms whose existence, like that of gravitational potential energy, we assume as connecting links between energy disappearing when the electric and magnetic conditions are produced, and energy appearing when those conditions change. Some form of energy disappears during the production of the conditions which we describe as electrification or magnetisation, and we find that the same or other forms of energy may be made to reappear by the electric and magnetic actions. We have no special senses which are affected by the electric or magnetic modifications of matter, and they are therefore in some degree more hypothetical than Kinetic energy, or Heat, or Light.

Chemical Energy.—Lastly, we have to recognise a form of energy in *Chemical Separation.* Heat is necessary to decompose many substances into their constituent elements, the heat disappearing with no other result than the separation of the elements; while, on the other hand, there are numerous familiar examples of the appearance of heat on the union of elements or the chemical combination of previously separated substances. In the voltaic cell, chemical combination occurs when the cell is in action, while in and round the circuit heat is produced, or magnetic actions occur, or light issues from sparks, and these energies we suppose to have come from the constituents in the cell through the intermediate forms of electric and magnetic energies. Conversely, the electric current decomposes substances—the electric and magnetic energies being transformed now into the energy of chemical separation.

The following list includes all the distinct forms of energy which have yet been recognised :—

1. Kinetic Energy.
2. Potential ,,
3. Heat ,,
4. Strain ,,

(1 and 4 are also combined in Sound.)

5. Light and Radiant Energy.
6. Electric ,,
7. Magnetic
8. Chemical ,,

We have already given many examples in which the disappearance of one of these forms is accompanied by the appearance of one or more of the other forms, and our observations and experiments justify us in regarding such accompaniment as the universal rule. There is no example in which one kind of energy is absolutely annihilated without the appearance of another, nor in which one kind of energy appears *de novo* without the loss of another. Indeed, so far convinced are we of this, that an apparent exception would lead us to suspect, not the truth of this statement, but the completeness of our list of energies. We should at once look out for a hitherto unrecognised form which appeared or disappeared, and endeavour to obtain some independent evidence of its existence and only after most careful research and failure to find such a form could we suppose that energy was annihilated or came into existence *de novo*. It is, of course, quite possible, that there are forms of energy yet unrecognised. If we happened to observe the disappearance of energy at one time or place, and the appearance in connection with it of energy at another time or place, and yet could not suppose the existence of any yet known form as connecting link between the two, we should be driven to suppose that some form existed, hitherto unknown, so that the disappearing energy took this form and that on its reappearance it emerged from it, and we should at once look out for the conditions implying the new form. If, for example, the cases of so-called "telepathy" were placed beyond question, we should probably have disappearance of energy from one individual accompanied by its appearance in another at a distance, and it is quite conceivable that we should have to suppose that the energy in the intervening space was in a form not yet known. But up to the present the evidence hardly warrants the assumption that our list is thus incomplete.

It is usual to go somewhat further than the statement that the forms of energy may replace each other—this being all that experiment alone would warrant us in saying—and to regard these replacements as rather transformations of one and the same thing—energy; which we suppose to be identical with itself, though varying in its phenomenal appearance. With this belief we may make the following statement:—

The Identity of Energy.—There is something which we term "energy," and which may be recognised in various forms. When it disappears in one form it appears in one or more other forms.*

The Conservation or Constancy of Energy.—So far we have considered energy only in its qualitative aspect, and have not considered any mode of measuring it except in the kinetic and potential forms. We now proceed to consider in what sense we can assert the constancy in quantity, as well as the permanence of existence of something which appears to us in various forms not directly comparable with each other, *i.e.* not directly measurable in terms of the same unit. An illustration may assist us here. If a man possesses some bank-notes, some gold, and some silver, he may use them all for purchase, since they are all money; and he may exchange any one form for one or more of the others. He may change a note for a mixture of gold and silver, and he may change silver for gold or notes. We may, therefore, assert qualitatively that money appears in various forms. But we may go further, and make a quantitative statement. There is a fixed rate of exchange between the various forms, so that he may exchange a £5 note for five sovereigns and each sovereign for twenty shillings. With this fixed rate of exchange, the total amount reckoned in terms of any one form is constant. If, for example, he begins with four £5 notes he may exchange all or any of them for gold or silver, or both, and the total quantity reckoned in terms of any one coin will be constant, and he will possess twenty pounds or four hundred shillings, even though the actual money may be a mixture of paper, gold, or silver. Were there no fixed rate of exchange, this could not be asserted. If, for instance, his money were partly in English, partly in French notes or coins, the amount reckoned in any one form would, of course, vary at different times according to the rate of exchange. The question then arises, does energy undergo its various transformations according to a fixed rate of exchange in each case? If so, and only on this supposition, the total quantity reckoned in terms of any one form is constant.

To answer this question, we have, in the first place, to consider how the various kinds of energy are to be measured. Kinetic energy we have defined as measurable by $\frac{mv^2}{2}$, in terms of a perfectly definite unit, and we know how the mass of any body may be found, at least in theory, so

* The belief in the identity of energy is no doubt metaphysical, as metaphysical as is our belief in the continued existence of any portion of matter, and its identity under various modifications. As, however, the metaphysical addition somewhat simplifies the form of the statement, and is never likely to lead us wrong in our experimental interpretation, we see no reason to exclude it. Were we to do so we should have to speak of the correlation of the energies, not of the constancy of energy. Instead of describing the conversion of, say, kinetic energy into heat, we should have to say that kinetic energy disappeared, and that at the same time heat energy appeared.

that this expression applies to any kind of matter moving with any velocity.

In the case of Heat, too, we can fix on a perfectly definite unit—say that raising 1 gm. of water from 15° to 16° C.—in terms of which may be expressed any quantity of heat in whatever kind of matter, and whatever the temperature of that matter. Common experience of the working of the steam-engine, or of other modes of transforming heat into kinetic energy, raises a strong probability that if heat energy is measured simply by the quantity of heat, that is, as so many calories, and not by any more complicated expression, then there is a fixed rate of exchange between it and kinetic energy; for the greater the amount of work to be done, the greater the amount of fuel required, and roughly in the same proportion. This probability is converted to certainty by the investigation on the rate of exchange between kinetic energy and quantity of heat, carried out by many physicists and especially by Joule, whose celebrated researches on the value of the rate of exchange we shall presently describe.

But in other forms of energy we have no such general units. In Light, for instance, we may compare the light given by one monochromatic red flame with that given by another red flame of the same quality, and we may, more or less easily, fix on a unit of red light of some particular quality. But another, say a green light, cannot be transformed to red to effect comparison with the unit, and so it cannot be expressed in terms of a red unit. But for each quality or wavelength of light, we may show that the quantity of illuminating power in terms of a unit of its own quality is proportional to the quantity of heat developed when the light is absorbed by a surface on which it falls. For both the heat developed on absorption in a given time, and the illuminating power vary together, inversely as the square of the distance from the source. Hence, each kind of light, which we may assume to be light energy, is transformed into heat according to a fixed rate of exchange. Again, in Chemical energy we have no general unit. We may say that in each special kind of chemical action, the amount of action is proportional to the quantity of substance formed; but we have no direct common unit connecting the action in one case with that in another, where quite a different compound is formed. But in each case, separately, the quantity of heat developed is proportional to the quantity of the substance formed, so that here also, if we assume that chemical energy is proportional to the amount of substance which may be formed by union, the rate of exchange between chemical energy and heat is fixed.

In the case of Strain energy, we may always think of the strain as produced by the transformation of kinetic energy, and since the kinetic energy can be regained on allowing the body to unstrain, with, in general, a small loss accounted for by the heat developed in the processes, we may fairly suppose that the strain energy gained is equal to the kinetic energy lost, and we usually measure it by the amount so lost. Hence, our mode of measurement, itself, assumes a fixed rate of exchange, as in the case of potential energy, and the fact that any discrepancy can be accounted for by other energies appearing, justifies the assumption. We might, perhaps, measure strain energy in terms of that possessed by a standard body with a standard strain, comparing the energy in any other strained body by transferring its energy in a suitable manner to the standard body, and noting the strain produced. Were this done without any consideration of work performed in the transfer, we should then have to measure the kinetic-energy-equivalent of the standard strain of the standard body. But this has not been as yet attempted, and we, therefore, content ourselves with considering the kinetic-energy-equivalent of each case separately.

Electric and Magnetic energies, as we have already remarked, give very little direct evidence of their existence. They are connecting links imagined to come between some recognised form of energy disappearing and others appearing under circumstances such that we cannot imagine a direct transfer. Hence we look for a fixed rate of exchange between the known forms lost on the one hand in electrification or magnetisation, and the known forms gained on the other hand when the electric and magnetic conditions cease. We shall see when we come to consider the energy relations of the electric current that this fixed rate of exchange is shown to exist. We suppose then that it holds also for the transformation into, and out of, the intermediate electric and magnetic conditions, which, indeed, we measure by the supposition of this fixed rate.

Since the other energies are ultimately measured on their transformation either into kinetic energy or into heat energy, and since we have fair evidence for fixed rates of transformation into these two it remains to examine the rate of exchange between heat and kinetic energy. We shall proceed to give an account of the experimental evidence, all of which goes to show that here also the rate is fixed. And anticipating this result we may sum up the whole discussion in the following

Statement of the Principle of the Conservation of Energy.—Energy is recognised in various forms, and when it disappears in one form it appears in others, and in each case according to a fixed rate of exchange. The total quantity of any energy, measured in terms of any one form, is therefore constant whatever forms it may assume.

The Rate of Exchange between Mechanical or Kinetic Energy and Heat Energy, or the Mechanical Equivalent of Heat.—The determination of the rate of exchange depends on the measurement of the work done on some system in which that work results only in a development of heat, and the simultaneous measurement of the heat so developed. The rate was first determined by calculation from the specific heats of air. This method was first set forth clearly by R. Mayer in a paper published in *Liebig's Annalen* in 1842. We shall therefore give it first, though a more direct method was shortly afterwards carried out by Joule.

Mayer's Calculation of the Mechanical Equivalent from the Specific Heats of Air at Constant Pressure and Constant Volume.—The Specific Heat of air, as of other gases, at constant pressure exceeds that at constant volume, and if we can assume that the excess is due entirely to the work done in pushing out the surrounding air in expanding, that is if we can assume that no energy is absorbed in merely separating the particles of air, that they possess no appreciable cohesion, this external work is the mechanical equivalent of the difference between the two Specific Heats. Making this assumption, let us suppose that the volume of air is 272·5 cc., which, according to Regnault, increases 1 cc.

for each 1° rise at constant pressure. Let this volume be at 0° and 760 mm. pressure in a vertical cylinder 1 sq. cm. section, and let the atmospheric pressure be represented by a piston loaded with a column of mercury 76 cm. high and 1 sq. cm. in section, and so weighing

$$76 \times 13{\cdot}596 = 1033 \text{ grammes weight.}$$

When the air is heated from 0° to 1° the volume expands 1 cc. and the piston is pushed out 1 cm., so that work is done equal to 1033 cm. gms.

Now let us turn to the heat measurements. The density of air at 0° and 76 cm. is, according to Regnault, 0·001293, so that the mass of air heated is

$$272{\cdot}5 \times 0{\cdot}001293 = 0{\cdot}3523 \text{ gm.}$$

At constant pressure the specific heat of air C_p is, according to Regnault, 0·2375, and according to E. Wiedemann, whose value we shall take, 0·2389, and this is the heat put in per gramme of air in raising its temperature from 0° to 1°.

But it is only the excess of this over specific heat at constant volume C_v, which is the equivalent of the work done. Now C_v may be found from C_p from the relation (chap. xviii.)

$$\frac{C_p}{C_v} = \frac{\text{Adiabatic Elasticity.}}{\text{Isothermal Elasticity.}}$$

This ratio has been determined in various ways, and we may take its value as very near 1·405,* whence

$$C_v = \frac{0{\cdot}2389}{1{\cdot}405} = 0{\cdot}1700.\dagger$$

Then $$C_p - C_v = 0{\cdot}2389 - 0{\cdot}1700 = {\cdot}0689.$$

Multiplying by the mass heated, we find

$$0{\cdot}0689 \times 0{\cdot}3523 = 0{\cdot}02427 \text{ calory}$$

as the heat equivalent of the 1033 cm. gms. of work done. Then the mechanical equivalent of the calory is $1033 \div 0{\cdot}02427 = 42560$ cm. gms., and this is the mechanical equivalent in centimetres and grammes weight. Mayer, using the data available in 1842, found 36500 cm. gms.

The calculation depends entirely on the assumption that no work is done in the mere separation of the particles of air. In a second paper published in 1845 ‡ Mayer supported the assumption by quoting an experiment by Gay-Lussac (*Mémoires d'Arcueil*, 1807 : *Gilbert's Annalen*, xxx., 1808, p. 249), which went to show that if a gas expands from one vessel into another equal vessel previously empty, the first loses just as much heat as the other gains. Mayer saw that the cooling in the first vessel is due to the work done by the remaining gas in pushing out that which passes into the second vessel, and the heating in this vessel is due

* Meyer, *Kinetic Theory of Gases*, p. 123.

† This is slightly less than Joly's direct determination (p. 85), which gives $C_v =$ 0·17154. The uncertainty in the Specific Heat of Air at constant pressure also makes the result uncertain.

‡ Helm, *Die Energetik*, p. 24.

to the work done on the gas already in it as successive portions enter and compress it. But since, if the heat developed in one were transferred to the other the original temperature would be regained, the experiment shows that mere expansion of a gas when no external work is done does not produce a change of temperature. Thus Mayer's mode of calculation was justified.

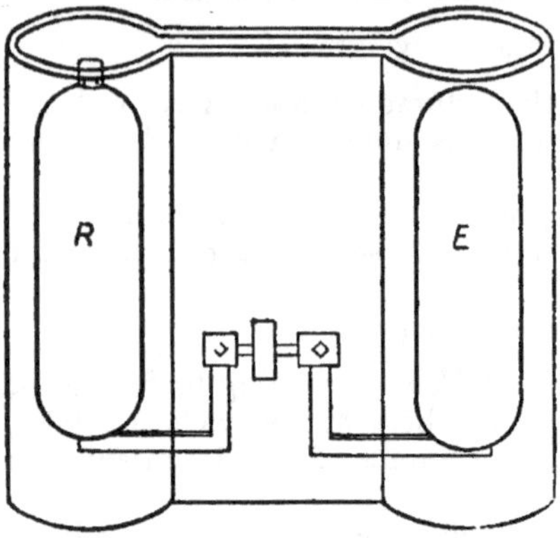

FIG. 73.—Joule's Expansion of Air Apparatus.

Joule's Researches.—Already Joule had begun his series of researches and sought to determine the mechanical equivalent by compressing or rarefying air and equating the heat developed or lost to the work done on or by the air. Here again the calculation depends on the absence of cohesion of the air, and Joule devised an experiment apparently independently, but on the same lines as Gay-Lussac's. Two copper vessels R and E (Fig. 73) were connected by a pipe provided with a stop-cock. R was filled with air at about 22 atmospheres pressure, E was exhausted, and the two were placed in a vessel containing water. The stop-cock was then opened to allow the air to expand and fill both vessels. On stirring the water and taking its temperature, no appreciable alteration was found.

When R and E were arranged as in Fig. 74 in separate calorimeters a notable cooling was observed in the vessel surrounding R and a very nearly equal heating in that surrounding E. This last form of experiment was almost identical with Gay-Lussac's.* The method is not capable of great exactness owing to the large heat capacity of the vessels and calorimeter compared with that of the contained air. But by a method devised later (chap. xviii.) Joule and Thomson showed that there are slight absorptions or evolutions of heat in a gas on mere change of volume, though the amounts are too small to affect the foregoing calculation of the mechanical equivalent.

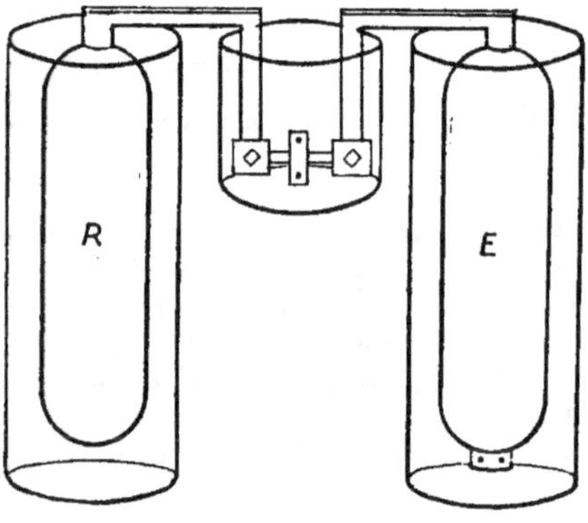

FIG. 74.—Modified Form of Joule's Apparatus.

Joule's Researches on the Mechanical Equivalent of Heat.—At the Cork meeting of the British Association in 1843, Joule gave an

* Gay-Lussac used two globes, each about 12 litres, and each containing a sensitive spirit thermometer. One was exhausted and the other contained air at atmospheric pressure. On opening the connecting-cock one thermometer rose 0·58°, the other fell 0·61°.

account of his earliest experiment on the mechanical equivalent (*Scientific Papers*, i. p. 123). In this he measured the work done in turning a small machine, which we should now call a dynamo, and he also measured the current produced. From his law of heating effect due to current, discovered shortly before, he was able to determine the total heat evolved in the circuit of the dynamo from a measurement of that evolved by a known current in a known resistance, and hence he found the mechanical equivalent to be 836 foot-lbs. per pound of water heated 1° F. In a postscript to the paper describing this experiment he says that by working a piston perforated by small holes, forming narrow tubes, in a cylindrical glass jar holding about 7 lb. of water, each lb. of water was heated 1° F. by the work equivalent to raising 770 lb. 1 foot. In the following year he gave as the results of experiments on the work done in the compression of air and the heat generated, 823 (*Scientific Papers*, i. p. 171), and in a continuation of this work he was led to devise the experiment on the expansion of air described above. The result of this continuation was 798. In 1845 he first described a method in which falling weights were employed to churn water in a calorimeter, and the mechanical energy lost by the weights was taken as equivalent to the heat developed in the water. The result was 890. He then proceeded to improve the conditions of this last and most direct experiment, and in 1850 a full account was published in the *Philosophical Transactions* (*Scientific Papers*, i. 298). The general nature of the apparatus will be seen from Fig. 75. Two masses, each either 10 lbs. or 29 lbs., were attached by strings each to the axle of a wheel and axle *wa*, *wa*. From the wheel strings passed horizontally to the drum *d*, attached to a spindle on which were fixed paddles of brass, 8 in number, revolving in a calorimeter C of copper and containing a known weight, about 7 or 8 lbs. of water. The drum *d* could be detached from the spindle so that the masses could be wound up to a height of 5 feet from the floor without rotating the paddles. When they were wound up the drum was re-attached to the spindle, and as the masses fell the paddles spun round. In the calorimeter were fixed four brass vanes, cut out like the wards of a lock. These allowed the paddles to pass through them, but prevented any continuous circulation of the water and therefore any permanent acquisition of kinetic energy by it. The water was only churned up by the motion of the paddles, and its kinetic energy was rapidly transformed to heat through fluid friction. The arrangement inside the calorimeter is shown by the horizontal and vertical sections in the figure. The masses had a terminal velocity on reaching the floor varying from 1·4 to 3·1 inches per second. This was noted and allowed for as, in effect, diminishing the height of fall. The wheels were mounted on bearings with as little friction as possible, but the residual friction was calculated by subsidiary experiments, and the amount of kinetic energy absorbed by it was allowed for. The calorimeter was on a wooden stand with transverse slits, in order that the calorimeter should rest on a few points of the wood only. Loss by conduction was thus reduced to a very small quantity.

In each experiment the masses were wound up and allowed to fall twenty times, the duration of an experiment being somewhat over half an hour. The rise of temperature of the water in the calorimeter was then noted, and the effect of radiation to or from the surroundings determined by subsidiary experiments. Knowing the capacity for heat of the calorimeter, the total quantity of heat generated could be determined.

The result obtained was that the kinetic energy due to the fall of 772 lbs. through 1 foot at Manchester would, on transformation to heat, raise 1 lb. of water between 55° and 60° F. through 1° F.

Joule also made experiments with mercury instead of water, using an iron calorimeter, and he obtained as a result 774.

In another series of experiments, he made the masses, in falling, turn

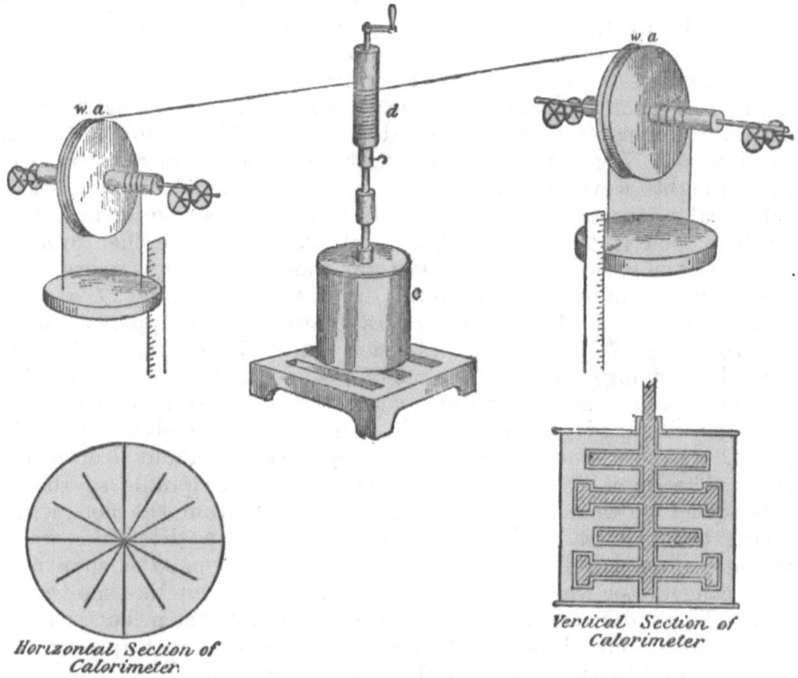

FIG. 75.—Joule's Water-Churning Apparatus for Determining the Mechanical Equivalent of Heat.

a bevelled iron wheel against another fixed bevelled iron wheel, both being placed in the calorimeter and surrounded by mercury. The result was nearly 775. It is worthy of note, as illustrating the care with which he experimented, that he allowed for the loss of energy by the sound given out by the vibration of the apparatus, estimating it by the work required to produce an equal sound as heard at the same distance from a violoncello. But neither of the modes of experiment with mercury was quite as satisfactory as that with the water calorimeter, and they were disregarded in favour of the water-friction experiment.

Later Repetition.—Many years after the publication of Joule's paper giving an account of this work, the British Association, in framing a system of electric units, took as the primary unit of resistance that in which unit rate of working would maintain unit current, and a committee of the Association prepared resistances determined in terms of this unit. Now, in this electro-magnetic system the unit rate of working is 1 erg per second, and the current, C, is measured in terms of dynes on unit pole at 1 cm. distance by 1 cm. length, *i.e.* the resistance, R, is really measured in mechanical units, and when we put for any circuit

$$\text{Rate of working} = C^2R,$$

the product C^2R is in mechanical units, if C and R are in terms of the above units.

But Joule showed that the heat developed in any resistance is proportional to C^2R, which leads us to suppose that the work required to maintain C in R is all transformed to heat in the wire having that resistance. If, then, we measure the heat in calories, the current in electro-magnetic measure, and the resistance in terms of the new unit, we shall have on the one hand the heat, and on the other its mechanical equivalent in C^2R. As soon as the new resistance standards were fixed, Joule himself made an experiment to obtain the mechanical equivalent by this method (*Scientific Papers*, i. p. 542). Converting to foot-lbs. and degrees Fahrenheit, the result was 782 instead of 772, so different a result that Joule was induced to undertake a repetition of his water-friction experiment to find whether the error lay in his work or in that of the committee. The result of his second great experiment was published in the *Phil. Trans.*, 1878, pt. ii.

The method adopted generally resembled that of his earlier experiment, but he employed a different contrivance for doing work, one which had already been used by Hirn. The set of paddles was rotated at a uniform speed by a handle. If the calorimeter had been free to move round the axis of the paddles, it would have spun with them, but it was kept fixed by applying a couple which could be measured. This couple was equal and opposite to that which was exerted on the water and calorimeter by the paddles, and equal to the couple applied to the handle since the rotation was steady. Now, the total work done by a couple G in n revolutions is

$$2\pi n G,$$

so that the value of the mechanical energy was known at once from n and G. For other details the original paper may be consulted (*Scientific Papers*, i. 632). Joule's final result was that 772·55 lbs., falling 1 foot at the sea level in the latitude of Greenwich, would acquire kinetic energy which, transformed to heat, would raise 1 lb. of water from 60° to 61°. This confirmed in a remarkable manner the accuracy of the earlier work, and showed that the determination of the electric standard of resistance was faulty. But the fault had been made plain from other experiments, and a redetermination showed an error of more than 1 per cent.

It will be convenient to express Joule's value of the mechanical equivalent here in various units, for the sake of comparison with later results. These later results are all somewhat larger than Joule's, chiefly through the adoption of the nitrogen or hydrogen scale of temperature, with slightly larger degrees about 15°, than those of the mercury-in-glass scale used by Joule.

Writing after the result the units in which it is expressed, we have

772·55 foot-lbs. at Greenwich per lb. of water per 1° F. at 60° F.;

multiplying by $\frac{9}{5}$ for the Centigrade scale,

1390·6 foot-lbs. at Greenwich per lb. of water per 1° C. at 15° C;

multiplying by 30·48 to convert to centimetres, and remembering that we may use any unit of mass, if we use it on both sides of the transaction:

42385 centimetre grammes at Greenwich per gramme of water per 1° C. at 15° C.;

multiplying by 981·17 the value of g at Greenwich,

$4{\cdot}155 \times 10^7$ ergs per gramme of water per 1° C. at 15° C.

Later Researches. — Joule's work has been followed by other researches on the value of the mechanical equivalent of heat. We shall give a brief account of some of the more important of these.

Rowland's Experiment.—In 1877–78 Rowland repeated Joule's water-friction experiment on a large scale at Baltimore (*Proc. Am. Ac.*, xv., 1879, p. 75). The calorimeter and its contents had a water equivalent about 9 kgm. The paddles were worked by a steam engine, and the spindle to which they were fixed passed up through the bottom of the calorimeter. The calorimeter was hung on a wire, which of course tended to twist when the paddles revolved, but a measured couple was put on to bring the calorimeter back to its undisturbed position. When the rate of working was about 2200 kgm. metres per minute, the rise in temperature was about 25° in 40 minutes. The temperature was measured by mercury-in-glass thermometers, which were standardised in terms of the air thermometer, and the air thermometer temperatures were then reduced to absolute temperatures by means of Thomson and Joule's experiment (chap. xviii.).

More recently Dr. Day has determined Rowland's temperatures in terms of the hydrogen scale, which is now generally used as the standard (*Phil. Mag.*, xlvi., 1898, p. 1). In the table below are given Rowland's values in ergs per gramme degree of water on the absolute scale, and the values corrected to the hydrogen scale, by Dr. Day:—

Absolute Temperature above 0° C.	Rowland's Values in ergs and absolute degrees.	Corrected ergs in degrees on hydrogen scale.
5	$4{\cdot}212 \times 10^7$	...
6	4·209	$4{\cdot}203 \times 10^7$
7	4·207	4·201
8	4·204	4·199
9	4·202	4·198
10	4·200	4·196
11	4·198	4·194
12	4·196	4·192
13	4·194	4·191
14	4·192	4·189
15	4·189	4·188

Absolute Temperature above 0° C.	Rowland's Values in ergs and absolute degrees.	Corrected ergs in degrees on hydrogen scale.
16	4·187	4·186
17	4·185	4·185
18	4·183	4·184
19	4·181	4·182
20	4·179	4·181
21	4·177	4·180
22	4·176	4·179
23	4·175	4·178
24	4·174	4·177
25	4·173	4·176
26	4·172	4·176
27	4·171	4·175
28	4·171	4·175
29	4·170	4·174
30	4·171	4·174
31	4·171	4·174
32	4·171	4·174
33	4·172	4·174
34	4·172	4·174
35	4·173	4·175
36	4·173	4·175

These values bring out very clearly that there is a variation in the work required to raise 1 gramme of water 1° at different parts of the scale—in other words, that there is a variation in the specific heat of water with temperature. They also point to the existence of a minimum value in the neighbourhood of 30°. Rowland's work was the first to indicate the minimum value, and his conclusion has since been confirmed by others.

Miculescu's Experiment.—In 1892 Miculescu presented to the Paris Faculty of Sciences a thesis, in which he gave an account of a water-friction experiment carried out on new lines. The stirring power was supplied by a 1 h.-p. electric motor which was balanced on knife edges in the horizontal line of the axis of the armature. The spindle of the armature was prolonged through a stuffing-box into the water-stirring vessel, where the paddles were attached to it. The couple exerted on the water was measured by hanging a weight on to a horizontal arm projecting at right angles to the axis of suspension of the motor sufficient to keep the motor from turning round. The water in the stirring-vessel was kept at a constant temperature by circulating round the outside of it a current of colder water passing at such a rate that it took heat from the calorimeter as fast as it was developed by the stirring. The heat was measured by the quantity of cooling water sent round and the difference of temperatures of entrance and exit. This difference was measured by a thermocouple which was graduated in terms of a Tonnelot thermometer, and ultimately in terms of the hydrogen scale. The temperature of the water varied between 10° and 13°. We may take the result as making $4{\cdot}187 \times 10^7$ ergs equivalent to the heat required to raise 1 gramme of water at 11·5° through 1° on the hydrogen scale.

Experiment of Reynolds and Morby.—Yet another mode of carrying out the water-friction experiment has been described by Osborne Reynolds and Morby (*Phil. Trans.*, A. 190, 1897, p. 301). Having at command a 100 h.-p. engine provided with a hydraulic brake, the idea occurred to Professor Reynolds that this might be used to determine the amount of work needed to raise the water in the brake from 32° F. to 212° F. The brake itself might be regarded as consisting of paddles working in a water-stirring vessel so arranged that the couple exerted in the stirring could be varied at will and measured at any value. The water was delivered into the brake at 32° F., and was raised in it to about 212° and then passed out, the rate of flow being regulated so that the rise should be through about 180°. The quantity flowing while a given amount of work was done was measured.

The time of running was 62 minutes with a speed of 300 revolutions per minute, and various horse-powers and various quantities of water were used, the total quantity of water rising in some experiments to nearly half a ton.

The final value obtained is that the mean specific heat of water between 32° F. and 212° F. measured in foot-lbs. at Manchester is 776·94. In ergs and degrees centigrade it is $4{\cdot}1832 \times 10^7$.

Griffith's Experiment.—In 1883 E. H. Griffiths gave an account of an experiment to determine the mechanical equivalent of heat by the method of electrical heating (*Phil. Trans.*, 184, A., 1893, p. 361). This research was carried out with the greatest care in every detail, and the original paper should be consulted for particulars and especially for the method of temperature regulation.

A coil was immersed in a calorimeter 8 cm. deep and 8 cm. wide, containing various quantities of water up to about 250 gms. The calorimeter was closed by an air-tight lid through which passed the thermometer, the stirrer, and the wires to the coil. It was suspended in an exhausted enclosure and the walls of this enclosure were double, the cavity between being filled with mercury. A graduated tube led out from the cavity so that it formed, practically, the bulb of a big thermometer in the middle of which the calorimeter was suspended. Any variations in the temperature of the enclosure could thus be detected. The double-walled vessel was surrounded by water, and was kept as nearly as possible at a uniform temperature in order that radiation loss could be exactly allowed for.

When a current was passed through the coil, if E was the fall of potential in it (determined by comparison with a Clark's cell), and if R was its resistance, both in electro-magnetic units, the rate of energy supply was E^2/R ergs per second. Measuring the heat developed in any time the number of ergs per calory could be determined.

The thermometer—mercury-in-glass—was compared with a platinum-resistance thermometer and the indications of this were found first in terms of the air thermometer and later in terms of the hydrogen scale. Different temperature ranges were used between 15° and 25°, and the final result corrected to the hydrogen scale was very nearly

$$4{\cdot}2 \times 10^7\{1 - {\cdot}000266(t - 15^\circ)\} \text{ ergs per gramme of water heated } 1^\circ \text{ of the hydrogen scale,}$$

where t is the temperature on the Centigrade scale.

Experiment of Schuster and Gannon.—Another determination by the method of electrical heating was described by Schuster and Gannon in the *Phil. Trans.*, 185, A., 1895, p. 415. They immersed a coil in a calorimeter containing about 1500 gms. of water, and passed a current through it. The energy supplied was E*Ct* ergs in time *t*. By passing the current through a silver voltameter, and using the known electrochemical equivalent of silver the value of C*t* was determined, while E was measured by comparison with a Clark's cell. The temperature was measured by a mercury thermometer, which was standardised in terms of the nitrogen and the hydrogen scales. It is interesting to observe the difference in the result expressed in ergs per calory at 19·1° on these three scales—

$4{\cdot}1804 \times 10^7$ on mercury thermometer of hard French glass.
$4{\cdot}1905 \times 10^7$ on nitrogen scale.
$4{\cdot}1917 \times 10^7$ on hydrogen scale.

The results below, taken from Schuster and Gannon's paper, give the

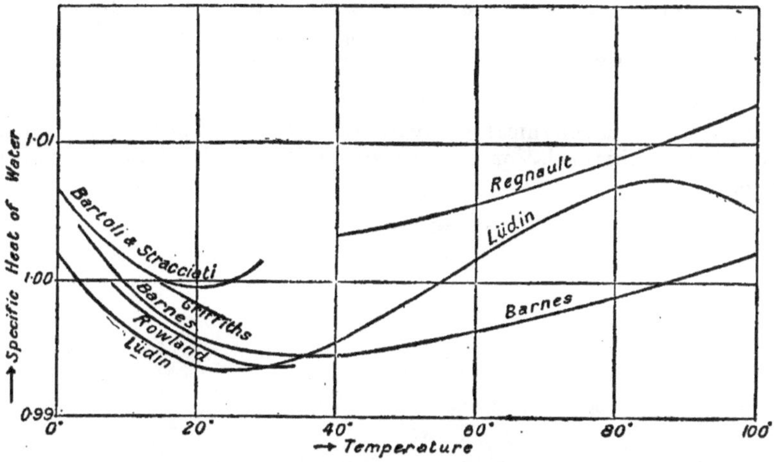

FIG. 60.—Results of Different Experiments on the Specific Heat of Water.

values obtained by different observers in foot-lbs. at Greenwich, and Fahrenheit degrees of the nitrogen thermometer at 15° C.; the values in ergs and Centigrade degrees of the same thermometer are added:—

	Foot-lbs. °F.	Ergs. °C.
Joule (1878)	775	$4{\cdot}178 \times 10^7$
Rowland	778·3	$4{\cdot}1895 \times 10^7$
Miculescu	776·6	$4{\cdot}180 \times 10^7$
Griffiths	780·2	$4{\cdot}199 \times 10^7$
Schuster and Gannon .	779·2	$4{\cdot}194 \times 10^7$

And in terms of the mean calory:

Reynolds and Morby . .	776·94	$4{\cdot}1832 \times 10^7$

The values given in terms of the nitrogen scale must be increased by 1 in 2500 to bring them to the hydrogen scale. In Fig. 60 the results obtained by different experimenters are exhibited in terms of the variation of the specific heat of water.

An inspection of these numbers shows that the results by the two electrical methods are somewhat closely in agreement, and higher than those given by water-friction methods. It appears probable that there is some error in the electrical relations assumed, and some recent experiments suggest that this is in the value of the E.M.F. of the Clark's cell. Meanwhile, Rowland's corrected value at 15° may be taken, and the first three figures of this may be put as $4{\cdot}19 \times 10^7$ ergs. Hence we have the mechanical equivalent—

$4{\cdot}19 \times 10^7$ ergs;
4·27 metre grammes;
778 foot-lbs at Greenwich;

in terms of degrees on the hydrogen thermometer about 15° C., the last, of course, on the Fahrenheit scale.

The First Law of Thermodynamics.—When exchange occurs between work and heat, the researches we have been describing all tend to show that the ratio of exchange is fixed. Hence if W is the work in ergs equivalent to heat H measured in calories, we have

$$W = JH$$

where J is the mechanical equivalent, taken provisionally as equal to $4{\cdot}19 \times 10^7$. This relation is known as the First Law of Thermodynamics.

CHAPTER IX.

THE KINETIC THEORY OF MATTER.

Atomic Hypotheses—Solids—Liquids—Gases—Kinetic Theory of Gases—Mean Value of the Square of the Velocity of Translation $\bar{V}^2$—Mixture of Gases—Relation between $\bar{V}$ and Temperature—Energy of Translation and Internal Energy—Joule's Approximate Method of Calculating the Velocity of Mean Square—Effusion or Transpiration through a small Orifice into a Vacuum—Thermal Transpiration—The Mean Free Path—The M.F.P. calculated from the Coefficient of Viscosity—Conduction of Heat in Gases—The Diameter of the Molecules and the number of Molecules per Cubic Centimetre—Forces acting on unequally heated Surfaces in High Vacua—The Gas Equation of Van der Waals

THE belief in the identity of energy throughout its apparent transformations naturally leads us to attempt to explain these various manifestations as really identical in form, though affecting our senses differently. As examples of hypotheses of this kind, we may instance the electro-magnetic theory of light, in which light (or rather radiant energy) is supposed to consist of a mixture of electric and magnetic energies; and the electrical theory of chemical energy, which identifies chemical action with electric energy. But we obtain our primary idea of energy from the power of doing work possessed by a moving body, and we are able to study most thoroughly the energy-transactions in a system which we need only consider mechanically—*i.e.* with regard to its kinetic and potential energies. Our aim, then, must be to frame hypotheses which shall reduce other energies to these forms, for such hypotheses can be most completely worked out, and their consequences most thoroughly compared with the facts of observation.

Atomic Hypotheses.—The basis of these attempts must almost necessarily be the hypothesis of an atomic or molecular constitution of matter. This hypothesis we owe to the Greek philosophers, who possibly arrived at it from considerations of the compressibility, the disintegration, and the diffusion of matter. If matter is composed of small particles separated by interspaces, it is easy to explain *compression* as an approach of these particles, *solution* as an entry of the particles of one substance into the spaces separating the particles of the other, *cleavage* as the forcing apart of the particles by the insertion of a wedge of some other substance, *diffusion* as the scattering of the particles, and so on. If we suppose that matter is continuous, then we must accept such facts as compression and solution as simple facts—*i.e.* facts incapable as yet of explanation.

Considered only as explaining the general and most obvious properties of matter, the atomic hypothesis is merely qualitative. It became quantitative when used by Dalton and other chemists to explain the laws of chemical combination, and the simplicity of this explanation has no doubt largely contributed to its universal acceptance. The hypothesis in its original form may be stated thus: A given mass of an element contains a definite number of exactly similar indestructible particles, termed atoms, these particles being exceedingly minute and exceedingly numerous in any volume which comes within the range of our senses. Each atom of a given element has a definite mass, which is always the same for the same element.

Though this form of the hypothesis is sufficient for many purposes, the relations which exist between the atomic weights and the other properties of the various elements lead to the supposition that the atoms of different elements are not simple indestructible particles unrelated to each other, but are built up of some common material, the differences being in the quantity and arrangement of this common material. Were an atom of iron, for instance, entirely unrelated to an atom of platinum, had they existed as they are through all time, it would be difficult to account for such a relation as the equality of their atomic heats. It is now generally believed that such relations may ultimately be deduced from the supposition that the atoms are complex bodies, containing still smaller particles or corpuscles, all alike in all elements, but different in number and arrangement in the different atoms. Recent experiments, especially certain experiments on the discharge of electricity in rarefied gases, have given great weight to this supposition, since they may be to some extent explained if the atoms are regarded as being split up in the discharge, giving out corpuscles, the same for different gases. In this chapter, however, we shall not go beyond the supposition that the atoms are simple bodies.

In chemical compounds the dissimilar atoms of the constituent elements are supposed to be grouped together to form molecules similar to each other, the masses of the components being in the same proportion in the molecules as in the compound as a whole, and the laws of chemical combination find a simple explanation by supposing that the mass of each kind of atom is definite.

The atoms, even in an elementary substance (in which, so far as we know, they are all similar), are probably to be regarded as grouped, perhaps in pairs, perhaps in greater numbers, and we may fairly extend the term " molecule " to describe a group so formed.

When we seek to explain the various forms of energy on this atomic or molecular hypothesis, we have further to suppose that the atoms are held together by forces, for work is required to effect chemical decomposition.

Work, too, is required to alter the volume of any liquid or solid body in its normal condition. Hence, we must regard the molecules also as held together by forces. We may have energy of motion both of atoms and of molecules, and energy of position both of atoms with regard to each other, and of molecules with regard to each other.

We will now consider the explanation which the atomic hypothesis gives of the phenomena of heat.

When two solid bodies are pressed together and rubbed, one on the other, work is done against friction and ordinary kinetic energy disappears as fast as it is supplied to the bodies. The surfaces in contact are found to rise in temperature. We may suppose that the friction means that the surface-particles of one body catch against the surface-particles of the other, and that there are forces between these particles against which the work is done. The collision will give the colliding particles greater motion than that which they possessed before, and so we get the idea that the heat developed is, at least in part, due to the increase in kinetic energy of these particles.

When a body is heated it generally expands—*i.e.* its molecules are on the average at a greater distance apart. This again implies that motion has been given to them to move them apart: but it also probably implies the existence of energy of position, the equivalent of the work done in increasing the separation of the molecules against the forces between them.

We also know that a hot body radiates energy to the surrounding bodies—energy which affects our sense of sight, if the source be hot enough. We know, too, that this energy is associated with waves, implying a vibratory motion at the source. The waves are so small, and the frequency of vibration producing them is so great, that we are led to suppose the individual molecules or atoms to be themselves the vibrating bodies. As to the mode of vibration we can only guess. It may be that each separate atom vibrates in itself as a bell when struck—or each atom may travel to and fro about some mean position, as a bell on a spring might do—or both modes may co-exist. We may therefore have energy, existing in various ways, associated with the atoms and molecules. We may have vibrating energy of the atom, energy of motion of the atom as a whole, energy of motion of the molecule as a whole, energy of separation of the atoms from each other, energy of separation of the molecules from each other, and we must now regard heat as a mixture of some or all of these. In other words, heat consists in the kinetic and potential energies of the ultimate particles of matter.

We must make further assumptions as to the forms and constitutions of the atoms and the actions between them if we wish to go further and work out their motions by ordinary mechanics. One attempt of this kind was made in the ring-vortex theory of Lord Kelvin,* which supposed that the atom was a ring-vortex of fluid existing in an infinite fluid. According to this theory, the energy of any system is entirely kinetic, either energy of motion of the vortices, or energy of motion of the surrounding fluid. It would therefore reduce all forms of energy to the single form of kinetic energy, and on this account it is worthy of mention, though recent electrical work has led to the imagination of a very different type of atom.†

Without entering into any details as to the construction of the atoms or molecules, we may still be able to form some general notions of their arrangements and motions in the various states of matter.

In solids, the atoms and molecules are probably only agitated about some mean position, for we know that solids keep their shape for indefinitely long periods if not exposed to external action. Carefully preserved jewels engraved by the ancients still possess all their original sharpness of outline. We cannot, then, suppose that the molecules are travelling about, but only moving to and fro to a very limited extent. The molecules

* Maxwell's *Scientific Papers*, vol. ii. p. 470; "Atom," from *Encyc. Brit.*, 9th ed.
† J. J. Thomson, *Electricity and Matter*, p. 90.

are sufficiently close together to act very considerably on each other, and work has to be done to alter their arrangements in any way, as is shown by the elasticity of solids.

In liquids, we must suppose that the molecules are not only agitated but are travelling about, though only progressing slowly, for the phenomena of liquid-diffusion show that the molecules take some time to travel any considerable distance in a liquid. We may suppose that the molecules possess on the average sufficient energy to do the work needed to get away from their neighbours, but that they are still so near together that they readily become entangled again. Their great resistance to compression and their cohesion show, too, that the molcules are near enough to act very considerably upon each other.

The existence of *viscous solids*, such as pitch, intermediate between solids and liquids—substances which will flow like liquids if a sufficiently long time is given to them—seems to show that in these solids there is a certain amount of travelling about of the molecules. We may, perhaps, suppose that a given molecule will, in its excursions about its mean position, come within the sphere of action of its neighbours in such a way as to receive from them continual supplies of kinetic energy which enables it to increase the extent of its excursion until it can break away and travel on to some new position of less constraint. It is quite possible that exactly the same process takes place in every liquid, only on a much greater scale, many of the liquid molecules vibrating about a mean position like solid molecules for a time, but ultimately, by the action of neighbouring molecules, becoming detached and travelling about till they become entangled by new surroundings.

We may probably explain, by the aid of this supposition, the viscosity of a liquid. If a liquid is in motion in one direction, but in such a way that each layer moves slightly faster than the one below it, then there is a tangential action between the layers proportional to the change of velocity per unit length perpendicular to the layers.

If, for example, the successive layers A,B,C,D, Fig. 76, are all moving in the direction AX, but A moves v per second more than D at a distance d from it, the tangential force per square centimetre exerted by each layer on the next is $\eta\frac{v}{d}$, where η is the coefficient of viscosity. Now each element of the liquid is being sheared by the relative motion of the successive layers, and we may suppose that for a short time after the shearing the element resists the shear just as a solid element would, but that the resistance rapidly dies away, owing to the breaking loose of the molecules from their old positions and their adjustment in new positions. The strain producing the tangential stress is not the whole strain of the element since the beginning of motion, but only that part of it in which the particles have not yet had time to rearrange themselves. It is easy to see that if rate of decay of strain is proportional to the strain, then if the relative velocity per unit length v is doubled, this effective strain is doubled, and

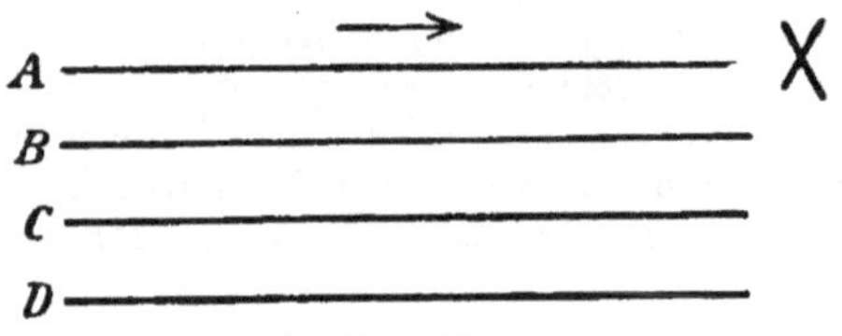

FIG. 76.

the force is doubled, or η is a constant. Anything which hurries the rearrangement will lessen the viscosity. For instance, we have every reason to suppose that rise of temperature increases the energy of the molecules, and therefore enables them to get free from each other more frequently. This agrees with our knowledge that the viscosity of a liquid decreases as the temperature rises.

Kinetic Theory of Gases.—In applying the kinetic theory to *gases* we can go into much greater detail. The molecules in a gas are much farther apart on the average than in solids or liquids. A cubic centimetre of water at 100° C. forms about 1600 cubic centimetres of steam at 100° C. and 760 mm. pressure. Hence, the molecules of the steam are about $\sqrt[3]{1600}$, say 12 times as far apart as those of the water. And this increased distance results in an almost complete absence of cohesion, as is shown by the experiments of Gay-Lussac and Joule (p. 119), in which a gas is allowed to expand without doing external work. The change of temperature is only infinitesimal, showing that practically no work is done against the mutual actions between the molecules. The extreme rapidity of gaseous diffusion shows that the molecules are in very rapid motion. But as there are an enormous number of molecules, even in a very small space, and as they are not mere points, but have some volume of their own, they must continually be colliding with each other, and under the term "collision" we must include every case in which two molecules come sufficiently within each other's sphere of action to influence each other's motion. We need not necessarily suppose that in a collision the force between two molecules is repulsive. We may illustrate this by the motion of a comet which comes into our system from outer space. Drawn by the sun's attraction it rushes inwards, travels round the sun, and then rushes away to outer space again. From our present point of view this is a "collision," and the collision of molecules may perhaps be similar. The collisions in gas at ordinary pressure cannot, however, occupy a very appreciable fraction of the time of travelling about, for this would be inconsistent with the absence of cohesion. When, in Joule's experiment, there are 22 times as many molecules in a given space as in ordinary air, there are many more collisions in a given time for each molecule; in fact, the number for a given gas per second is nearly proportional to its density. If these collisions occupied an appreciable fraction of the time in ordinary air they would, therefore, occupy a still larger fraction in the compressed air. At any one instant, then, an appreciable fraction of the total number of molecules would be in collision, and hence work would have to be done to separate them, that is, to lessen the number within each other's sphere of action. In other words, there would be cohesion.

We shall, in the first place therefore, regard the collisions as instantaneously altering the velocities and directions of motion of the colliding molecules. We shall not attempt to inquire what goes on in a collision, and in speaking of the velocities of the molecules we shall regard only the velocities when free from each other's actions.

When a collision occurs, the velocities and directions of motion of the two colliding bodies are changed; but we shall suppose that the total energy of the motion of translation remains the same after the collision is finished as it was before it began. We know that the energy of translation does not in reality always remain constant, for a glowing gas, when cooling, loses some of its energy of translation, and loses it in part by radiation. This implies that in some cases some of the energy of motion is converted into vibratory energy of the molecules and atoms in the collision, and is then radiated out. In other cases the collisions may convert energy of vibration into energy of translation, for radiant energy is also being absorbed from surrounding bodies to some extent. When, therefore, the temperature is maintained constant we must suppose that there is a balance between loss and gain, and that the mean kinetic energy of the whole group remains the same, even though it may change for individual pairs of colliding molecules.

The assumptions that the time occupied in collision is very small, and that the number of molecules in even a small space is very large, taken together, imply that the dimensions of the molecules are exceedingly small compared with their distance apart.

The most characteristic features of a gas are its diffusibility throughout any vessel in which it is contained, however the volume of the vessel may be increased, and the uniformity of the pressure which it exerts over the containing-walls in accordance with Boyle's Law. On our theory, the diffusion is due to the rushing of the molecules into any space open to them, while the pressure on the containing-walls is due to the cannonade of the molecules against them. Each molecule as it comes up against the walls rebounds, and imparts momentum to the walls equal and opposite to that which it receives. Though on an area comparable with the dimensions of a molecule the cannonade in a very small time may vary, on any sensible area an enormous number of impacts will occur in any sensible time, and the average will be practically constant. If some molecules lose energy in the collision, others gain, and on the average we must suppose the loss and gain equal. Hence the average momentum imparted per second (that is, the pressure exerted) will be the same everywhere. We can see, too, in a general way how Boyle's Law is explained, for doubling the number of molecules in a space doubles the number colliding in a given time against the containing-walls, and so doubles the momentum imparted per second. In other words, the pressure is proportional to the density.

The Mean Value of the Square of the Velocity of Translation.—Knowing the pressure and the density of a gas, it is possible to calculate the mean value of the squares of the velocities with which the molecules are moving.

If we follow any one molecule in imagination, its velocity will be continually changing through collisions; but if we consider a large number of molecules, say those in a cubic centimetre, we may safely assume that, so long as the conditions exhibited by the whole are the same, the velocities are distributed in such a manner that a definite and constant fraction of the whole will be moving with a given velocity—or with a velocity within narrow given limits, though the individuals may be continually changing. This assumption * is justified by our

* More advanced theory than we can give here shows how the velocities of the molecules must be distributed in order that the collisions may not affect that distribution. We may refer the reader to O. E. Meyer's *Kinetic Theory of Gases*, from which much of the theory given in the text is derived.

experience of statistics of population dealing with large groups of individuals. Whenever the group as a whole shows constant features, we find that it can be subdivided into smaller groups, also showing constant features, even though the individuals in these groups change; for example, in a large town where the circumstances remain pretty nearly the same, the percentage of the population whose age lies between given limits will remain constant, though fresh individuals are coming into the group and others are moving out of it. Or, to take another illustration more nearly resembling the case of a gas. With similar external circumstances as to day, hour, and weather, probably a certain fraction of the people in a town will be in the streets on different days, with so many moving at four miles per hour, so many at three miles per hour, so many in collision, stopping to talk to each other. The individuals forming each of these groups will change, but the number in the group is probably nearly constant.

Similarly, if we consider a sufficient number of molecules, we may assume that a constant number move with a given velocity in a given direction.

Let us take, for simplicity, a vessel whose interior is cubical and one centimetre each way, and let ABC be three perpendicular faces meeting at an angle. Consider a molecule impinging with velocity V_1 in a certain direction against the face A of the vessel, and resolve this velocity into three components u_1, v_1, w_1, perpendicular respectively to A, B, and C.

Then $$V_1^2 = u_1^2 + v_1^2 + w_1^2$$

We shall suppose that the walls of the vessel are perfectly plane and with coefficient of restitution unity, so that a molecule impinging on a side has its velocity perpendicular to that side exactly reversed, while the other components are unaffected. This, as we have seen, is probably not true for individual impacts—in some there may be a gain of energy, in others a loss; but, so long as the gas and wall are at the same temperature, the average energy is the same before and after impact, and the number of molecules moving away with a given velocity is the same as if in each impact the above supposition were true. That this is the right view is shown by considering the case in which it is no longer true, that in which the temperature of the gas differs from that of the wall. Then the energy of the gas is different after impact, and from this difference, as we shall see later, such motions as that of the radiometer can be explained.

If we take the mass of the molecule as m, it moves up with the velocity u_1 perpendicular to the wall, and has this changed to $-u_1$, or there is a total change of momentum, $2mu_1$. This, therefore, is the momentum given to the wall by the impact.

Through the velocity u_1 the molecule would move to and fro u_1 times per second if there were no collisions, and impinge against A $\frac{u_1}{2}$ times. Now, though no actual molecule does this, the effect is the same as if it did, for when it has its velocity altered by a collision, some other molecule at the same distance from A takes its place, and moves on with the same velocity, and no appreciable time is, by our supposition, lost by the collisions. In one second then the total momentum imparted to A by one molecule will be $2mu_1 \times \frac{u_1}{2} = mu_1^2$. If we have in all n molecules, their velocities perpendicular to A being $u_1, u_2 \ldots u_n$ and if the total momentum imparted per second to A is equal to p,

$$p = m(u_1^2 + u_2^2 + \ldots\ldots + u_n^2).$$

Considering the face B we similarly obtain

$$p = m(v_1^2 + v_2^2 + \ldots\ldots + v_n^2)$$

and on C

$$p = m(w_1^2 + w_2^2 + \ldots\ldots + w_n^2)$$

Adding these three together,

$$3p = m(u_1^2 + v_1^2 + w_1^2) + m(u_2^2 + v_2^2 + w_2^2) + \ldots\ldots + m(u_n^2 + v_n^2 + w_n^2)$$
$$= m(V_1^2 + V_2^2 + \ldots\ldots + V_n^2)$$

If $\overline{V}^2$ is the mean of the squares of the velocities, then we may write this in the form

$$3p = mn\overline{V}^2$$

But mn is the total mass of gas in 1 cc., that is, is equal to the density ρ.

Hence $$3p = \rho\overline{V}^2 \text{ or } \overline{V}^2 = \frac{3p}{\rho}$$

$\overline{V}$ is not the mean velocity, but the square root of the mean of the squares of the velocities. It is termed "the velocity of mean square," and it may be shown that it is somewhat greater than the mean velocity.

Maxwell investigated the law of distribution of velocities about the mean which would justify the supposition of constancy of distribution, and he showed that it is of "exactly the same mathematical form as the distribution of observations according to the magnitude of their errors, as described in the theory of errors of observation. The distribution of bullet-holes in a target according to their distances from the point aimed at is found to be of the same form, provided a great many shots are fired by persons of the same degree of skill" (Maxwell's *Theory of Heat*, p. 309, ed. 5). It can be shown that the mean velocity U is given by

$$U = \sqrt{\frac{8}{3\pi}}\overline{V}$$

or very nearly $$U = \frac{12}{13}\overline{V}$$

If the gas is hydrogen at 0°, then if $p = 1014000$ (the value of 76 cm. of mercury in dynes per sq. cm.) ρ is very nearly ·00009.

Whence $$\overline{V} = 184000 \text{ cm./sec. nearly;}$$

and $$U = 170000 \text{ cm./sec. nearly.}$$

If we take any other gas at the same temperature and pressure the velocity is inversely as the square root of the density. Hence for oxygen we must divide by 4, and

$$\overline{V} = 46000 \text{ nearly,}$$
$$U = 42500 \text{ nearly.}$$

For nitrogen we must divide by $\sqrt{14}$, and

$$\overline{V} = 49000 \text{ nearly,}$$

$$U = 45400 \text{ nearly.}$$

If we assume that $\overline{V}^2$ is the same for a given temperature, whatever the pressure—and this is the natural assumption, seeing that the temperature of a gas only alters very slightly if its volume is changed without doing external work—then at once we have $\frac{p}{\rho} = \text{constant}$ for a given temperature, or Boyle's Law follows.

It must, however, be observed that our investigation is based on the assumption that the collisions of each molecule take up only a negligible fraction of the time. If through a crowding up of the molecules this ceases to be true, we can no longer assume that a molecule of type having velocity u perpendicular to a face of the cube will travel u cm. per sec., for some of the time will be wasted in collisions. The more time thus wasted the fewer returns will the molecules make to the wall, and it will contribute to the pressure a less amount than that calculated. But we shall return to this point later and consider how we may obtain a more correct relation between pressure and volume than that given by Boyle's Law.

Mixture of Gases.—Maxwell showed that in a mixture of gases the different kinds of molecules will exchange energy with each other till the average kinetic energy of a single molecule of each kind is the same, and that this is the condition for steadiness of distribution.

Now, assuming this, suppose that we have equal volumes of two different gases at equal pressures and of densities ρ_1 and ρ_2, the equality of pressure gives

$$\rho_1 \overline{V}_1^2 = \rho_2 \overline{V}_2^2$$

If the two gases be allowed to mix there is no work done externally, and no change of temperature, and the pressure remains the same. This is accounted for if we suppose that the velocity of mean square of each gas is the same after mixture as before.

But if m_1 is the mass of a molecule of the first gas, and m_2 that of a molecule of the second gas, Maxwell's investigation shows that

$$m_1 \overline{V}_1^2 = m_2 \overline{V}_2^2$$

or

$$\frac{\rho_1}{m_1} = \frac{\rho_2}{m_2}$$

But if N_1 N_2 be the numbers of the two kinds of molecules in unit volume before mixture,

$$N_1 m_1 = \rho_1 \qquad N_2 m_2 = \rho_2$$

whence

$$N_1 = N_2$$

or two different gases at the same temperature and pressure contain the same number of molecules per c.c. This is known as **Avogadro's Law.**

Relation between $\overline{V}$ and Temperature.—If we take a volume of a single gas and keep its density constant while altering its temperature, we have the relation between the pressures at t° and at 0° C.

$$p_t = p_o(1 + at)$$

But $$\frac{\overline{V}_t^2}{\overline{V}_o^2} = \frac{p_t}{p_o} \text{ if } \rho \text{ is constant.}$$

Therefore, $$\overline{V}_t^2 = \overline{V}_o^2(1 + at) \quad \text{or} \quad \overline{V}_\theta^2 = \overline{V}_o^2 a\theta,$$

where θ is the temperature on the gas scale. Hence the energy of translation of the molecules in a c.c., $\rho\frac{\overline{V}^2}{2}$, is proportional to the gas temperature.

Energy of Translation and Internal Energy.—If our investigation applies to real gases it is easy to show that the energy of translation is not for most gases the only energy possessed by the molecules. When a gas is heated we must suppose that in general some of the energy goes to increase molecular potential or molecular vibrational energy. For the energy of translation of 1 c.c. at 0° C. is

$$\frac{\rho\overline{V}_o^2}{2} = \frac{3p_o}{2}$$

and if the volume is constant the increase in translational energy, fcr a rise of 1°, is

$$\frac{3ap_o}{2}$$

where a is the coefficient of pressure increase.

The total increase of energy is K_v, the work measure of the specific heat at constant volume.

The difference of the specific heats at constant pressure and constant volume is given by $K_p - K_v = ap_o$, the work done in expansion.

Putting $$\frac{K_p}{K_v} = \gamma, \text{ we have } (\gamma - 1)K_v = ap_o$$

Hence, $$\frac{\text{increase in translational energy}}{\text{increase in total energy}} = \frac{3ap_o}{2K_v} = \frac{3(\gamma-1)K_v}{2K_v} = \frac{3}{2}(\gamma - 1)$$

This is unity, or the total energy given is converted wholly into translational energy when and only when

$$\gamma - 1 = \frac{2}{3} \text{ or } \gamma = \frac{5}{3} = 1{\cdot}66$$

This is the value of γ found by experiment for mercury vapour, argon, and helium. For these gases, then, we must suppose the collisions to be of such kind that the internal conditions of the molecules are not appreciably affected by the collisions; in fact, that there is no interchange between the energy of translation and the energy of position or vibration of the constituent parts of a molecule.

According to a theorem due to Boltzmann, if a molecule be regarded as a mechanical system having n "degrees of freedom" or such different modes of motion that it requires the knowledge of n different quantities to specify its position and configuration at any instant, the energy must be equally shared between the different modes when the distribution of velocities and of internal energy is permanent. A purely translational motion has three degrees of freedom, say the motions parallel to three perpendicular axes. If there are other degrees of freedom implying possibility of change of internal arrangement, making with the three translational degrees n in all, we have

$$\frac{\text{translational energy}}{\text{total energy}}=\frac{3}{n}=\frac{3}{2}(\gamma-1), \text{ whence } \gamma=1+\frac{2}{n}$$

If $n=3$ we have the case of mercury vapour, helium, and argon. We may, merely for illustration, picture the molecules in these gases as small perfect spheres, perfectly smooth if they come in contact at collision, or else in an encounter never actually touching. Then their mutual actions always pass through their centres, so that there never can be any interchange of energy between the translational and the rotational forms, and we need not consider the co-ordinates necessary to specify the configuration of the sphere. Of course this is a very crude illustration, but it serves to show that it is conceivable that there may be degrees of freedom like those expressing the rotation of the spherical molecules which may be omitted from consideration, since the forces producing or destroying that rotation do not come into play in encounters.

For oxygen, hydrogen, and nitrogen,

$$\gamma=1\cdot 4 \text{ very nearly.}$$

Then

$$n=5$$

We may here picture the molecules as pairs of atoms rigidly attached and forming, as it were, dumb-bells. If the actions at encounters always pass through the axis of symmetry the co-ordinates expressing the orientation round that axis may be omitted and the position and configuration of a molecule is sufficiently given by the three co-ordinates of its centre of gravity and the two angles giving the direction of the axis of symmetry, thus accounting for $n=5$.

If the distance between the atomic pair is variable, $n=6$ and $\gamma=1\cdot 33$, a value possessed by some gases.

As n increases, γ approaches 1, and observation shows that for more complex gases this is true.

It may be observed that if the omission of any degree of freedom is admissible through its force not coming into play at collision, then if motion in that degree is produced by the absorption of radiation of a given wave-length passing through the gas, such radiation will not affect the energy of translation but only the molecular configurations. Thus we may suppose that the pressure of argon against the sides of the containing vessel will not be appreciably altered if it is exposed to radiation which it can absorb.

But the questions here discussed or rather indicated are still open. It is held by many that Boltzmann's theorem does not really apply to molecular systems; that, in fact, the conditions assumed by Boltzmann in the proof of his theorem are not realised in actual molecular systems.

Joule's Approximate Method of Calculating the Velocity of Mean Square.—The first calculation of the velocity of the molecules was made by Joule by a method which is obviously only approximately correct, but which is valuable in that it enables us to obtain easily results of the right order and which gives us at any rate insight into the principles of the theory. Considering, say, a cubic centimetre, let us think of the molecules as divided into six streams moving perpendicularly to the six faces of the cube, one stream towards each face, and let us omit all consideration of the collisions between molecules. Then the stream moving towards one face at any instant has mass $\frac{\rho}{6}$ where ρ is the density of the gas. Let the velocity of the stream be V. Then the total mass moving up to the face in one second is that which would be contained in V c.c., or is $\frac{\rho V}{6}$. But its velocity is changed by impact against the face from $+V$ to $-V$, so that the momentum imparted to the face in one second is $\frac{\rho V}{6} \times 2V = \frac{\rho V^2}{3}$. Equating this to the pressure, we get $V^2 = \frac{3p}{\rho}$.

Effusion or Transpiration through a small Orifice into a Vacuum.—The phenomena of "transpiration" through a small orifice may be generally explained by the aid of Joule's method, though, of course, the method cannot be expected to give a complete account. If a gas of density ρ is contained in a vessel with a small orifice, area S, we may think of mass $\frac{\rho V}{6}$ as moving up per second towards the face containing S, and the mass escaping through S will be $\rho\frac{VS}{6}$ per second.

The mass escaping when the distribution of velocities according to Maxwell's Law is taken into account can be shown to be

$$\rho\frac{US}{4} = \frac{12\rho VS}{13 \times 4} = \frac{3}{13}\rho VS$$

where U is the mean velocity and V that of mean square, so that the approximation in Joule's method gives the numerical coefficient too small in the ratio 13 : 18, and this example may serve to show the kind of error introduced by that approximation.

If we have two vessels containing different gases at equal pressures escaping through equal orifices, the masses escaping per second will be in the ratio $\rho_1 V_1 : \rho_2 V_2$, and the volumes escaping in the ratio $V_1 : V_2$; or since $V_1^2 : V_2^2 = \rho_2 : \rho_1$, when the pressures are the same, the volumes are as

$$\sqrt{\rho_2} : \sqrt{\rho_1}.$$

Hence the times of efflux of equal volumes are as

$$\sqrt{\rho_1} : \sqrt{\rho_2}.$$

This formula is in accordance with Graham's experiments. Below are given some of his results (Meyer, *Kinetic Theory*, p. 84):

Gas.	$\sqrt{\text{Specific Gravity}}$.	Time of Efflux of Given Volume through	
		Drawn-out Glass Tube.	Perforated Brass Plate.
Air	1	1	1
Hydrogen	0·263	0·277	0·276
Oxygen	1·051	1·053	1·050
Nitrogen	0·986	0·984	0·984
Carbonic Acid	1·237	1·218	1·197

If the gas flows out, not into a vacuum but into a space containing another gas, the rate of efflux is still nearly the same so long as the pressure of the outside gas is small, say less than half that of the inside gas. The stream of issuing gas is to be regarded as all moving in the same direction, sweeping the external gas away, and its velocity will depend chiefly on the velocity of the molecules within the vessel.

But if the pressure outside is nearly equal to that inside, the external molecules will collide with those issuing, and the time of efflux will be lengthened out. Still, with the two gases issuing under the same difference of pressure, the masses moving up to a very short orifice will be given by $\frac{\rho VS}{6}$ and the same fraction of each will issue if the external gas is at the same pressure in each case, and still the ratio of the times of efflux will be as

$$\sqrt{\rho_1} : \sqrt{\rho_2}.$$

Thermal Transpiration.—If we have the same gas, but at two different temperatures, θ_1 and θ_2, on the two sides of a small orifice, the masses moving up to the orifice on the two sides will be proportional to $\rho_1 V_1$ and $\rho_2 V_2$ respectively. If the pressures are equal to begin with,

$$\rho_1 : \rho_2 = \theta_2 : \theta_1$$

and

$$V_1 : V_2 = \sqrt{\theta_1} : \sqrt{\theta_2}$$

whence

$$\frac{\rho_1 V_1}{\rho_2 V_2} = \sqrt{\frac{\theta_2}{\theta_1}}$$

or if $\theta_2 > \theta_1$, $\rho_1 V_1 > \rho_2 V_2$, and more gas moves in from the cold side than moves out from the hot side and the pressure of the hot gas increases, the increase will go on until there is a balance between the two streams or until

$$\rho_1 V_1 = \rho_2 V_2$$

or putting

$$V^2 = \frac{3p}{\rho}$$

until

$$p_1 \rho_1 = p_2 \rho_2.$$

But
$$\frac{p_1}{p_2}=\frac{\rho_1\theta_1}{\rho_2\theta_2}$$

therefore the balance is attained when

$$\frac{p_1^2}{p_2^2}=\frac{\theta_1}{\theta_2}\text{ or }\frac{p_1-p_2}{p_2}=\frac{\sqrt{\theta_1}-\sqrt{\theta_2}}{\sqrt{\theta_2}}$$

The pressures are then as the square roots of the temperatures on the gas scale.

The phenomena of thermal transpiration have been investigated by Osborne Reynolds (*Phil. Trans.*, part ii. 1879). He maintained the gas at constant temperatures of 100° C. and 8° C. respectively, in chambers on the two sides of a plate of biscuit, meerschaum, or stucco, and determined the difference of pressure in the steady state. He found that at low pressures the formula is nearly verified. As the pressure on both sides increases, however, the difference of pressure is nearly inversely proportional to the mean pressure. In his paper he gives a full investigation of the theory, which agrees with his observations.

The Mean Free Path.—The very great velocity of the molecules as calculated on p. 136 might lead us to expect that gaseous diffusion would be extremely rapid, so that if, for instance, a gas-tap were turned on in a room the coal-gas, with a molecular speed comparable with half a mile a second, would almost instantly spread all over the room. But this does not agree with observation. If the air in the room is free from draughts it may be quite a considerable time before the coal-gas is in sufficient quantity to be perceived, even a few feet from the tap. The diffusion is hindered by the collisions of the molecules with each other. If we could follow a given molecule we should see it continually colliding with, or being interfered with by its neighbours, pursuing a given direction only for a very short distance and a very small time, then colliding with another molecule and changing its direction of motion for another short distance, then colliding again, and so on. The mean distance travelled between successive collisions is termed **the mean free path.** We shall denote this by M.F.P. The number of collisions per second made by a molecule is termed the **Collision Frequency.** Evidently the collision frequency × the M.F.P. is equal to the mean velocity.

We shall see later that we may estimate the M.F.P. in air at atmospheric pressure as of the order of a hundred-thousandth of a centimetre, and the collision frequency as more than a thousand million. During a second, then, a molecule changes its direction of motion thousands of millions of times, moving now forward, now backward, now up, now down, now to this side, now to that. The different displacements will to a very large extent neutralise each other, so that at the end of a second a molecule will generally only be a very short distance from the point it occupied at the beginning of the second.

Molecular Dimensions. Molecular Sphere of Action.—We cannot at present form a working hypothesis, useful for the kinetic theory, of the structure of the molecule, or of the field of force around it. We cannot, therefore, say how the molecules act upon each other when they approach and are in collision. We must be content to take what is at the best merely an approximate representation, by supposing that the centres of the molecules approach till they are, on an average, a distance s from each other, and that then they recede. It is often convenient, for calculation, to think of a sphere of action surrounding one molecule of a colliding pair, and to concentrate our attention on the centre only of the other molecule. We then regard the centre of the first molecule as surrounded by a sphere of radius s, within which the centre of the second molecule cannot penetrate, and we term s the radius of molecular action. We ought really to picture two spheres of action, one round each molecule, and each of radius $s/2$, but the result is the same. Since the molecular centres do not get within distance s, we may regard s as the diameter of each molecular system. If the gas is reduced to the liquid or the solid condition, we think of each molecule as being just within the spheres of action of its neighbours all the time, and we therefore regard s as indicating approximately the distance of a solid or liquid molecule from its immediate neighbours.

Dependence of the M.F.P. on Molecular Dimensions and on the Density of the Gas.—If the molecules were mere points, and if they exerted forces upon each other only at infinitely small range, the M.F.P. would be infinitely great so long as the number of molecules in a finite space was finite. For consider a single molecule projected from a point. The spheres of action of the surrounding molecules within any finite distance would fill up an infinitely small fraction of what we may term its field of vision, since the total solid angle subtended by any finite number of molecules at the point would be infinitely small. If then a line were drawn in any assigned direction the chance that it went through another molecular point within a finite distance would be infinitely small. But if we assign a finite value to the radius of molecular action and now think of a molecule as projected from a point, in whatever direction we draw the line of projection it is practically certain that within some finite distance it will impinge on the sphere of action of another molecule. This is illustrated by a shower of rain which has only to be of sufficient breadth to hide entirely objects beyond it.

We may form a mental picture of the M.F.P. by imagining that all the molecules but one are fixed in the configuration which they have at a given instant. We may then project the one from its position in turn in all directions till it comes into collision with another molecule, and take the average distance traversed before collision as equal to the M.F.P. The fixing of the molecules, which we have assumed for simplicity, gives us, as Clausius and Maxwell showed, too great a value for the M.F.P. It is easy to see that the motion of the molecules increases the chance of collision, for imagine a spherical shell $2s$ thick drawn at a distance from the point of projection. The projected molecule will travel through this in time $2s/V$, and meanwhile a molecule within the shell, and having the same velocity, will travel a distance $V \times 2s/V = 2s$, and sweep out an area $4s^2$. The effective area subtended at the centre of projection by the molecule will be $\pi s^2 +$ a fraction of $4s^2$, the fraction depending on the inclination of the motion to the direction of travel of the projected molecule. Thus the chance of collision is increased. It can be shown to be $\sqrt{2}$ times as great as the chance when only one molecule moves.

If the number of molecules per c.c. remains the same, while the cross section of each is increased, evidently the chance of hitting one within a short distance is increased. We shall show below that, as we might perhaps expect, the collision frequency with a given velocity is proportional to the cross section of the molecular systems. If the cross section remains the same while the number of molecules is increased, the frequency of collision increases in the same proportion. This again might perhaps be expected. For if we put into a given space a second equal number of molecules, we might expect that the projected molecule would collide with the added molecules as often as with those previously in the space if the packing was so open that the second set were not appreciably screened by the first, and thus it would have double the collision frequency. These results may be obtained as follows. Imagine a straight line AF drawn from a point occupied by a molecule A, and passing in succession through the spheres of action

of molecules at B, C, D, &c. Then the average of the lengths AB, BC, CD, &c., is the M.F.P. For it is the average distance which a molecule projected in the direction AF will travel before it collides with another molecule, and this average distance will be the same whether we project always in one direction or whether we project in all directions in turn. Now let us suppose that round the molecule at A is a sphere of action of radius s and cross section πs^2, and let us represent the other molecules at B, C, D, &c., by points. Let πs^2 sweep forward in the direction AF

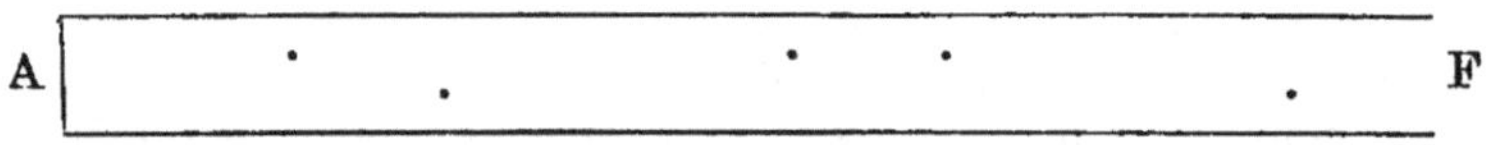

through l cm. If it impinges on n molecular points in this distance the M.F.P., which we shall denote by L, is equal to l/n. But the volume swept out by πs^2 is $\pi s^2 l$. Let the number of molecules per c.c. be N. Then if $\pi s^2 l$ is large enough the number of molecules in it will be $\pi s^2 l\text{N}$. Equating this to n or l/L, we have $\pi s^2 l\text{N} = l/\text{L}$, or $\text{L} = 1/\pi s^2\text{N} = m/\pi s^2\rho$ where $\text{N}m = \rho$, m being the mass of one molecule and ρ the gas-density. L, then, is inversely as the molecular cross section and inversely as the density, if πs^2 is constant, and we may probably assume that it is constant, at a given temperature.

The Mean Free Path calculated from the Coefficient of Viscosity of a Gas.—If a gas is moving in a given direction, but faster on one side of a given plane containing that direction than on the other side, then the slower-moving gas exerts a dragging action on the faster-moving gas, which in turn tends to hurry on the slower-moving gas. This tangential force is termed the force of viscosity. It is explained in the kinetic theory as due to the interchange of molecules between the two portions of gas across the plane. If we think of the plane as horizontal and the upper part of the gas as moving the faster, then the molecules moving downwards through the plane have on the average a greater momentum in the given direction than those moving upwards through the plane to replace them, and therefore the upper portion tends to lose momentum in the given direction and the lower portion tends to gain it. In other words, there is a force on the lower gas parallel to the plane and in the given direction and an opposite force on the upper gas.

Let us consider a gas contained between two plane parallel boundary walls AB, CD (Fig. 77) a distance d apart, the lower plane CD being fixed and the upper plane AB moving with constant velocity v from left to right. We shall assume that the layer of gas in contact with each wall has no motion relative to that wall, and that the velocity increases uniformly as we pass up from CD to AB, so that at a distance x from CD it is $\frac{xv}{d}$. Let the viscous tangential force of one layer on the next layer per square centimetre in the direction of motion be F. The motion of each layer relative to the one below it being uniform, the force F is the same on each layer and ultimately acts on each boundary.

Direct experiments on the vibration of a plane disc close to another plane disc fixed parallel to it show that for a given gas in a given condition F is proportional to $\frac{v}{d}$, and so tend to justify our assumptions. It may be mentioned, however, that when the pressure of the gas is

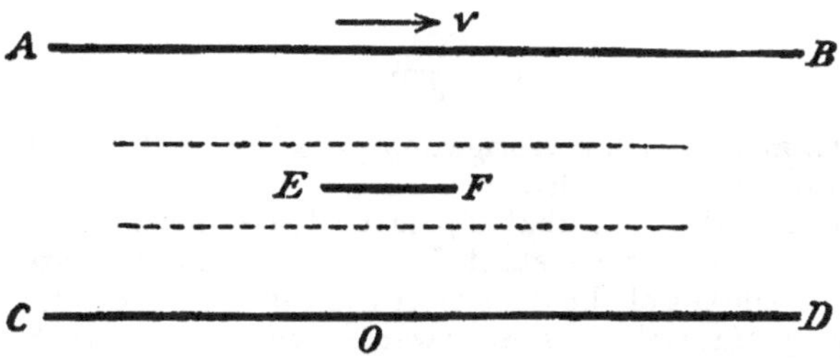

FIG. 77.

very much reduced we are no longer able to assume that the layers in contact with the boundary planes are fixed relative to them. There is, in fact, side slip. This, however, is inversely proportional to the pressure of the gas, and is negligible at ordinary pressures.

Let us now imagine a plane EF, 1 cm. square, parallel to the boundary planes and consider the transfer of molecules across it. Let u be the average velocity of the layer at EF from left to right. The molecules moving down through EF will come from various distances and so carry with them various amounts of momentum parallel to u. Let us suppose that they carry on the average the momentum possessed by the molecules in the plane indicated by the upper dotted line and distant from EF by the M.F.P. = L say. According to Joule's method, the mass moving down through EF in one second is $\rho\frac{V}{6}$.

The velocity parallel to u, at the distance L above EF, is $u+\frac{Lv}{d}$.

Hence the momentum parallel to u transferred across the plane in one second is

$$\rho\frac{V}{6}\left(u+\frac{Lv}{d}\right)$$

But an equal mass $\rho\frac{V}{6}$ comes up through EF from below, and if we suppose that this has on the average the momentum of the layer indicated by the lower dotted line, distant L below EF, the momentum parallel to u, brought up through EF, is

$$\rho\frac{V}{6}\left(u-\frac{Lv}{d}\right)$$

Hence the gas above loses momentum equal to the difference of these, viz.:—

$$\rho\frac{V}{3}\cdot\frac{Lv}{d}$$

while the gas below gains an equal amount. But this transfer measures the tangential force F. Then

$$F=\frac{\rho VL}{3}\cdot\frac{v}{d}$$

But we also have $F=\eta\frac{v}{d}$, where η is the coefficient of viscosity,

whence $\eta=\frac{\rho VL}{3}.$

Maxwell's more exact investigation gives $\eta = \cdot 30967\rho LU$, where U is the mean value of the velocities.

We have already seen that $L\rho$ is probably constant for a constant temperature. V is also constant for a constant temperature, so that η should be independent of the density of the gas. This result, first obtained from theory by Maxwell, was afterwards verified by direct experiment. Since V increases with the temperature, η should also increase, a result borne out by experiment.

Maxwell's method of obtaining η consisted in principle in allowing a horizontal circular disc suspended by a wire to vibrate about its axis, another horizontal disc being fixed close below and parallel to it. The layer of gas between the two discs was, therefore, sheared by the motion, and the viscous resistance of the gas gradually "damped" the vibrations of the moving disc. From this damping η could be calculated, and it was found to be independent of the pressure within a wide range. If, however, the pressure is sufficiently reduced the method fails through slip of the gas on the surface of the discs. And even if there is no slip, it the pressure is so far reduced that the M.F.P. is comparable with the distance between the discs, evidently the method of investigation is no longer admissible.

The following values of η at 0° C. are taken from the extensive tables given by Meyer (*Kinetic Theory of Gases*):—

Air	·00017
Hydrogen	·00008
Oxygen	·00019
Water vapour	·00009

While the viscosity increases, as the theory indicates, with the temperature, it does not increase merely in proportion to V or in proportion to the square root of the absolute temperature, but more rapidly, apparently indicating that at a fixed density the free path also increases with the temperature (Meyer, *l.c.*, p. 216).

In the equation $\eta = \frac{\rho VL}{3}$, substituting for V from the equation $\rho V^2 = 3p$, we obtain

$$L = \eta\sqrt{\frac{3}{p\rho}}$$

whence L can be found when η is known.

Since the number of collisions per second, or the collision frequency, is equal to mean speed/mean free path, we can also calculate the collision frequency, *i.e.*, the number of collisions per second,

$$\frac{V}{L} = \frac{\rho V^2}{3\eta}$$

Below we give the values obtained for several gases at 760 mm. pressure from our approximate numbers, which are sufficiently near the truth to show the order of the magnitudes involved. We give the value for water vapour on the supposition that it could be compressed to I atmosphere at 0° C. Air is regarded as a simple gas.

Gas.	Density at 0° and 760.	Coefficient of viscosity η.	M.F.P. $L = \eta\sqrt{\frac{3}{p\rho}}$	Collision frequency $V/L = p/\eta$.
Air . .	$129 \div 10^5$	$17 \div 10^5$	$82 \div 10^7$ cm.	59×10^8
H . . .	$9 \div 10^5$	$8 \div 10^5$	$146 \div 10^7$,,	125×10^8
O . . .	$143 \div 10^5$	$19 \div 10^5$	$87 \div 10^7$,,	53×10^8
H_2O . .	$81 \div 10^5$	$9 \div 10^5$	$55 \div 10^7$,,	111×10^8

Had we used the exact formulæ we should have obtained values for the M.F.P. about $\frac{1}{8}$ greater, and of course the collision frequencies would be proportionately reduced.

Conduction of Heat in Gases.—The conduction of heat in gases can be explained in a manner similar to that in which viscosity has been explained. If there is a temperature slope in a gas, there is a continual passage of more energetic molecules from the hot side across any given plane, and a continual passage of less energetic molecules from the cold side, with the net result that there is a transfer of energy down the slope. The following investigation, though very incomplete, gives an estimate of the amount of conduction.

Let EF, Fig. 78, be 1 square cm. in a layer of which the temperature is θ, and let the slope of temperature perpendicular to EF be $\frac{d\theta}{dx}$. Let us take mass $\rho\frac{V}{6}$ as passing through EF per second from the upper side and as having on the average the temperature of the layer AB a distance from EF equal to L, the M.F.P.—*i.e.* a temperature $\theta + L\frac{d\theta}{dx}$.

In the opposite direction let mass $\frac{\rho V}{6}$ pass upwards at the temperature of the layer CD a distance L below EF, that is, at temperature $\theta - L\frac{d\theta}{dx}$.

The excess of energy carried through EF from above over that returned from below is therefore the same as if mass $\frac{\rho V}{6}$ were cooled through $2L\frac{d\theta}{dx}$, and if κ_v is the specific heat of constant volume, this will be equal to

$$\kappa_v\frac{\rho V}{6} \times 2L\frac{d\theta}{dx} = \frac{\kappa_v \rho V L}{3}\ \frac{d\theta}{dx}$$

If K is the conductivity for heat of the gas, the quantity of heat passing down per second is $K\ \frac{d\theta}{dx}$, whence $K = \frac{\rho V L}{3}\kappa_v$, or putting $\frac{\rho V L}{3} = \eta$ the coefficient of viscosity, $K = \eta\kappa_v$, which is independent of the density

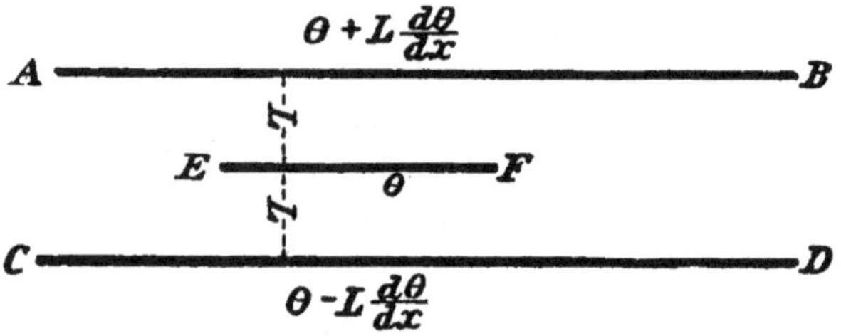

FIG. 78.

of the gas, a result predicted from the kinetic theory by Maxwell and subsequently verified by experiment.

The Diameter of the Molecules and the Number of Molecules per Cubic Centimetre.—The value of the molecular diameter, or rather the radius of the sphere of action, was first calculated by Loschmidt in 1865.

Taking s as the radius of the sphere of action as defined on p. 143, $\frac{s}{2}$ may be taken as the radius of each molecular system when two molecules are at their nearest approach with their centres s apart. Then the volume of a singular molecular system is of the order

$$\frac{4}{3}\pi\left(\frac{s}{2}\right)^3 = \frac{\pi s^3}{6}$$

The total volume of the molecules in 1 c.c. is therefore $\frac{N\pi s^3}{6}$.

If we could suddenly destroy the translatory motion of the molecules in 1 c.c., still keeping the molecular systems unchanged in themselves, they would simply fall to the bottom of the space and occupy a volume of the order $\frac{N\pi s^3}{6}$. If we imagined them all exact spheres, piled up like shot, the volume would be $3\sqrt{2}/\pi$ times greater. Probably we have an approach to such a condensation in the liquid and solid states, where the molecules may be regarded as each within the spheres of action of its next neighbours all the time, and we shall therefore assume, as an approximation of the right order, that the volume of a gas containing N molecules is $\frac{N\pi s^3}{6}$ when liquefied.

If, then, 1 c.c. of gas at 0° and 760, and density Δ, condenses to volume ν of liquid of density δ,

$$\nu = \frac{\Delta}{\delta} = \frac{N\pi s^3}{6} \qquad (1)$$

But our equation for the free path is

$$N\pi s^2 L = 1 \qquad (2)$$

From (1) and (2) we have $s = 6\nu L$.

We know ν and L in a number of cases, and can therefore find s. Then, substituting this value of s in (2), we can find N. The following table shows the results obtained for hydrogen, oxygen, and water vapour. It is hardly necessary to point out that, seeing the assumptions made, the results are only to be taken as indicating the order of magnitude involved. The value of N, for instance, differs for different gases, whereas by Avogadro's Law it should be the same.

Gas.	Gas Density Δ.	Liquid Density δ.	$\nu=\frac{\Delta}{\delta}$.	M.F. Path. L.	$s=6\nu L$	N in 1 c.c.	Mass of one Molecule.
H	$9\div10^5$	$7\div10^2$	$130\div10^6$	$146\div10^7$	$11{\cdot}4\div10^8$	$3{\cdot}4\times10^{18}$	$2{\cdot}7\div10^{23}$
O	$143\div10^5$	1·24	$115\div10^5$	$87\div10^7$	$6\div10^8$	$10{\cdot}2\times10^{18}$	$14\div10^{23}$
H_2O	$81\div10^5$	1	$81\div10^5$	$55\div10^7$	$2{\cdot}7\div10^8$	8×10^{18}	$10\div10^{23}$

Using the more correct formulæ, Maxwell calculated that N for gases at 0° and 760 should be 19×10^{18}, say 2×10^{19} (*Papers*, ii. p. 372), whence m for hydrogen comes out $4{\cdot}5\div10^{24}$. The value of N can also be found by an entirely different method—from the consideration of the electrical properties of gases. The value so found is $3{\cdot}9\times10^{19}$ (Thomson, *Conduction of Electricity through Gases*, p. 130). Maxwell further obtained for s with hydrogen the value $5\div10^8$. If we assume that this is about the value of the molecular diameter in other cases, an assumption perhaps warranted from the nearness of the values of ν for different substances, then this would imply that in solids and liquids the centres of the molecules are a distance apart comparable with $5\div10^8$.

Forces Acting on unequally heated Surfaces in High Vacua.—The motions of unequally heated surfaces in rarefied gases were first investigated by Crookes, who was led to the invention of the radiometer—a beautiful illustration of such motions. In a simple form, it consists of four mica vanes, each lampblacked on one face and attached to the four arms of a cross, pivoted to move with very slight friction about the centre (Fig. 79). The lampblacked faces are so placed as all to move round forward or all backward.

The vanes are enclosed in a glass bulb, which is exhausted by a mercury-pump till the pressure is exceedingly small, and then sealed. On exposing the bulb to a source of radiation, the lampblacked surfaces of the vanes are more heated than the bare surfaces and move away from the source, and so rapid rotation results.

FIG. 79.

The complete explanation of the action is due to Osborne Reynolds ("On Certain Dimensional Properties of Matter in the Gaseous State," *Phil. Trans.*, part ii., 1879). The action was also sufficiently explained very shortly after by Maxwell ("On Stresses in Rarefied Gases arising from Inequalities of Temperature," *Phil. Trans.*, part ii., 1879). The theory is altogether beyond our scope, but the following account of what occurs may give some idea of the action. It is to be remembered that it is an account, and not an explanation. *

Let us imagine that a plane is suddenly introduced into a gas, one side of the plane being hotter than the gas, while the other side is at the same temperature with it. Consider a small area on the plane far from the edges. The molecules which come up on the hot side are raised in temperature by contact with the plane and rebound with a greater velocity than that with which they arrived, while those on the cold side go off with the same velocity. For a moment after the introduction of the plane the number coming up to the area is the same on the two sides, since it is conditioned by the temperature and pressure of the surrounding gas, and these are not yet affected by the presence of the plane. Hence the extra kick off of the molecules rebounding from the hot side implies a greater pressure on that side. But very quickly this excess of pressure will fall off, the rebounding molecules on the hot side sharing their extra energy with the molecules with which they collide, and as soon as a uniform temperature slope outwards is established, the decrease in density, and therefore the decrease in the number of molecules coming up to the hot side, compensates for the extra impulse at each collision, and so the pressure falls to the same value as on the cold side.

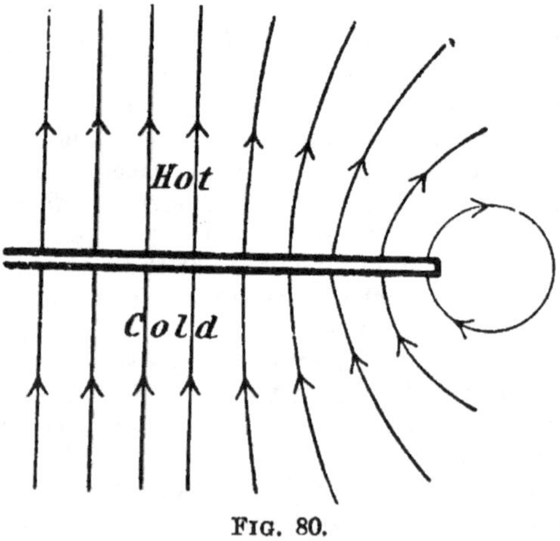

FIG. 80.

* Lord Rayleigh (*Nature*, July 15, 1909, p. 69) considers the extreme case of a

But near the edges this compensation will not be complete. For a distance inwards comparable with the M.F.P. the lines of flow of heat will diverge from each other as shown in Fig. 80, and the temperature falls at an increasing rate from the hot plate outwards. The density at the average distance from which we may suppose the molecules to come to the plane is greater than well inside the edge, and the change in temperature and in velocity of the molecules at their impacts is greater. This results in a greater pressure against the area near the edge on the hot side, which continues as long as that side is hotter. On the cold side there will be a similar edge effect, for the hotter molecules will to some extent come round the edge and carry heat into the cold surface. The temperature at the average distance from which we may suppose the molecules to come will be higher and the density less than well inside the edge on the cold side. There results a defect of pressure near the edge. Reynolds and Maxwell both showed that the excess of normal pressure on the surface is proportional to

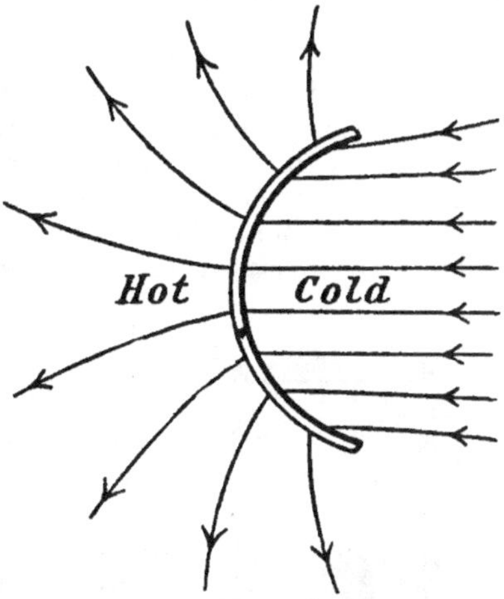

FIG. 81.

$$\frac{\text{rate of increase of temperature slope outwards}}{\text{pressure}}.$$

Hence, since the temperature slope with a plane decreases outwards only near the edges, and for a width of surface comparable with the M.F.P., both the numerator and denominator contribute to make the total force greater the less the pressure of the gas; hence the necessity for the very high vacua with vanes of the ordinary size.

If the vanes are curved, as in Fig. 81, the convex side being hot, the lines of flow are somewhat as represented in Fig. 79, and the divergence on the hot side is more marked, while it is hardly noticeable on the cold side. In this case the excess of pressure on the hot side will be greater than with a plane surface.

At ordinary pressures the edge acted on by the pressure excess is exceedingly narrow—of the order of the M.F.P., which, as we have seen, is comparable with 10^{-5} cm., while the pressure excess can be shown by the theory to be inversely as the pressure. Hence the total force is very minute. But if the body acted on is reduced to exceedingly narrow dimensions, and its mass correspondingly reduced, the effect may still be noticeable at ordinary pressures. This was pointed out by Reynolds, and he succeeded in detecting the action on spider lines and silk fibres at pressures comparable to the atmospheric pressure, one side of the fibre being exposed to radiation (*l.c.*, p. 768).

We have, perhaps, an example of the action, which we may call "radiometer action," on exceedingly small surfaces, at ordinary pressures, in the dust-free region surrounding a heated body.

If a thick copper rod is placed so as to project into an enclosure filled with dusty air or smoke, on heating the part outside so that the rod inside the enclosure becomes heated by conduction, the rod looked at endwise, when the light is properly directed, may be seen to be surrounded by a dust free space as in Fig. 82, where the rod is supposed to come end-on against a window in the side of the enclosure, the light being thrown through the space towards the observer.

Lodge explains this by supposing that the heated body heats the nearer face of the dust particles, and that there is a backward movement if the dimensions of the particles are comparable with the mean free path. They retreat on all sides from the heated body, leaving a space of clear air. This ascends by convection, as illustrated in the figure, being replaced by fresh air, which in turn is cleared of its dust.

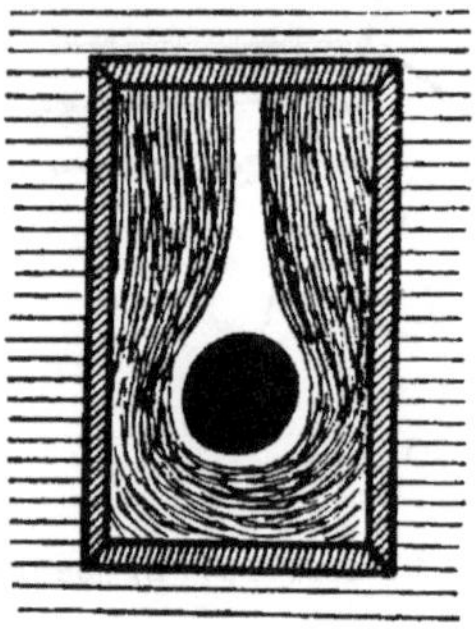

FIG. 82.

If the body be cooler than the surroundings, the converse happens, the dust being forced on to the cooler surface. Many illustrations of the deposit of dust and smoke on cooler surfaces may be found. Plaster ceilings very frequently show the course of the laths and rafters behind the plaster. Where the plaster is backed by wood, it is probably kept warmer, and the dust is not so freely deposited as on the neighbouring cooler parts. Walls above hot-water pipes are often very soon blackened, the hot dusty air depositing its dust against the cooler wall.

The Gas Equation of Van der Waals.—The gas equation, expressing the laws of Boyle and of Charles, viz.:

$$pv = \mathrm{R}\theta$$

is only approximately in agreement with observation. In a celebrated paper "On the Continuity of the Liquid and Gaseous States of Matter" (English translation, *Physical Memoirs of the London Physical Society*, vol. i. pt. 3), Van der Waals deduced from theory the equation

$$\left(p + \frac{a}{v^2}\right)(v - b) = \mathrm{R}\theta$$

where a and b are certain constants. This equation, though still not in agreement with observation, represents the relation between pv and θ much more exactly than the original gas equation. The following method of obtaining the equation may serve to show how the kinetic theory accounts for the failure of Boyle's Law.

Returning to the investigation of the pressure exerted by a gas on the sides of the containing vessel, let us now take into account (1) the size of the molecules as affecting their length of path; (2) their cohesive forces at collision as lengthening out the time of collision. These are really two aspects of one transaction, but it is convenient to consider them separately.

Let us take first the size of the molecules. Suppose a molecule to start normally from one face of a unit cube vessel. If it meets another molecule in direct collision, moving with equal and opposite velocity, there will be simply an interchange of velocities, and the second molecule will take its place on the journey across the cube, and we may now fix attention on the second molecule. If this comes into direct collision with, and interchanges velocities with a third molecule, this third molecule takes the place of the second, and so on. But at each collision the substituted molecule starts with its centre a distance s farther on, where s is the radius of molecular action or the diameter of a single molecular system. If, then, there are ν collisions per second of this kind, the total distance covered will be, not V but, $V+\nu s$, and the molecule, or its representative, will return to the original starting-point oftener than if the molecules were mere points in the ratio $V+\nu s : V$, and on this account the pressure will be greater than that originally calculated in the same ratio. But this is on the supposition that all the collisions are direct, whereas they must be regarded as of all degrees of obliquity.

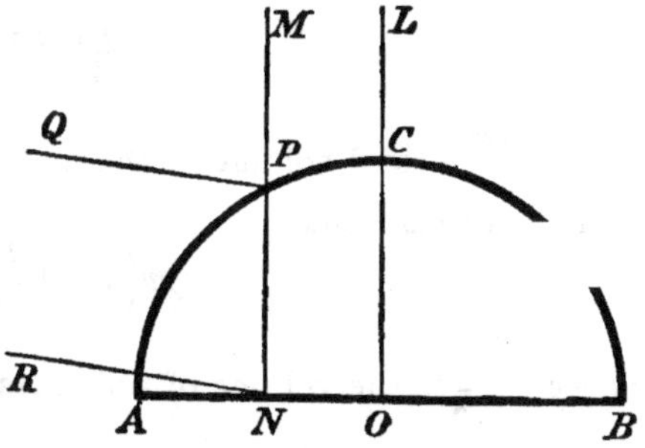

FIG. 83.

To find the average increase of path, let us suppose one of the pair of molecules in collision to have its centre at O, Fig. 83. Let ACB be a section of the hemisphere of molecular action with radius $OC=s$. If the second molecule comes along LC, in direct collision, the path omitted is $CO=s$; but if the molecule comes along MP, and rebounds along PQ, producing MP to N, the path omitted is PN. For, if the radius s were indefinitely small, the second molecule with the same obliquity of collision would move off along NR parallel to PQ.

Now the number of molecules making collision of a given type over a given area of the hemisphere will be proportional to the projection of the area on the diametral plane through AOB, for we may suppose the numbers moving up towards that plane evenly distributed over it. If the number moving up in given time to unit area is n, the number moving towards a given ring of radius $ON=r$ and breadth dr is $n\times 2\pi r dr$, and these each add path PN, or total path

$$n \times 2\pi r dr \times \mathrm{PN}.$$

Then all the molecules coming up in unit time add path $n\Sigma 2\pi r dr \times \mathrm{PN}$, where we sum up for the whole hemisphere. But $\Sigma 2\pi r dr \times \mathrm{PN} = \frac{2}{3}\pi s^3$, the volume of the hemisphere, and the total path added by all the $n\times\pi s^2$ molecules coming to the hemisphere is

$$n\frac{2\pi s^3}{3}$$

The average path added is, therefore,

$$n\frac{2\pi s^3}{3} \div n\pi s^2 = \frac{2}{3}s$$

and we must use this instead of s.

In place of
$$p=\frac{\rho}{3}V^2$$

we shall, therefore, put
$$p=\frac{\rho}{3}V^2\frac{V+\frac{2}{3}\nu s}{V}$$

where ν is the number of collisions per second.

But
$$\nu=\frac{V}{L}=VN\pi s^2$$

whence
$$p=\frac{\rho}{3}V^2\left(1+\frac{2}{3}N\pi s^3\right)$$
$$=\frac{\rho}{3}V^2\left(1+4\frac{N\pi s^3}{6}\right),$$

Here $\frac{N\pi s^3}{6}$ = total volume of the molecular systems per c.c., and since this is small we may put
$$pv\left(1-4\frac{N\pi s^3}{6}\right)=\frac{V^2}{3}$$

where $v=\frac{1}{\rho}$ is the volume of unit mass.

If we put
$$4\frac{N\pi s^3}{6}v=b \text{ we get } p\,(v-b)=\frac{V^2}{3}.$$

Now let us turn to the second consideration, that of time lost in collision. We may think of a slackening of the velocity on close approach and then an acceleration, so that in each collision there is a time greater on the average by τ than if one free path had instantaneously changed to the next. If ν is the number of collisions in one second, the time occupied in traversing a given path will be greater than that obtained on the supposition of instantaneous change in the ratio

$$1+\nu\tau:1,$$

and a molecule or its representative will return less often to a face of the containing vessel in the ratio $1:1+\nu\tau$. On this account then, the pressure will be reduced to

$$p=\frac{1}{v-b}\,\frac{V^2}{3}\,\frac{1}{1+\nu\tau}$$

or, instead of p, we must put $p+p\nu\tau$.

Now $\nu=\frac{V}{L}=VN\pi s^2$ and $Nm=\rho$ where m = mass of a molecule.

therefore
$$\nu=V\rho\frac{\pi s^2}{m}.$$

Putting for p its approximate value $\frac{\rho V^2}{3}$ in the small quantity $p\nu\tau$ we get

$$p + p\nu\tau = p + \frac{\rho^2 V^3 \pi s^2 \tau}{3m}$$

$$= p + \frac{a}{v^2}$$

where $$v = \frac{1}{\rho} \text{ and } a = \frac{V^3 \pi s^2 \tau}{3m}$$

and the equation becomes

$$\left(p + \frac{a}{v^2}\right)(v - b) = \frac{V^2}{3} = R\theta.$$

The assumption that a is constant implies that $V^3\pi s^2\tau$ is constant. We have no data to justify this, and herein lies a weakness in the investigation.

It may be mentioned that Van der Waals has calculated b from observations on the compression of gases. This gives another method of finding s, and the results are of the same order as those already found.

The equation of Van der Waals has been modified in various ways to make it represent the facts of observation more closely. But the modifications are probably suggested rather by the lack of accordance with fact than by the kinetic theory, and must be regarded as somewhat empirical. For an account of some of these modified equations we refer the reader to Meyer's *Kinetic Theory of Gases*, p. 100. Another modification has been proposed by Callendar (*Proc. R. S.*, lxvii., 1900, p. 266). We shall consider the equation of Van der Waals further in chap. xi.

We have already seen that some of the energy given to a gas on raising its temperature goes, in all probability, to increase the internal energy of the molecule. This brings us to the point where the theory for the present stops. No satisfactory hypothesis as to the construction of the molecules, and the mode in which they possess energy, has as yet been devised.

In order to find out what goes on in a molecule we should, of course, like to isolate and study it in the method by which alone we could do so, viz., by examining the nature of the energy it gives out. But this is impossible. The nearest approach to isolation is in a rare gas, where the molecules have a long free path, so that, for a great part of their time, they are not interfering with one another. When such a gas is heated under certain conditions, it sends out waves of light of definite refrangibilities, showing that in all probability the molecules or their parts are vibrating, but we are quite unable to do more than guess at the mode of vibration giving rise to these light waves. It may be that the different atoms in the molecule are vibrating about the common centre of gravity, as the earth and moon vibrate about their common centre of gravity; or, it may be that the separate atoms vibrate as a bell vibrates. If the latter is the mode of vibration giving rise to light-waves, then we must suppose that the contiguity of other dissimilar atoms alters the character of the vibrations, for we know that the light sent out by incandescent compounds is different in character from that sent out by the constituents before combination. As a heated gas, which is emitting a line spectrum, is compressed the character of the light sent out is found gradually to change. On examining the spectrum it is found to consist at first of a greater or less number of bright lines. As compression goes on these first widen out into bands, and then gradually a continuous spectrum appears, showing that light of every refrangibility is being sent out. This is explained by the gradual shortening of the free path. The molecules interfere more and more with each other, their time for uninterrupted natural vibration becomes less and less, while the times of clashing and general disturbance in collisions become greater and greater, and this general disturbance gives rise to the continuous spectrum. If, instead of a gas, we take a glowing liquid or solid, we cease to have any appearance of bright lines or bands. There is merely a continuous spectrum, showing, apparently, that the molecules interfere with each other too much to allow the natural vibrations to have play. We might, perhaps, give an illustration of this. If a number of bells are hung in a room by strings from a ceiling, and are set swinging, they will occasionally collide, and the energy of translation will partly be converted into energy of vibration. If the bells are far apart the notes given out will be the natural notes of the bells, for the free paths will be long and the time of free vibration long, compared with the time of clashing. But the closer the bells are together the more prominent will the clashing be. If they are packed loosely in a box and rattled about the natural vibrations will be quite overpowered by the clashing.

Recent researches on the emission of radiation by gases appear to show that each gas only emits a line spectrum when chemical change is proceeding.* Thus sodium vapour when raised to a high temperature in a neutral gas, one with which it does not combine, does not give the characteristic D lines, but probably a continuous spectrum. But if it combines with the surrounding gas the D lines appear. This would appear to show that the waves corresponding to the D lines are not emitted by the molecules as a whole, but by the parts which are changing places in the chemical actions proceeding. At present, however, our hypotheses as to the nature of these actions are hardly definite enough to warrant us in giving an account of the origin of the radiations.

* An account of our present knowledge will be found in the *Reports of the Congrès International de Physique*, 1900, vol. ii., p. 100, by E. Pringsheim.

CHAPTER X.

CHANGE OF STATE—LIQUID—VAPOUR.

General Account of Evaporation—Vapour-Pressure—Boiling—Delayed Boiling—Condensation on Nuclei—Measurements of Vapour-Pressure—Determination of Vapour Density—Density of Saturated Vapour—Measurements of Latent Heat of Vapours—Specific Heat of Saturated Vapour—Spheroidal State.

Change of State.—We are accustomed to find each substance occurring most commonly in one particular state of matter. Iron, for example, is usually solid, oil liquid, and air gaseous. But we are also familiar with the change of the same kind of matter from one state to another, which is effected by supplying or withdrawing heat, and experiment leads us to believe that, unless chemical change intervenes, every substance may be made to assume any one of the three states, even though our present experimental arrangements may be insufficient to accomplish this. The change as we add heat is, in general, from solid to liquid, and from liquid to gas, though it may be from solid to gas without the intervening step. There are also apparent exceptions, as in the case of the preparation of solid red phosphorus by heating ordinary molten phosphorus. But we probably have here an absorption of energy accompanying quite a different arrangement of the molecules—a change of state of another kind. We may take it, therefore, as a rule that the three states—solid, liquid, gas—are in ascending order as regards the quantity of energy possessed, and the quantity required to effect the change from one state to the next is usually large. If, for example, we take a quantity of ice below 0° C. and supply heat to it, the temperature rises steadily to 0°. There is then a pause while melting takes place, a very considerable quantity of heat being absorbed merely to effect the change from ice to water while the temperature remains steady. This heat is said to be *latent*, a term which was given on the supposition that the general effect of heat was to raise the temperature of bodies, whereas this heat is not affecting the temperature. From this point of view the term is a very good one, though, of course, the heat is not latent in any other sense, as its presence is quite evident in change of physical state.

The ice being melted, the water again rises steadily in temperature till it begins to boil, turning rapidly into steam or water-gas, when there is another pause, and a still larger quantity of "latent" heat is required merely to effect the change from water to steam without rise of temperature. But besides this rapid change at boiling, when the water is in an open vessel a gradual change into steam takes place at the upper surface even at ordinary temperatures, a change which is more rapid as the temperature rises. This change, which is termed *evaporation*, even takes place from the surface of ice—a fact which is familiar to us in the disappearance of snow in a dry east wind, though the temperature does not rise to the melting point. Evaporation also absorbs heat, a fact which may be easily illustrated by pouring a few drops of ether on to the hand, when the cooling of the skin as the ether evaporates shows that heat is being rapidly absorbed.

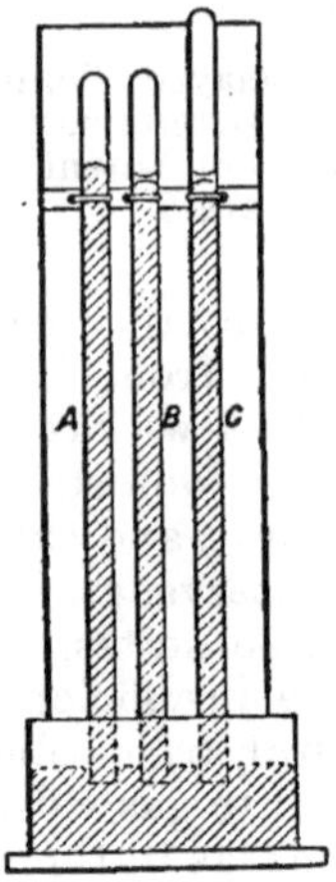

FIG. 84.—A, ordinary barometer; B and C, barometers with small quantities of water above the mercury.

In studying change of state in detail, we shall consider first the change into the gaseous condition and the converse, because, with the aid of the kinetic theory of gases, we are able to give a more complete account of it than is as yet possible in the other cases.

Liquid-Gas Change.—Let us suppose that an ordinary barometer-tube, held with the closed end downwards, is nearly filled with mercury, the remaining space being filled with water. On inverting the tube in the ordinary way, and unclosing the end in a cistern of mercury, the water floats up to the top, and some of it evaporates into the Torricellian vacuum. Allowing for the pressure of the small quantity of water, and comparing the height of the column thus corrected with that of a barometer made with dry mercury, it is found to be somewhat less, or a certain pressure is exerted on the top of the water resting on the mercury column. This pressure is due to the water-gas, or, as it is usually termed under such circumstances, the water-vapour, present in the space which would otherwise be a Torricellian vacuum. If the temperature is maintained constant, and if there is sufficient water, this pressure rapidly attains a definite value, quite independent of the volume of the tube above the mercury. This may easily be shown by raising this tube or depressing it in the cistern, or by tilting it, or by using different tubes. Thus in Fig. 84 A is an ordinary barometer, B and C other barometers, each with a small quantity of water at the top. The height of the columns in B and C is the same, and less than that in A.

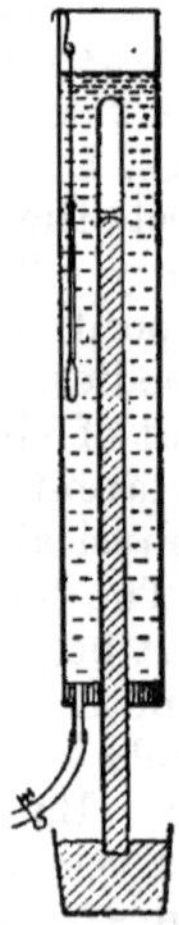

FIG. 85.—Barometer with Water above the Mercury and surrounded by a Heating Bath.

In an ordinary tube, the evaporation is very rapid, so that the steady pressure of the vapour is soon attained. In this state the space above the mercury is said to be *saturated* with vapour, and the pressure exerted is termed the "maximum vapour-pressure" or "vapour-tension" for the given temperature; maximum, because any decrease in the volume only leads to condensation of the vapour, not to increase in its pressure.

Use of the Terms "Gas" and "Vapour."—In ordinary language a gas is usually described as vapour when thought of as given off by a liquid or solid, especially when it is not far from the maximum pressure at which it begins to condense. We shall see later that there is probably for every gas a temperature below which it can always be condensed to liquid by sufficient pressure, and above which such condensation is impossible. It has been suggested that the gas should be described as a vapour below this temperature, and as a gas above it. This restriction of the term "vapour" is in accordance with general usage, but it is probably more convenient to use the term "gas" in the general sense, including "vapour" as a particular case.

If the barometer-tube B of Fig. 84, containing the water be surrounded with a water-bath, as in Fig. 85, so that its temperature can be raised, then it is found that the pressure of the vapour rises with the temperature, more and more of the water evaporating into the space above; and at each temperature there is a definite maximum pressure, which is more or less rapidly attained.

In the case of water, the following are some of the values of the maximum pressure. We shall describe later how the accurate values are obtained.

Temperatures.	Pressure in mms. of mercury.
0°	4·6
20°	17·4
50°	92
100°	760
150°	3581·2
200°	11689

For other substances the values are very different.

Temperatures.	Pressures in mms. of mercury.		
	Alcohol.	Ether.	Mercury.
0°	12·7	184·4	·020
20°	44·5	432·8	...
34·97°	...	760·0	...
50°	219·9	1264·8	...
79·7°	760·0	...	...
100°	1697·6	4953·3	·746
200°	...	...	19·9
300°	...	...	242·15
357·25°	...	...	760
400°	...	...	1588

If the space into which the evaporation takes place contains air or any other gas the evaporation goes on, but at a diminished rate. The pressure exercised by the vapour and, therefore, the total vapour present in a given space when the steady state is attained, is still very nearly the same, though it appears to be always slightly less than in a space containing only the vapour. The practical equality may be shown by an arrangement similar to that in Fig. 86.

A is a piece of wide glass-tubing, several inches long, closed at its upper end by a tap, above which is a small funnel. A is connected to the open tube B by a piece of strong flexible tubing. The tap *t* is opened, and mercury is poured into B until it reaches a certain mark in A, standing, of course, at the same level in B. The tap is then closed, and ether is poured into the funnel. B is lowered so as to reduce the pressure in A, and when *t* is opened, a small quantity of the liquid is pressed into A without the escape of any air; *t* is now closed, and when the steady state is reached—an excess of liquid being still present—B is raised till the level of the mercury in A is at the original mark. The air occupying its original volume, still exerts the atmospheric pressure. The difference of levels of the mercury in the two tubes is, therefore, due to the pressure of the ether-vapour, and this is found to be equal to its value when the air is absent. The same arrangement may easily be used to find this value by opening *t* and raising B till all the air is driven out of A, a small quantity of liquid being still left in the tube above the mercury and below the tap. If we close *t* and lower B, then as soon as the surface in B is a depth below that in A equal to the difference between the atmospheric pressure and the pressure of ether vapour, evaporation takes place. When this begins, even if B is slowly lowered still further, A also falls so that the difference between the levels in A and B will remain constant.

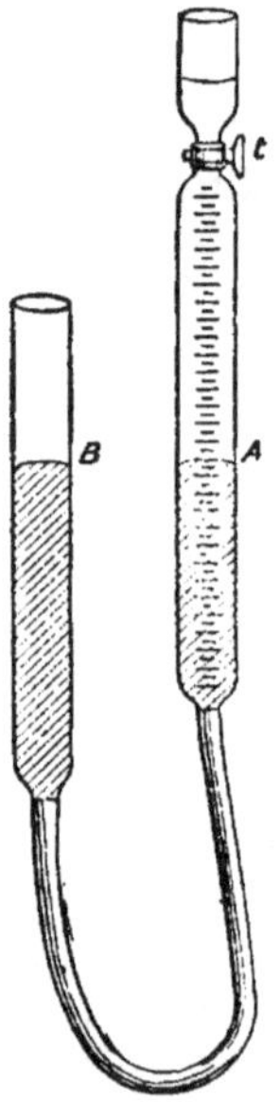

FIG. 86.—Evaporation into an Air Space.

This evaporation into a space already containing air is, of course, continually occurring from water on the surface of the earth. The pressure of the vapour present in the air, in fine weather, is always less, below the cloud level, than the maximum for the existing temperature. Evaporation, therefore, takes place from the surface of any water, or from any damp material present, and tends to bring about saturation. But fresh drier air is continually being brought over the water or damp material. This may take place by winds, or it may be brought about by convection, for the water-vapour is lighter than the air it replaces, and the vapour-charged air tends to rise, colder air from above, containing less vapour, taking its place. The evaporation, therefore, continues, but probably owing to continual renewals of the air, the maximum pressure is never produced by evaporation from the surface of the earth alone.

As a converse to evaporation, we continually have condensation occurring in the atmosphere in the formation of clouds and fogs. These are, probably, always produced by cooling. A mass of air containing water-vapour, not very far from the maximum pressure, becomes cooled, and normally as soon as the temperature falls to that for which the maximum pressure of water-vapour is equal to the existing pressure, condensation begins in the form of cloud, which is well described as water-dust.

Distillation.—Condensation is used in the process of distillation to obtain pure water.

The simplest form of still is one in which the steam rises from a vessel, leaving the impurities behind in the water. The steam is conveyed through a jacket, through which cold water is kept circulating, so that the temperature of the steam falls below the condensing-point, and the condensed water trickles down into a receiver, as in Fig. 86*a*.

We may summarise the results already described in the following statement:—

A liquid with a free surface and a space above it will evaporate, or change into gas, into that space, until a pressure is reached definite for each temperature. This limiting pressure is called the maximum vapour-pressure or vapour-tension. Its value is only very slightly affected by the presence of other gases, though these retard evaporation. It rises

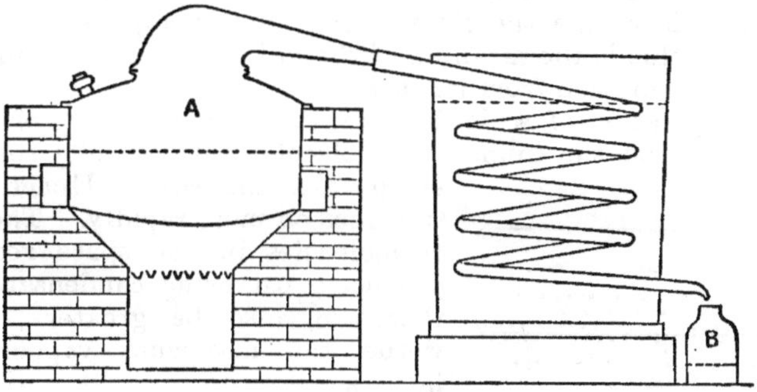

FIG. 86*a*.—Apparatus for Distillation. A, still; B, receiver.

with the temperature, and differs greatly for different liquids at the same temperature.

It may assist us in arranging these facts in our minds, if we seek to give an explanation of them on the kinetic theory of matter.

Let us first consider the case of a liquid partially filling a closed vessel, the space above being a vacuum to begin with.

The molecules in a liquid are, in general, within each other's spheres of action, and are entangled with each other; but there are a great many of the molecules which have sufficient kinetic energy to escape from the groups which they may be near at a given instant, and these move off to become entangled with other groups. We might, perhaps, represent the liquid at any instant as a network of attached molecules, forming a loose solid with numbers of gas molecules moving about in the interspaces. But the members composing the network are continually changing places with the freely moving molecules, so that the network is not permanently solid. If, now, one of the freely moving molecules happens to be close to the surface, and moving from the general body of the liquid, it will entirely escape and move into the space above the liquid, that is, it will evaporate. Escapes will take place all over the surface of the liquid in the same way, and very soon the space above will contain a great number of these molecules, which from their kinetic energy, will behave to each other as gas molecules, colliding, and escaping again from each other's spheres of action, cannonading the sides of the vessel, and so exerting pressure. Through their collisions with each other and with the sides, many of the molecules will have their direction of motion reversed, and will return to the liquid. Some of these will become entangled, and again become liquid molecules. The greater the number of molecules in the space, the greater is the number thus returning to the liquid, and as evaporation continues a point is at last reached when the number returning is equal to the number escaping. This corresponds to the "maximum vapour-pressure," the steady state being due—not to a cessation of evaporation—but to a balance between evaporation and condensation. If, through diminution of volume, the pressure tends to exceed the maximum vapour-pressure, the condensation exceeds the rarefaction until the steady state is again arrived at.

We may note that the molecules escaping are the most energetic in the liquid; their escape therefore lessens the average energy of those remaining, and this is the meaning of the fall of temperature in a liquid, produced by evaporation from its surface.

If the temperature of a liquid rises, the average energy of the molecules increases, and the number of molecules with velocity sufficient to escape also increases. Hence evaporation goes on more rapidly. The number of molecules in the space required to produce a balancing condensation must, therefore, also be greater — in other words, the maximum vapour-pressure increases.

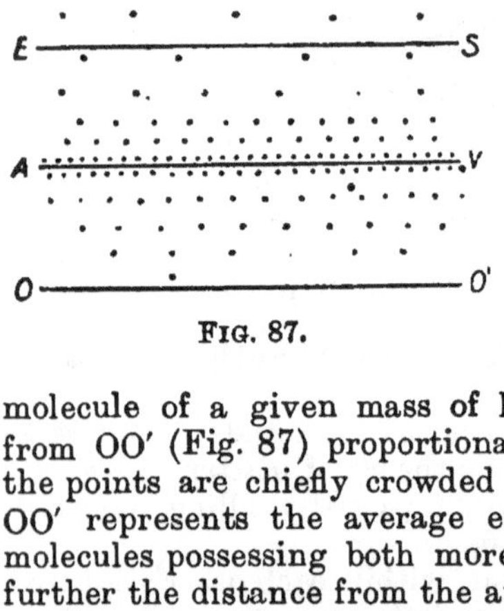

FIG. 87.

The rapidly rising rate of increase may, perhaps, be explained, or at least illustrated, as follows:—Let us plot on a diagram the energy possessed by each molecule of a given mass of liquid by putting a point at a distance from OO′ (Fig. 87) proportional to its energy. We may suppose that the points are chiefly crowded about the line AV, whose distance from OO′ represents the average energy. But there will be numbers of molecules possessing both more and less than the average, though the further the distance from the average, the less the number of points.

Let the energy which a molecule must possess in order to escape, be represented by the distance between ES and OO′. All molecules, therefore, represented by points above ES will escape if they have the opportunity.

As the temperature rises, the average energy rises, so that the whole diagram of points may be supposed to be stretched upwards and the number above ES increases. But the Specific Heat being nearly constant, the total energy, and therefore the average energy, increases by nearly the same amount for each rise of temperature of 1°, that is, AV approaches ES by nearly equal steps. But as the points are crowded more and more, the nearer we approach to AV, each successive degree-rise brings a greater and greater number above ES. The number able to escape and the vapour-tension, therefore, increase much more than in proportion to the rise of temperature.

So far we have supposed the space above the liquid to contain only the vapour of the liquid. But if some other gas—say air—is present as well, its chief effect is to lengthen out the process by which the steady state is reached. The air molecules will hinder the evaporation by knocking back into the liquid many of the molecules trying to escape, but they will equally hinder the condensation by getting in the way of those returning to the liquid. The two processes being equally interfered with, the same number of molecules must be present in the space for a balance, whether the air is there or not. We may, in fact, compare the effect of the air to a screen perforated with holes and laid upon the surface. Such a screen would equally interfere with both processes, and merely lengthen out the time required to attain the maximum vapour-pressure.

Boiling.—Besides the quiet transformation from liquid to gas, which we have hitherto considered, there is the more rapid conversion which occurs in boiling. If we heat some water in an open glass vessel, so that we can notice what occurs, evaporation of course goes on from the top-surface. This evaporation is indicated by the clouds forming where the vapour mixes with the colder air, and condenses through cooling below the temperature at which its pressure is the vapour-pressure. As the temperature rises, bubbles, which gradually increase in size, appear on the sides of the vessel. These consist, partly, of gases driven from solution—the higher the temperature, the less the quantity of gas which a liquid will dissolve—and, partly, of water-vapour, for evaporation takes place into the bubbles. Perhaps these bubbles have been formed by the swelling out of bubbles already existing, but too small to be seen; or they may have been formed in cavities free from liquid, for however smooth the surface is, there are still probably minute irregularities. These cavities may serve as nuclei into which the expelled gas and the water vapour can pass. As the bubbles increase in size, their upward buoyancy at last detaches them and they float to the top, and with this stage is associated the well-known "singing," which occurs shortly before boiling. Soon afterwards, boiling commences, that is, bubbles of steam form at points on the containing-walls and rise up to the surface. These bubbles always rise from definite points, each point supplying a constant succession of bubbles, and as the boiling continues these points diminish in number. The bubbles are never formed in the middle of the liquid. Even if they appear to be thus formed, closer examination always shows a particle of foreign matter forming a boundary to the liquid. The temperature now ceases to rise and it is found that the boiling-point is that *temperature at which the maximum pressure of the vapour is equal to the atmospheric pressure.* This may be illustrated by the barometer

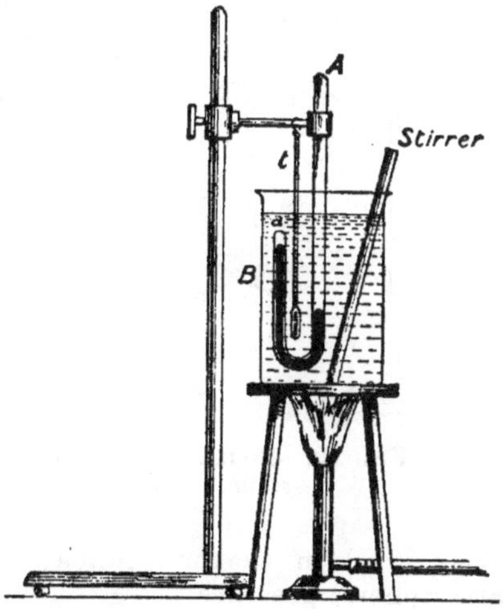

FIG. 88.—Vapour Pressure equals Atmospheric Pressure at the Boiling-Point.

experiment of Fig. 85, p. 158. For if the barometer column be jacketed throughout with steam, the pressure of the vapour of the water above the mercury just depresses the top of the column to the level of the cistern outside, that is, just exerts the atmospheric pressure. Or if a U tube A B Fig. 88, closed at the top of the short limb and open at the other limb, be filled with mercury round the bend to *a* as in the figure, and if then a small quantity of liquid—say, for convenience alcohol boiling at a lower temperature than water—be passed round the bend to the top of the mercury, if the tube is immersed in a water bath and gradually heated, the liquid begins to evaporate when its boiling-point is closely approached, and when it is just reached the mercury stands at the same level in the two limbs; that is, the vapour just balances the atmospheric pressure.

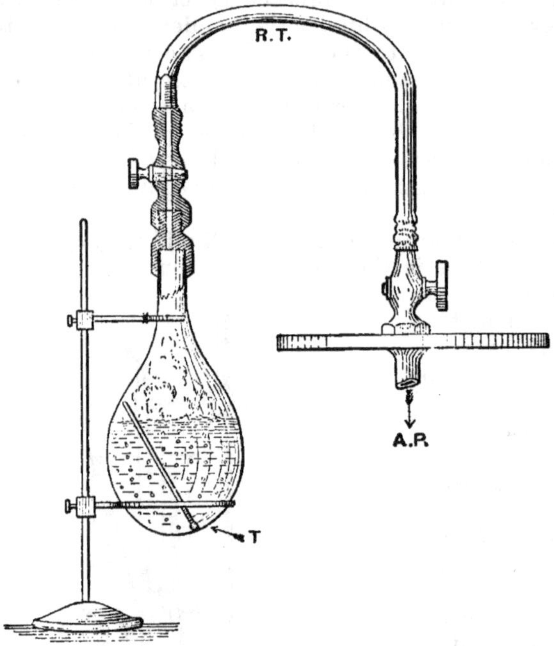

FIG. 89.—Reduction of Boiling-Point by reduction of pressure

If the external pressure be diminished, boiling may take place at a lower than the ordinary boiling temperature, for the vapour-pressure will equal the atmospheric pressure at a lower temperature. This may be illustrated by connecting a flask of hot water to an air-pump (Fig. 89). After a few strokes of the pump, boiling begins. Or, on boiling water in a flask till all the air is expelled, corking the flask up, and holding it under a stream of cold water (Fig. 90), the cooling is accompanied by condensation, and therefore diminution of the pressure on the water and rapid boiling sets in. Frequently the flask collapses in this process, owing to the excess of external pressure.

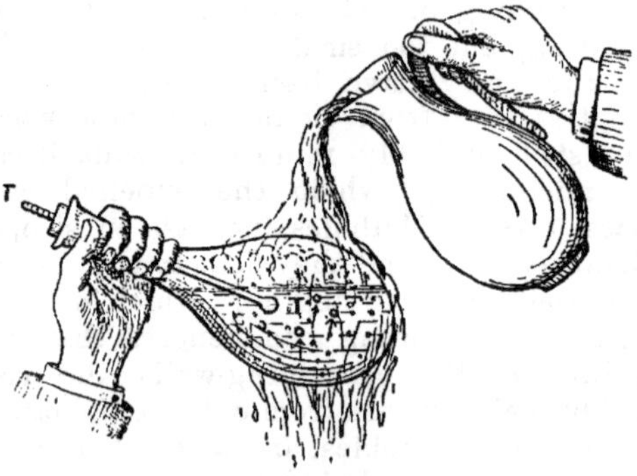

FIG. 90.—Boiling water by cooling it.

On high mountain levels the low atmospheric pressure produces a serious lowering of the boiling-point, that on the summit of Mt. Blanc being about 84° C. At such a temperature, boiling water has not its usual cooking qualities.

On the other hand, increase of pressure is accompanied by rise of boiling-point. In high pressure steam-boilers, the boiling-point rises many degrees above the normal. This rise of boiling-point is employed to obtain the gelatinous matter from bones, water above 100° dissolving this matter more freely than water at or below 100°. The bones are, therefore, heated in a closed vessel of water, fitted with a safety-valve, so that the water is exposed to the pressure of its own vapour, and this increases as the temperature rises. There is therefore no definite boiling-point short of that corresponding to the pressure at which the safety-valve is forced open. Such an arrangement is called a Papin's digester from its inventor, the discoverer of the rise of boiling-point with pressure.

Delayed Boiling.—When boiling has been going on for some time in a glass vessel, the temperature of the water begins to rise above the normal boiling-point, and at the same time the steam is given off in larger bubbles and from fewer points. If the vessel be very clean (it may be cleaned by rinsing with hydrofluoric acid and then with water) and the water very pure, by alternate boiling and cooling it is quite easy to raise the temperature of the water ultimately to 105° or 106° C. With still greater precautions, the temperature may be raised many degrees higher. In this state when boiling does occur it is almost explosive in its character, and the phenomenon is termed "boiling with bumping." On putting into the vessel sand, iron filings, or any rough material containing crevices or air bubbles, boiling again becomes normal at the normal temperature.

A still more striking experiment was made by Dufour, who prepared a mixture of oils of about the density of water and of high boiling-point, and then placed in it small drops of water, which remained suspended. They could be heated without boiling to at least 178° C.

The observation that normal boiling is resumed when air bubbles are present gives the clue to the nature of boiling. It is probably always associated with the presence of bubbles or cavities.[1]

We have seen that as the temperature rises towards the boiling-point, the dissolved gases are expelled, and collect in bubbles. As these are detached, each bubble probably leaves a small portion behind, just as a drop of water in breaking off from a surface and falling down, always forms a neck at which rupture takes place and leaves behind the part above the neck as in Fig. 91. These minute bubbles do not grow so largely as before, since most of the dissolved gas has been expelled in their first formation. Evaporation, however, goes on into them, and their size will be such that the vapour-pressure for the existing temperature + the pressure of gas or air in

FIG. 91.—Formation of a drop by pinching in at a neck.

[1] An excellent account of the phenomenon of melting and boiling, with experiments illustrating the explanation, is given by Aitken in the *Transactions of the Royal Scottish Society of Arts*, vol. ix, 1874–75.

them = atmospheric pressure. Now, when the temperature rises to the point at which the pressure of the vapour equals the atmospheric pressure, there can no longer be equilibrium, since the internal pressure exceeds the external by the pressure of the gas or air in them. The bubbles grow, their buoyancy increases, and finally they break away and float to the top.

The small portion of each still remaining serves as a fresh nucleus, and the process is repeated indefinitely, as we see from the constant stream of steam-bubbles from the same point in the containing vessel. The heat supplied to the liquid is taken up as the latent heat of the steam formed.

If the heat is supplied very rapidly the temperature of the liquid tends to rise above the normal boiling-point, and the evaporation into any bubble present tends to increase. Thus the bubble grows rapidly. It is true that evaporation and condensation are always going on together. But suppose that we are considering water at 101°. The vapour-pressure at 101° is about 787 mm., and only with that pressure of vapour in a bubble would the evaporation and condensation balance. The growth of the bubble, however, keeps the pressure within down at 760 mm., so that the rate of condensation is hardly affected by the rise of temperature, while the rate of evaporation has grown considerably. The unbalanced evaporation increasing the size of the bubble increases the evaporating surface. Steam is more readily formed, more latent heat is taken up, and if the evaporating surface is sufficient the temperature may be brought down to the normal boiling-point. If, however, the points at which steam is formed decrease in number—if there are not sufficient bubbles—the steam given off may not be sufficient to carry away all the heat supplied, and the temperature may rise appreciably above the normal boiling-point. In fact, with glass vessels some rise above it almost always takes place.

The presence in steam bubbles of gas other than steam—an observation due to Grove—supports this explanation.

After a time, through the removal of the foreign gases, the portions of the bubbles left behind probably get smaller—so small that their surface tension seriously affects the pressure within them.

The surface tension alone exerts a pressure $p = T\left(\frac{1}{r} + \frac{1}{r'}\right)$, where r and r' are the principal radii of curvature of the stretched surface, and T is the tension per centimetre. In order, therefore, that a bubble may grow, we must have (the vapour-pressure + the pressure of the contained gas) greater than (p + the atmospheric pressure), and if p becomes sensible, through the diminution of the residual bubble, the temperature must rise sensibly above the normal boiling-point before this condition will hold. When the bubble once begins to grow, r and r' increase, and p diminishes, so that the pressure within the bubble diminishes; but the evaporation into the bubble is still at the rate corresponding to the higher temperature of the liquid, while the condensation is only at the rate corresponding to the diminished pressure, now tending rapidly towards the atmospheric pressure. The bubble, therefore, grows with very great rapidity, almost explosively; much latent heat is taken up, and the temperature of the liquid falls, though not necessarily to the normal boiling-point. As the bubble rises up, the process is repeated with the nucleus left behind, the growth of the new bubble being delayed till the condition for its growth obtains, when there is another sudden swelling out. The introduction of sand or iron filings puts an end to this state of delayed boiling, since it presents a large number of air-bubbles as nuclei for evaporation, and then ordinary boiling occurs.

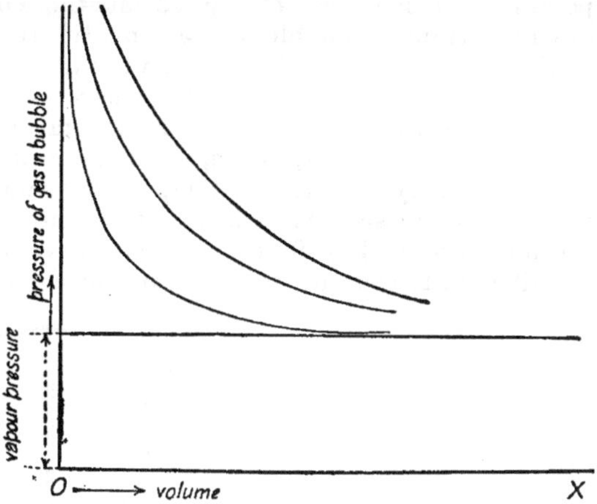

FIG. 92.—Relation between Volume and Pressure of Gas contained in a Bubble in a Liquid.

Stability of Bubbles in a Liquid.—The stability of bubbles formed on the side of a vessel may be discussed by the aid of diagrams. The pressure within a bubble is that of the vapour and that of the contained gas. This may be represented by Fig. 92, the different hyperbolas representing the relation between pressure and volume for different quantities of contained gas. Lifting the curves above the zero pressure a distance equal to the vapour pressure, then the relation between the internal pressure and the volume of a bubble containing a given quantity of gas will be represented by the abscissa and ordinate of the corresponding curve.

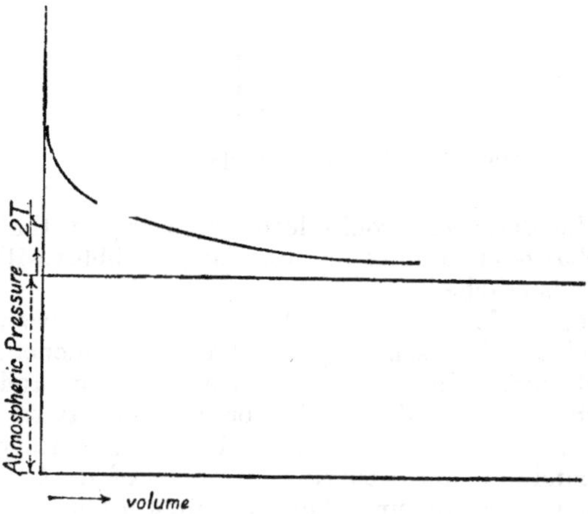

FIG. 93.—Relation between Volume and External Pressure on a Bubble. The surface tension curve slopes down more gradually than the hyperbolas in Fig. 92.

The external pressure is that of the atmosphere + the hydrostatic pressure due to depth below the surface + that due to surface tension $\frac{2T}{r}$. We shall include the hydrostatic in the atmospheric pressure. Supposing the bubble spherical, the latter term is inversely proportional to the cube root of the volume. Hence, for instance, to halve this part of the pressure, we increase the volume eight times. Then the pressure curve representing $\frac{2T}{r}$ slopes down much more gradually than any of the hyperbolas in

Fig. 92. Fig. 93 represents the curves raised a distance equal to the atmospheric pressure above the line of zero pressure.

Superposing the two figures, so that *ox* is the same for each, the points where the curve of Fig. 93 cuts those of Fig. 92 represent positions of equilibrium for bubbles containing different quantities of gas. As long as the vapour-pressure is less than the atmospheric pressure, the curve of Fig. 93 cuts the curves of Fig. 92, where they have a greater slope than it, and any increase of volume, therefore, means an excess of external over internal pressure, and, hence, equilibrium is stable. The bubbles, therefore, grow only with rise of temperature and addition of gas. But when the vapour-pressure exceeds the atmospheric it is possible for the surface tension curve of Fig. 93 to cut a gas curve of Fig. 92 at a less slope, and equilibrium is then unstable, as in boiling with "bumping." We leave

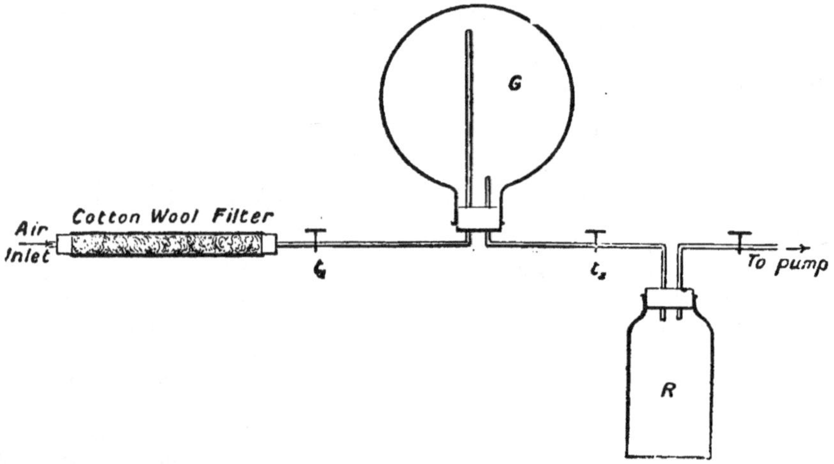

FIG. 94.—Apparatus for obtaining a Dust-Free Space.

the reader to consider how the diagrams would lead us to expect that after the first "singing" discharge of bubbles no more large bubbles will be found till the boiling-point is reached.

Condensation on Nuclei.—We have seen that the change from the liquid to the gaseous condition is essentially a surface phenomenon. It takes place at the top level surface in ordinary quiet evaporation, and in all probability at the surface of already existing bubbles or cavities in ebullition. The converse change in condensation is also generally a surface phenomenon, the condensation occurring on nuclei of dust or suspended matter other than air or vapour. This was first discovered by Coulier (*Journal de Pharmacie et de Chemie*, xxii., pp. 165 and 254, 1875), and later again by Aitken (*Trans. R.S.E.*, xxx., pt. 1, p. 337), to whom we owe extensive investigations on the subject.

When a beam of sunlight crosses a room, the visibility of its track is due to particles of "dust," many perhaps large enough to be seen, others quite beyond our range of sight. If the air be filtered through cotton-wool this dust is filtered out. The clean air can then be quite appreciably supersaturated with vapour without formation of cloud or fog, that is without condensation.

To show this a large clean glass globe may be arranged as in Fig. 94.

The globe G is, to begin with, filled with the air of the room, and contains some water which can be washed round the sides so as to ensure complete saturation of the air.

The taps t_1 t_2 being turned off, the pump exhausts R. If now t_2 is turned on for a moment, the air in G expands, cools, and becomes supersaturated at the new temperature. The excess of moisture at once condenses in the form of fog which, usually evident enough, is still more evident if a light, such as a candle flame, be viewed through the globe when beautiful diffraction rings may be seen round the flame, the rings being larger the smaller the drops. Then turning on both t_1 and t_2 let clean dust-free air be slowly drawn through the cotton-wool filter. After a short time t_1 and t_2 may be both turned off, and the air in G be again expanded and cooled by turning on t_2 for a moment. There will now be a much smaller number of dust nuclei, and as the vapour condenses on these alone, the drops are much larger and the fog is less dense. For though the same amount of water may be present as liquid in the air its concentration into larger drops diminishes its surface, and there is much less hindrance to the passage of light. The diffraction rings round the flame at the same time grow less.

If this process be repeated several times the drops soon become easily visible, like those in a Scotch mist, and move at an appreciable speed downwards in a shower of fine rain. When all the drops have fallen down the air is clear and a fresh small expansion does not produce a cloud or rain at all.

Aitken (*Nature*, March 1, 1888, p. 429, and Feb. 27, 1890, p. 394), has devised a "dust counter," an apparatus to count the number of dust particles or, at any rate, condensing nuclei per c.c. in any specimen of air. This consists essentially of a small chamber 1 cm. deep with a glass floor ruled in square millimetres and a glass top through which the glass floor can be seen. The chamber is connected to a pump, and by an inlet either the surrounding air can be introduced or any known proportion of filtered dust-free air. The air in the chamber is kept saturated. In general it is necessary to dilute the air to be tested with a large proportion of clean air to reduce the number of dust particles, and so increase the size of the drops. Suppose, for instance, that nine-tenths of the air is dust-free. If an expansion is suddenly made, drops are formed on the dust particles, and these should be heavy enough to fall at once on to the ruled floor. There they are evident as little specks. The numbers on several square millimetres are counted by the aid of a lens and averaged. Say that the average is 5. Then on a square centimetre there would be 500. This is the total number falling down out of a cubic centimetre. But the air has been diluted to one-tenth its original dustiness. Therefore, in its original condition it contained 5000 dust particles.

Aitken has examined the air in various localities, and finds that the number of dust particles per c.c. is greatest in rooms, greater in dry town air than in the country, greater at lower altitudes than on mountains, and that there is a great diminution after rain.

Thus (*Nature*, March 1, 1888) Aitken found the following numbers:—

Source of Air.	No. per c.c.
Outside air—raining . . .	32,000
,, ,, fair	130,000
4 feet from floor of room in which gas was burning . . .	1,860,000
Near ceiling	5,420,000
Air above Bunsen flame . .	30,000,000

Again (*Nature*, Feb. 27, 1890) he found at Kingairloch, on the shore of Loch Linnhe, numbers ranging from 205 to 4000 per c.c.; at the top of Ben Nevis, 335 to 473 per c.c. In London and Paris, the numbers were counted by the hundred thousand. Fridlander (*Quarterly Journal of the Royal Meteorological Society*, xxii., July 1896) has tested various specimens of air in a journey round the world, the numbers at sea varying from 200 per c.c. in the Indian Ocean to 4000 in the Atlantic.

The explanation of the thickness of town fogs as compared with those in the country or at sea is now evident. The number of dust particles is always far greater in town air, and if the smoke keeps near the ground, as it does when the upper air is much warmer than the ground air, the number may be enormously greater than in country air. When condensation occurs the fog is, therefore, exceedingly "dense" through the minute subdivision of the water deposited.

The action of the nuclei in condensing the vapour probably depends on a principle first pointed out by Lord Kelvin, to which we shall return later (chap. xix.). The principle is that the equilibrium or saturation vapour pressure of a space in contact with a liquid surface depends on the curvature of the surface. If P be the normal saturation pressure at a given temperature, it can be shown that in contact with a spherical surface with radius of curvature r, for moderate values of r the saturation pressure is

$$p = \mathrm{P} + \frac{2\mathrm{T}\sigma}{(\rho - \sigma)r} \qquad (1)$$

where T is the surface tension; ρ the density of the liquid, and σ that of the vapour.

If the surface is convex then r is negative.

When r is exceedingly minute so that p/P differs largely from 1

it can be shown that $$\log p/\mathrm{P} = \frac{2\mathrm{T}\sigma}{p\rho r} \qquad (2)$$

if we assume that Boyle's law will hold for the vapour.

It follows that small drops in a space just saturated over a flat surface will find the space under-saturated for their own curved surface, and will evaporate; while if, by any accident, there is a concave surface formed the space will be over-saturated and condensation will go on.

Now suppose that in some saturated dust-free air a few vapour molecules collide, and with so little energy that they form practically a liquid particle. The radius is so minute that the saturation pressure for the drop is far above the pressure of the vapour round it, and the drop instantly evaporates. But if dust nuclei are present, a liquid layer may form on a nucleus with radius sensibly equal to that of the nucleus, which may be so large that the saturation vapour pressure is only very slightly in excess of the normal pressure, and a very slight supersaturation may make the liquid increase. There is another way in which soluble dust may aid condensation. We shall see later that the saturation vapour pressure over a solution is less than that over a pure liquid, so that a normally saturated space may condense on to a liquid surface if the liquid contains salt in solution. If, for instance, a particle of common salt is floating in vapour-laden air, any accidental formation of a liquid particle on it may lead to solution and the formation of a drop with lowered vapour pressure, which may tend to grow even if the space is not quite saturated for a normal plane surface of pure water. Aitken believes that such condensation frequently occurs and largely accounts for the haze in air.

The conditions of condensation of water-vapour in air and other gases have been very carefully investigated by C. T. R. Wilson (*Phil. Trans.*, A., 1897, p. 265, and A., 1899, p. 403). It is sufficient here to consider the case of air. He found that if all dust be filtered out, then the saturated air at any temperature about 20° may be suddenly expanded in the ratio 1 : 1·25 without condensation. But that, if the expansion exceeds this, condensation on a few nuclei (not more than a few hundred per c.c.) does occur up to an expansion of 1 : 1·375. After this a dense fog appears, the denser the greater the expansion. Now this means that if we start with air, say at 20°, the sudden adiabatic expansion 1 : 1·25 cools it to − 6°, when the density of the vapour saturating the air at 20° is about 4·2 times the density of the vapour saturating the air at − 6°. The expansion 1 : 1·375 cools it to − 16°, when the density is about 7·9 times the density of vapour saturating the air at − 16°. Then Wilson's experiments show that there are always nuclei present in small numbers sufficient to condense the vapour when it has about four times its normal saturation density, and in large numbers if it has about eight times its normal saturation density. We can hardly suppose that the nuclei are of foreign matter. Wilson shows that drops of radius of the order 10^{-8} cm. would, by the formula (2) above, be in equilibrium at the 8-fold density, and such drops would contain but few molecules according to the calculations of molecular dimensions in chap. ix. We may suppose, then, that in the molecular collisions water particles of such order are continually being formed, and if the vapour density is 8 times the normal they will tend to grow instead of evaporating. As to the formation at the 4-fold density, Wilson supposes that there is chemical action occurring; for instance, in some of the collisions the oxygen, nitrogen, and water might combine to form nitric acid. The vapour at 4-fold its normal saturating density might be saturated for even the minutest particle of acid formed by collision, and so the particle would grow. Such combination may always be occurring to some slight extent, the compounds being made and unmade in successive collisions.

Wilson also found that ultra-violet light is active in producing condensation nuclei and that condensation occurs under its action even with slight expansions, the fog produced being of a blue colour. Tyndall had previously found this formation of blue fog in many other vapours denser than water, under the action of ultra-violet light. The colour implies that the drops formed are comparable in size with the waves of blue light and so scatter that constituent when white light falls on them.

Exact Measurements of Vapour-Pressures.—Various observers have made researches on the vapour-pressures of liquids at different temperatures, the most complete as well as the most important being those in the case of water.

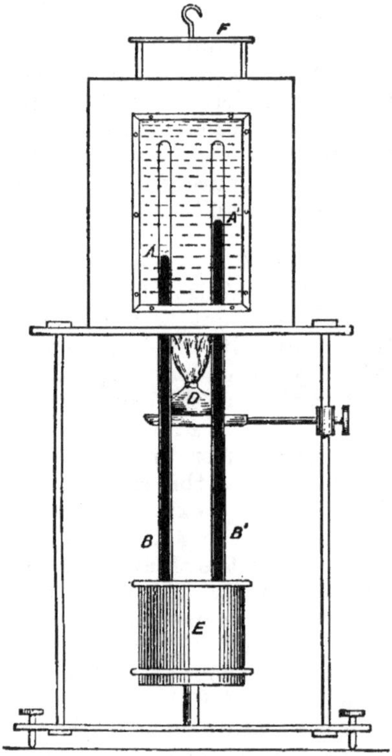

FIG. 95.—Vapour-Pressure of Water from 0° to 50°.

Two methods have been employed, the "*statical*" method, which is illustrated in its simplest form by the experiment described on p. 158, in which the pressure of the vapour depresses the column of mercury in a barometer, and the "*dynamical*" method, in which advantage is taken of the fact that a liquid normally boils when the pressure of its vapour is equal to the external pressure.

In the first method, the temperature is varied at will, and the corresponding maximum pressure measured; and in the second method, the pressure is varied at will, and the corresponding maximum temperature, that is the corresponding boiling-point, is observed.

These methods were carried out in Regnault's classical researches on the vapour-pressure of water, which may be taken as typical. He adopted three arrangements, corresponding to three ranges of temperature:—(*a*) From 0° to 50° C.; (*b*) below, and up to 0° C.; (*c*) from 50° upwards.

(*a*) *From* 0° *to* 50°.—Two barometers, AB, A'B', with a common cistern E (Fig. 95) are placed side by side—their upper ends being in a vessel containing water which serves as a constant-temperature bath. The front of this vessel is glazed, so that the levels of the mercury in the two tubes may be observed by a cathetometer. D is a spirit-lamp to raise the temperature of the water when required, and F is a stirrer to keep the temperature uniform throughout. A small quantity of water having been introduced into the tube AB, the vapour from it depresses the column, and after correcting for the pressure due to the small excess of water, for the effect of the water on the capillary depression, and for the temperature of the mercury, the difference in level in the two tubes gives the vapour-pressure at the temperature of the bath. Since the vapour-pressure at 50° is only about 92 mm., the bath was not very large and it was found possible to keep its contents at a uniform temperature. But above 50° a much larger bath would have been required, and the difficulty of maintaining a uniform temperature in a large mass of water is so great that Regnault abandoned this method for the dynamical.

(*b*) *For temperatures below* 0° he still employed two barometers, side by side (Fig. 96); but one barometer was bent round at the upper end, and terminated in a bulb E, containing ice. This was surrounded by a "freezing mixture," the temperature of which was taken by a thermometer. The temperature of the rest of the apparatus being above that of the freezing mixture, the pressure exercised by the vapour on the mercury corresponded to the temperature of the freezing mixture—for any tendency to rise above this pressure would be checked by condensation in E. The freezing mixture was liquid, and by stirring, its temperature was kept uniform throughout. Regnault was unaware of the fact that the vapour-pressure of ice below 0° differs from that of water below 0° at the same temperature, a fact deduced subsequently from theory by Kirchhoff, and later verified experimentally by Ramsay and Young. He, therefore, made no attempt to measure the vapour-pressure of water below 0°. In order to construct tables giving the most probable values of the vapour-pressure for each degree of temperature, Regnault employed the graphic method, plotting his results on a temperature and pressure diagram, and then getting rid of irregularities or discrepancies between different series by drawing a continuous curve most nearly representing the results. The curve, as drawn by Regnault, goes continuously through 0°, where the evaporating substance changes from water to ice. This throws doubt on the accuracy of his numbers about and below 0°. The water-vapour curve would no doubt be continuous, but the ice-vapour curve meets it at a small angle at 0° as represented in Fig. 97, theory showing that the ice-vapour curve diverges from the water-vapour curve about $\frac{1}{23}$ mm. per degree. Ramsay and Young (*Phil. Trans.*, ii., 1884, p. 461) have proved that the difference exists by maintaining ice and water in two separate connected vessels at a very low constant pressure. The temperatures of the two vessels became steady at the points at which their vapour-pressures were equal to this constant pressure. The ice was found to be at a higher temperature than the water.

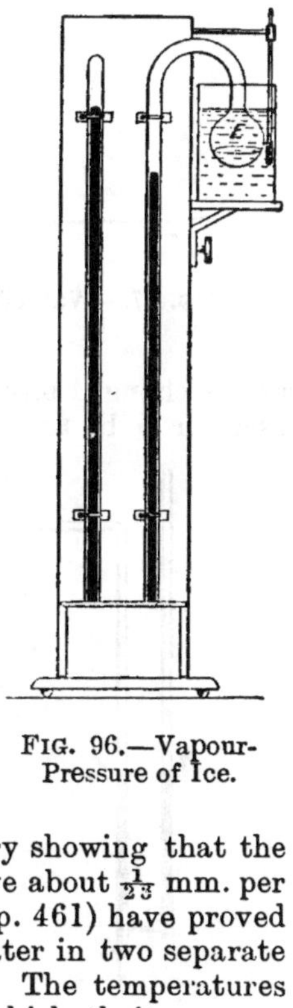

FIG. 96.—Vapour-Pressure of Ice.

(*c*) *For temperatures above* 50° Regnault employed the dynamical method. The arrangement of the apparatus will be seen from Fig. 98. A copper boiler, heated by a small furnace, is in connection with the reservoir of air G, the pressure of which is indicated by the manometer HK. Round the connecting tube at A is a wider tube, in which water circulates as in an ordinary still. The boiler is heated till the water in it boils, and the steam rises into the tube A. Here it is condensed, and the water trickles back into the boiler, so that there is no diminution of its contents. The pressure to which the water in the boiler is subjected

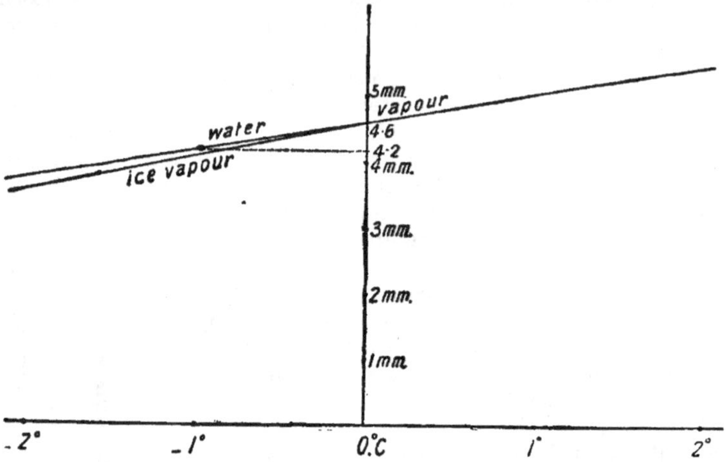

FIG. 97.—Water-Vapour Pressure and Ice-Vapour Pressure. They differ by 0·04 mm. at −1°.

can be changed either by withdrawing air from, or forcing air into, the reservoir G by the pipe F. Four iron tubes closed below were inserted

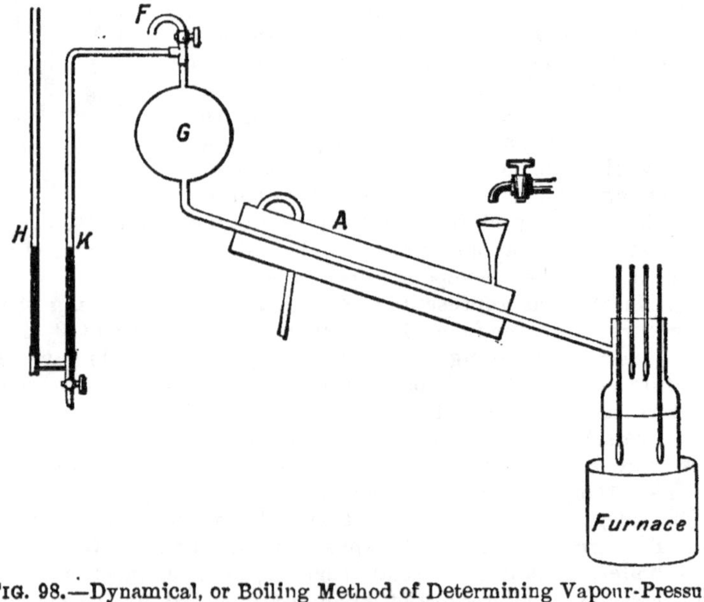

FIG. 98.—Dynamical, or Boiling Method of Determining Vapour-Pressure of Water above 50°.

through the top of the boiler, two reaching nearly to the bottom of the water, and two shorter tubes being surrounded by the steam. These tubes were filled with oil or mercury, and thermometers were inserted which gave the temperature. As the indications of the four agreed, the boiling was normal and not "with bumping."

It was found that on altering the pressure in G, the temperature soon reached a steady state, the boiling-point at the pressure indicated by the manometer. The manometer, therefore, gave the pressure of the vapour at the observed temperature.

For temperatures above 150°, the apparatus employed was exactly the same in principle, but much larger and stronger, and Regnault was able to work with it up to 230° at a pressure above 27 atmospheres.

The vapour-pressures of various other liquids have also been determined for a series of temperatures by Regnault and other observers using similar methods.

An account of the various methods will be found in Winkelmann's *Handbuch der Physik*, 2nd ed., vol. iii. p. 903. The values obtained for the vapours of water and various other liquids will be found in Landolt-Börnstein, *Tabellen*, p. 119 *et seq.*

Determination of Boiling-Points.—The boiling-points of liquids are often required, the knowledge of the vapour-pressure at other temperatures not being needed.

If the liquid whose boiling-point is in question is plentiful, it may be heated to boiling in a flask, or the apparatus used in determining the 100° point on the thermometers may be used, care being taken that the boiling is normal. A very simple arrangement, using the statical method, is that illustrated in Fig. 99.

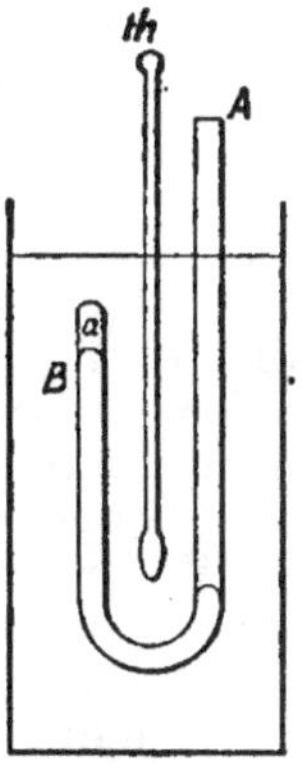

FIG. 99.

AB is a U tube. The shorter limb is closed, and is filled with mercury, which extends just round the bend. A small quantity of liquid is then introduced, and by careful manipulation may be floated up through the mercury without any air, so as to occupy the end *a* of the shorter limb. The tube is then immersed in a bath, which is heated until the vapour forms and depresses the mercury. When the level is the same in both limbs, the vapour-pressure is equal to the atmospheric pressure, and the temperature gives the boiling-point. This method, of course, admits of simple correction for variation of atmospheric pressure. The following are a few boiling-points:—

	°C.
Ether	34·87
Alcohol	78·4
Water	100·0
Mercury	358·5

Determination of Vapour-Density.—The method of determining the density of gases at 0° and 760 mm. pressure consists in filling a globe with the gas to be experimented on, at the atmospheric pressure H, while it is surrounded with melting ice; then closing it and weighing it. The globe is then put into the ice again, exhausted to a low pressure h, closed, and again weighed. Suppose the diminution in weight is W.

Assuming that Boyle's law holds, the weight of a volume of gas at 0° and 760 mm. equal to that of the globe is

$$\frac{W \times 760}{H - h}$$

Various corrections are of course needed, and we may refer the reader to Regnault's experiments, of which a short account is given in Jamin, *Cours de Physique*, tom. ii., or to the series of papers describing the experiments which led Lord Rayleigh to the discovery of argon (*Proc. R.S.*, Feb. 1888, Feb. 1892, March 1893; *Phil. Trans.*, A., 1895, p. 187).

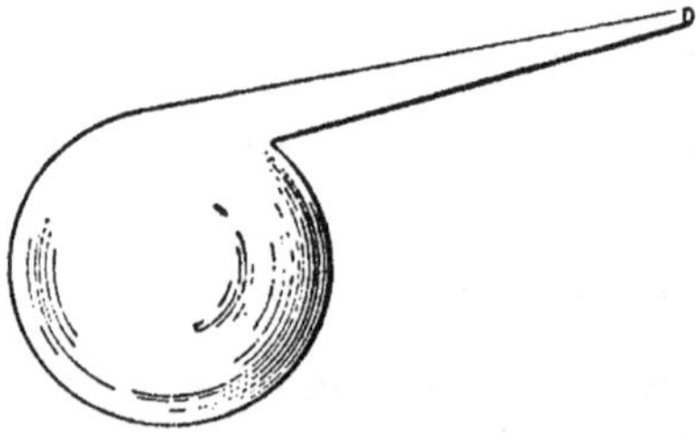

FIG. 100.—Vapour-Density Bulb: Dumas' Method.

Of course, this method is not applicable to substances which are liquid at ordinary temperatures and pressures. In their case, the weight of a known volume of the vapour must be determined at a known pressure, and at such a temperature that there is no condensation at that pressure, and the ratio of this to the weight of the same volume of air at the same temperature and pressure is termed the "Vapour-Density."

Various arrangements have been adopted for the purpose. In Dumas' method, a bulb (Fig. 100) ending in a fine open tube, is first filled with dry air and weighed, the temperature and pressure being observed. Some of the substance is then put into the bulb in the liquid state, and the bulb is immersed in a water- or oil-bath, the temperature of which is increased somewhat above the boiling-point of the liquid, the open tube projecting above the liquid in the bath. The liquid boils, and, if a sufficient quantity is present, its vapour drives out the air. When all the liquid has evaporated, and the vapour is at the temperature of the bath, the end of the fine tube is sealed. The bulb is now taken out, and when cool is weighed again. The point of the tube is now broker under mercury, the vapour condenses, and the bulb is practically filled with mercury. From its weight so filled, its volume can be determined. From this we may find the weight of the air filling it in the first weighing, and thence

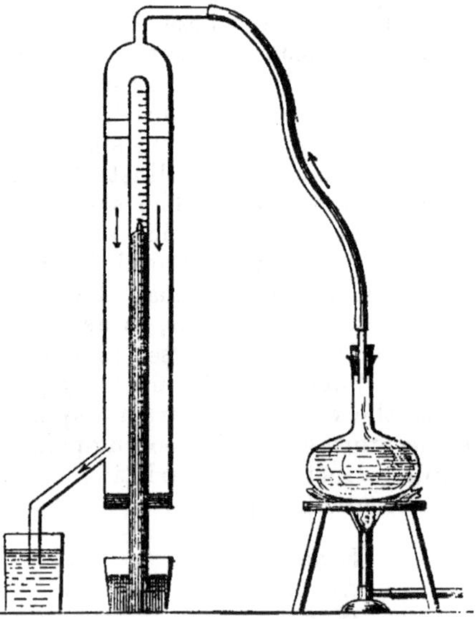

FIG. 101.—Vapour-Density Determination by Gay Lussac and Hoffmann's Method.

the weight of the bulb empty. Subtracting this from the weight in the second weighing, we have the weight of the contained vapour at the temperature of the bath and the atmospheric pressure. Finding the ratio of this to the weight of the same volume of air at the same temperature and pressure, we have the vapour-density.

In Gay-Lussac's method as modified by Hoffmann, a graduated barometer is enclosed in a wider tube (Fig. 101) through which circulates steam or other suitable vapour from a boiler, the barometer thus being maintained at a known temperature. A very small stoppered bottle is then quite filled with the liquid, and weighed, so that the weight of the liquid used is known. This is floated up through the mercury column, and on arriving in the Torricellian vacuum, the stopper is forced out, and the liquid evaporates. Such a quantity of liquid is taken, and such a temperature is maintained, that the whole of the liquid evaporates, its pressure, of course, depressing the mercury column. The volume occupied by the vapour is observed, and its pressure is given by the difference between the original and the new height of the barometer column. Hence we have the data for determining the vapour-density.

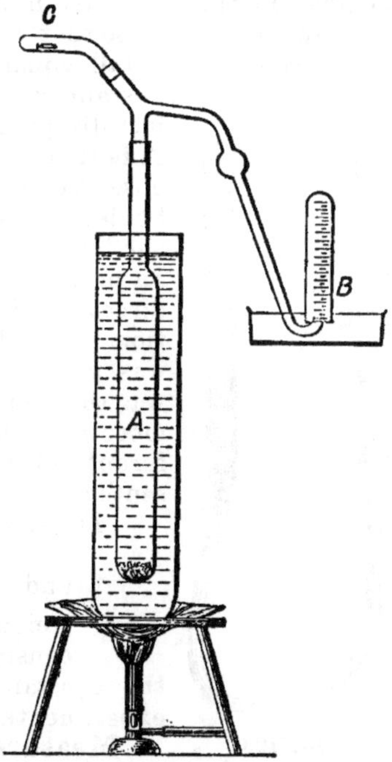

FIG. 102.—Victor Meyer's Method for Vapour-Density Determination.

Two methods have been devised by Victor Meyer. In that most commonly used, which we shall describe, a known weight of liquid is introduced in a small stoppered bottle into a heated flask containing air. The vapour formed displaces its own volume of air, and this is collected and measured. The method is applicable to vapours denser than air. One form is shown in Fig. 102.

The flask A contains air, and is heated in a constant-temperature bath. The neck rises above the bath, and a narrow side-tube leads down to the trough B, where the expelled air may be collected. The neck is closed by the slanting hollow stopper C, in which is placed the small bottle containing the liquid. When the temperature of the different parts of the apparatus is steady, C is turned round and the bottle falls down to the bottom of the flask on to some asbestos packing, put to prevent breakage. The stopper is forced out, and evaporation takes place, the vapour driving the air upwards. If the displaced air is raised sufficiently slowly, as each layer rises it takes the temperature and, therefore, the volume of that which it displaces. Hence the weight of the air forced out and collected at B is equal to that of the air displaced by the vapour in A. The vapour-density, as compared with that of air at the same temperature and pressure is, therefore, the weight of liquid introduced divided by the weight of air expelled.

Density of Saturated Vapour.—The densities of various saturated vapours, that is, of vapours in equilibrium with their liquids, have been studied by Hering. The principle of his method will be understood from the following imaginary experiment. Let A (Fig. 103) be a graduated tube closed at the top, and connected by a flexible tube with B open to the air at the top. Let A be entirely filled with mercury, which passes round the bend into B. Let a small weighed quantity of the substance be floated round through the mercury into A. By depressing B sufficiently, the pressure in A may be so reduced that the liquid begins to evaporate. At first the vapour in A is saturated, and exerts the vapour-pressure so that, as B is lowered still farther, the two mercury levels continue to fall equally. The vapour in A merely increases in quantity at the constant vapour-pressure. But when all the liquid has just disappeared, the volume occupied by the vapour in A is that of a known weight of vapour saturated at the observed temperature. The point of disappearance may be determined from the fact that, as soon as it is passed, the level in B begins to fall more quickly than in A. In practice it was found better to work in the opposite direction, since, through adhesion to the glass, evaporation was not complete till after the saturation-pressure was passed. The point at which condensation begins was therefore found. Hering used a much better but more complicated manometer. He found that for different substances the density might be nearly represented by

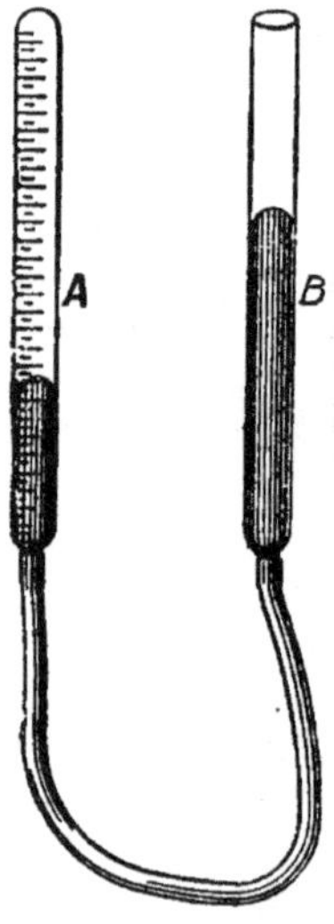

FIG. 103.

Density of saturated-vapour = Density as compared with that of hydrogen at the same temperature and pressure, calculated from the molecular weight $\times \sqrt{\text{temperature measured from } -273^\circ} \times$ constant,

the constant being the same for all the substances experimented on.*

Measurements of Latent Heat.—We have already seen that, when a liquid boils, its temperature normally remains constant, the heat supplied going to change the liquid into gas without rise of temperature, and we have called this "latent heat." But the term has a precise signification which may be expressed as follows:—

The latent heat of a vapour at a given temperature is the quantity of heat required to convert 1 gramme of liquid into vapour at its maximum pressure at that same temperature. Thus the latent heat of steam at 100° is the heat required to convert 1 gm. of water at 100° into 1 gm. of steam at 100°, and 760 mm. pressure.

The general principle of the method of measurement of latent heat will be best understood by considering a rough experiment with steam:—

A flask containing water, with a tube passing through the cork and nearly reaching to the bottom, is inverted, as in Fig. 104, and the water

* A series of researches has been made by Fairbairn and Tate on the density of saturated steam. For an account of their method the reader is referred to Baynes's *Thermodynamics*, p. 181.

boiled by a ring-burner, or otherwise. The steam issues from the lower end of the tube. A calorimeter, containing a known weight of water at a known temperature, is then brought under the flask, so that the steam passes into the water and is condensed there. Allowing the condensation to proceed for a short time, the calorimeter is then withdrawn and the temperature noted. Its increase in weight gives the quantity of steam condensed. Then, on the one hand, we have the gain of heat by the calorimeter and its original contents, and, on the other hand, the latent heat given up by each gramme of steam in condensing, together with the heat given up in falling from the temperature of boiling to the final temperature of the calorimeter.

Equating these, we can determine the latent heat.

For example, a calorimeter whose equivalent was 10 gms. contained 140 gms. of water at 15° temperature. Steam at 100° was passed into it until the temperature rose to 75°, that is, through 60°. The contents of the calorimeter weighed now 157 gms., so that the total quantity of steam condensed was 17 gms. If the latent heat given up by each gramme is L, it may be regarded as giving up L and then 25 more calories in falling to 75°. We therefore have the equation

$$150 \times 60 = (L + 25)17,$$

whence $$L = 504.$$

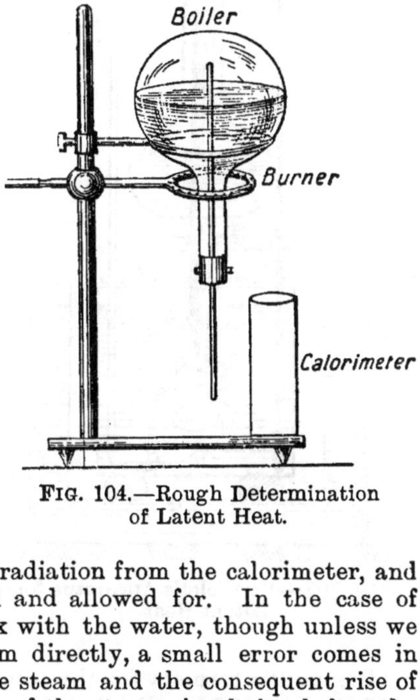

FIG. 104.—Rough Determination of Latent Heat.

This experiment, of course, is affected by very serious errors. Heat is introduced not only by the steam but also by the conducting pipe, and the amount so introduced must be determined and allowed for. Heat is, on the other hand, lost by conduction or radiation from the calorimeter, and its amount must also be determined and allowed for. In the case of water, we may allow the steam to mix with the water, though unless we measure the temperature of the steam directly, a small error comes in through the additional pressure of the steam and the consequent rise of the boiling-point, owing to the mouth of the steam-pipe being below the level of the water in the calorimeter. But in the case of other liquids, we cannot have mixture with the water in the calorimeter. It is better, therefore, to have a spiral condensing worm in the calorimeter, ending in an enlargement to collect the condensed liquid. The arrangement adopted by Berthelot (Fig 105) will illustrate this.

F is the vessel containing the boiling liquid, T the vapour-pipe, OSR the condensing worm, *l* a ring-burner under a gauze, the calorimeter being protected against radiation from the flame by a non-conducting cover faced with metal *nn*. With this apparatus Berthelot found the latent heat of steam at 100° C. to be 536·2 (the mean of three determinations, 535·2, 537·2, 536·2), agreeing nearly with Regnault's determination,

536·6, and therefore he concluded that the results obtained from it with other liquids were trustworthy.

Regnault made a series of determinations of the latent heat of steam at various temperatures, using at higher temperatures a boiler in which the pressure could be maintained at any desired value, so that the boiling-point could be regulated. From this the steam, still at the same temperature, was led into a spiral in a calorimeter, where it was condensed and the heat given out was measured (Jamin, ii. 243). By subsidiary experiments he determined the corrections for conduction and radiation. At lower temperatures the water was contained in a spiral immersed in a calorimeter, and the pressure was reduced to the point at which the water boiled. The vapour was led off and condensed in a vessel surrounded by ice, and the heat given up was there measured. Regnault put his results in the following form:—Calling the heat required to raise 1 gm. of water, from 0° to $t°$, and then to convert it into saturated steam at $t°$ the "total" heat of the steam at $t°$, and denoting it by Q, Regnault found that

$$Q = 606·5 + 0·305t.$$

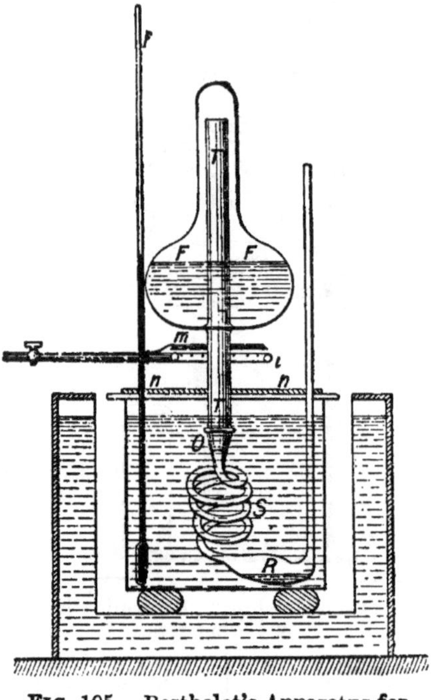

FIG. 105.—Berthelot's Apparatus for Latent Heat.

Taking no account of the alteration in the specific heat of water with rise of temperature, we have

$$Q = t + L$$

Hence, $$L = 606·5 + 0·305t - t$$
$$= 606·5 - ·695t.$$

From this we see that the latent heat decreases as the temperature rises, another indication that the liquid and gaseous conditions are approaching each other.

Regnault's formula for the total heat of steam gives the latent heat at 0° C., $L_0 = 606·5$. Winkelmann (*Wied. Ann.*, 9, 1880), re-examining Regnault's work, showed that the values obtained from the formula at low temperatures are probably higher than his experimental results warrant, and Dieterici (*Wied. Ann.*, 37, 1889) made a direct determination of L_0 by boiling water at 0° in a Bunsen calorimeter and measuring the volume of ice formed by the subtraction of a given weight of vapour. He obtained $L_0 = 596·8$.

Griffiths (*Phil. Trans.*, A., 1895, Part I., p. 261) devised a new method of determining the latent heat of steam, and used it for the two temperatures of 30° and 40° C. A small glass tube, open at one end and containing a known weight of water, was fixed in an exhausted silver flask immersed in a calorimeter containing oil. The water issued from the tube drop by drop, and, falling on to the silver, evaporated at the pressure corresponding to the vapour pressure for the temperature of the calorimeter. The vapour was pumped out as fast as it was formed, its latent heat being supplied by the oil in the calorimeter. The oil, of course, tended to fall in temperature, but its temperature was maintained constant by an electric heating coil and by very rapid stirring. The heat equivalent of the energy given to the oil by the current and by the stirring was determined, and this gave the latent heat of the steam formed. Griffiths' results at the two temperatures agree with the formula

$$L = 596{\cdot}73 - {\cdot}601t,$$

where the unit of heat adopted is the 15° C. calory, which agrees very nearly with the mean calory from 0° C. to 100° C. This gives $L_0 = 596{\cdot}73$, closely agreeing with Dieterici's value, and $L_{100} = 536{\cdot}6$, closely agreeing with Regnault's value $L_{100} = 537$, and at this higher temperature Regnault's work is probably very accurate. Griffiths' value for the total heat of steam is

$$Q = 596{\cdot}73 + {\cdot}399t.$$

Henning (*Ann. d. Physik*, xxi., 5, 1906, p. 849) measured the steam generated from water kept boiling continuously at constant temperature by an electric heater, in which the heat supply was measured. Over the range used, 30° to 100° C., $L = 598{\cdot}8 - 0{\cdot}5994t$ fairly represents the results, but $L = 94{\cdot}21(365 - t)^{0{\cdot}31249}$ gives a closer agreement on the whole. The 15° calory was taken as equal to 4·188 joules.

By superheating the vapour, that is, raising its temperature above the condensing-point, while maintaining the pressure the same, Regnault used the higher temperature apparatus devised for the determination of latent heat of a vapour to find also its specific heat at constant pressure. For, suppose that the latent heat of the vapour at $t°$ is L, the specific heat σ, and let the vapour be superheated in two different cases to $t + \theta_1$ and $t + \theta_2$. Then the heat given up in condensing is $L + \theta_1\sigma$ and $L + \theta_2\sigma$. We may determine both of these quantities, and knowing θ_1 and θ_2, we may obtain both L and σ.

Regnault found in this way that the specific heat of steam at constant pressure is constant within the limits of errors of observation, and the value he found is ·4805· He also determined the specific heat of other vapours. But it is to be remarked that in the expression $L + \theta_1\sigma$, L is much the larger quantity, and the value of $L + \theta_1\sigma$ will be affected by the errors of L. Hence, σ cannot be determined in this way with very great accuracy.

It is important to distinguish this specific heat of vapour at constant pressure from the specific heat of saturated vapour. In the former, the pressure, of course, remains constant during the rise of 1°. In the latter, the pressure varies, being always that at which the vapour is just on the point of condensation. For instance, the specific heat of saturated steam at 100° is the heat required to raise the temperature from 100° to 101°, the pressure being increased meanwhile from 760 mm. to 787·63 mm., the latter being the vapour-pressure at 101°.

It is easily shown from the second latent heat equation (chap. xix.), that the specific heat of saturated steam at 100° is negative, but we may see how it is so without any appeal to that equation. If we take a quantity of steam at 100° and 760 mm., and gradually alter the pressure to 787·63, not allowing any heat to escape, the work done in the compression goes to heat the gas, and its specific heat is such that the temperature rises above 101°. To bring it back to 101°, heat must be abstracted. Or, putting it in another way, when in the compression the temperature is 101°, the pressure is less than 787·63, the condensing pressure. During the rest of the compression, to keep the temperature at 101°, we must let heat escape. The specific heat, therefore, of saturated steam is negative. The total energy supplied is not negative, for the positive work done on each gramme is in excess of the mechanical equivalent of the negative specific heat.

If, therefore, we have a quantity of steam near its condensing point, and we compress it without allowing heat to escape, its temperature rises, and it gets further and further away from the condensing point. On the contrary, if we reduce the pressure it comes nearer to its condensing point, and, with sufficient reduction, we have a cloud formed. This cooling accompanying diminution of pressure explains a common formation of clouds in the atmosphere. This mode of cloud-forming is especially noticeable near the summits of mountains, when a current of air blows up the mountain side into regions of diminished pressure. It there expands, parting with energy to push out the surrounding air, and so cools. The water-vapour in the air at the same time expands, cools, and, at last, condenses. The same effect is also seen during the formation of thunder clouds. Large masses of damp and, therefore, light air rise, and cool through expansion against diminished pressure. The temperature falls below that for which the new pressure of the vapour is the maximum, and condensation occurs, forming the large cumulus masses high up in the air so characteristic of thunder clouds.

Other saturated vapours, among them alcohol, agree with water in having this so-called negative specific heat, so that condensation will occur with rarefaction. Some saturated vapours, such as that of ether, possess a positive specific heat, and these on sudden compression tend to condense. For, taking saturated ether-vapour at t°, and raising the pressure to that corresponding to saturation at $t+1^\circ$, the temperature is not so high as $t+1^\circ$, but say $t+\theta^\circ$, and heat is necessary to raise the temperature from $t+\theta$ to $t+1$. But if this heat is not supplied, the pressure is above the maximum at $t+\theta$, and condensation occurs and continues until the latent heat yielded raises the whole to $t+1^\circ$.

Spheroidal State.

When a metal plate is heated nearly to rednesss, and a few drops of water are thrown upon it, the drops rolls about on the plate without spreading out, and without boiling. They are then said to be in the *spheroidal* state. If the plate is a flat one, by looking at it edgeways it is possible to see that the drops are not in contact with the plate.

It is easy to show this also by bringing a wire nearly in contact with the plate, and connecting the wire with the plate through a battery and galvanometer as in Fig. 106, where the plate is the top of an inverted platinum crucible. Allowing a little liquid to trickle down the wire it assumes the spheroidal condition when it approaches the plate, and the galvanometer remains unaffected. On removing the source of heat, and allowing the plate to cool, a point is reached at which contact takes place, the galvanometer is deflected and the liquid boils violently. When in the spheroidal state, the liquid never reaches the temperature of boiling, the highest temperature for water being at ordinary pressure about 98°. The plate must be above 140°.

Perhaps we may give a general explanation of this somewhat as follows:—If the two surfaces were both non-volatilising solids, the air between them would tend to get into a steady state at a pressure equal to the atmospheric, and the upper solid would settle down into contact more or less rapidly. But even here some little time would be taken in the adjustment. This may be illustrated by allowing a small, very hot plate of glass to fall flat on a smooth, cold surface, when it moves freely about for a short time, evidently on a cushion of hot air. The layer of air between the two becomes heated, its pressure is increased, and it only slowly escapes out through the narrow space round the edge of the heated glass. The excess of pressure, meanwhile, sustains the weight of the glass.

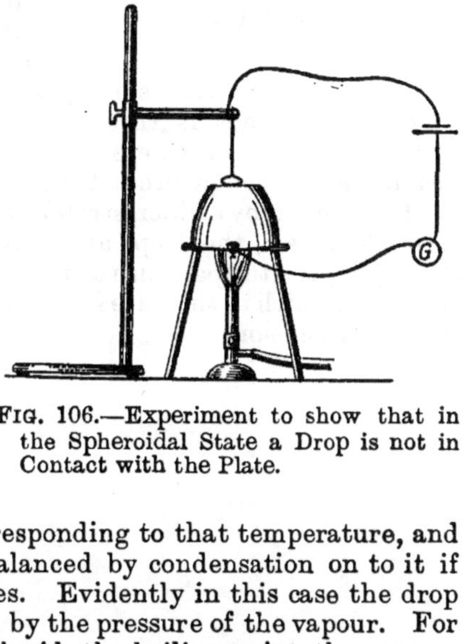

FIG. 106.—Experiment to show that in the Spheroidal State a Drop is not in Contact with the Plate.

In the case of the drop, evaporation comes in to maintain the excess of pressure. For by the heat received from the plate the surface of the drop is rapidly heated, and evaporation takes place at the rate corresponding to this higher temperature. Let us imagine that the drop is a large flat one, and that it is held in position in some way a short distance above the plate. First suppose both at the same temperature. The space will tend to be filled with vapour at the pressure corresponding to that temperature, and evaporation from the drop will be balanced by condensation on to it if we neglect the escape round the edges. Evidently in this case the drop will have to be held up otherwise than by the pressure of the vapour. For at the maximum temperature of the liquid, the boiling-point, the vapour-pressure only equals that of the atmosphere. But now make the plate much hotter than the boiling-point of the liquid. Let us suppose that the pressure of the vapour between it and the drop is still the vapour-pressure at the temperature of the liquid. The contact of the vapour with the hot plate superheats it, *i.e.* increases the momentum of the molecules, and if they are as a whole still exerting the same pressure on their return to the liquid there must be a diminution in the number returning to compensate the more violent impacts. Hence the number condensing is lessened, or the value of the pressure is not an equilibrium value, since the condensation is not equal to the evaporation. The pressure will, therefore, increase until the escape round the edges balances the excess of evaporation over condensation. Now this pressure will, in general, be above the atmospheric pressure, for the drop is itself near the boiling-point, where the normal vapour pressure is atmospheric, so that the excess may suffice to sustain the weight of the drop and we need not think of it as held up by any outside mechanism. If the pressure does not suffice, the drop comes lower down, nearer the source of heat, the superheating is greater, while the escape round the edges is less, and the pressure rises till the drop is sustained.

A curious case of the spheroidal state may often be noticed when a stream of water falls on a water surface. Of the drops which splash up and fall down again many will be seen to remain some little time as drops without coalescing with the general body of the liquid. The effect often occurs when, in rowing on still water, the oars are held steady, and the drops are allowed to fall on the hitherto undisturbed surface. And it is beautifully illustrated when a fiddle bow is drawn across the edge of a round glass vessel containing methylated spirit. As soon as the vessel vibrates regularly a shower of drops is thrown by the vibrating segments of the glass towards the centre, where they remain on the surface for a short time, forming a most exquisite pattern.

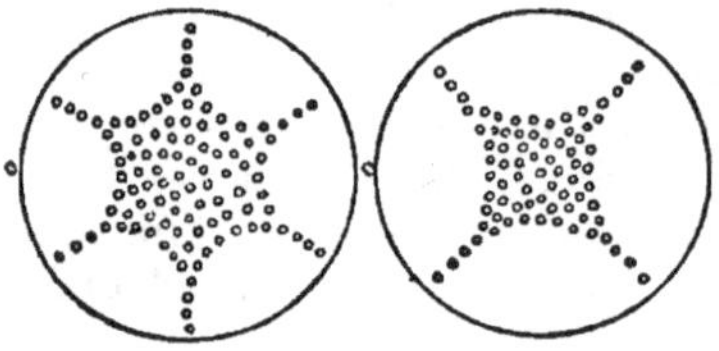

FIG. 107.

But though by sudden stretch of surface, and by evaporation in passage through the air, the drops are doubtless colder than the general body of the liquid, the temperature difference can hardly be enough to account for the effect, which in this case is probably due, in some way not yet explained, to surface tension.

CHAPTER XI.

CHANGE OF STATE. LIQUID—VAPOUR (continued).

Indicator Diagram—Critical Point—Critical Constants—Equation of Van der Waals—Liquefaction of Gases.

The Indicator Diagram for Water-Steam.—The indicator pressure volume diagram gives us a convenient graphic method of representing many of the facts already mentioned in connection with the water-steam change of state, and it also serves as an introduction to other facts

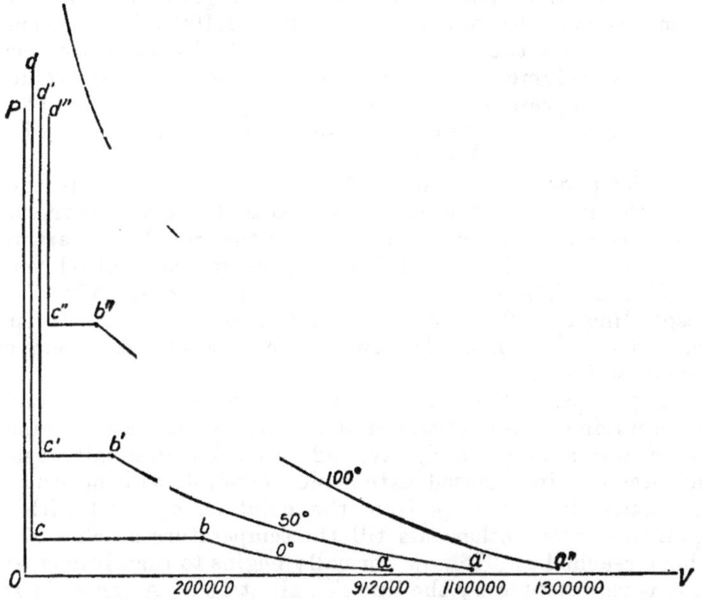

FIG. 108.—Indicator Diagram for Water-Steam.

now to be described. In this diagram volumes are measured from left to right, and pressures from below upwards, starting from two lines OP, OV at right angles, Fig. 108.

Suppose we take 1 gramme of steam at 0°, and, say, 1 mm. mercury pressure, it will have a volume of rather more than 900,000 c.c. This will be represented by a point *a* on the diagram, a distance proportional to rather more than 900,000 from OP, and a distance proportional to 1 mm. from OV. If we gradually increase the pressure, decreasing the volume, but keeping the temperature at 0°, the steam will be represented by successive points along a curve *ab*, which is nearly a hyperbola, since the pressure multiplied by the volume is nearly constant. But when the pressure reaches 4·6 mm., and the volume is 200,000 c.c., the state being represented by *b*, condensation normally begins and goes on at the same pressure till all is condensed to water at 0°. The curve, therefore, changes at *b* into a horizontal straight line *bc*, points on this line representing different proportions of the mixture of steam and water; *c* will represent the volume of the condensed water, just over 1 c.c. The pressure may now be increased, but the water diminishes only very slightly in volume, so that subsequent points lie along a line *cd*, only slightly leaning towards OP. In the figure it is impossible to represent the curve on proper scale. It is, therefore, only drawn so as to show its general nature. If we now start again with 1 gramme of steam at 50° and 1 mm. pressure, its volume will be about 1,100,000 c.c., represented by the point *a′*. As the pressure increases, the temperature remaining the same, the volume diminishes, the relation between volume and pressure being represented by the nearly hyperbolic curve *a′b′*, till at 92 mm. pressure and 12,000 c.c. volume condensation begins and goes on at constant pressure till all is water. This change is represented by the line *b′c′*. The point *c′* will represent about 1·01 c.c., and is, therefore, very near the line *cd*. Subsequent increase of pressure corresponds to the line *c′d′*, this again only slightly leaning towards OP. Starting anew with the steam at 100° and at 1 mm. pressure, the initial volume is about 1,300,000. The subsequent changes are represented by *a″ b″ c″ d″*, *b″* corresponding to 760 mm. pressure and 1700 c.c. volume, *c″* representing about 1·04 c.c. volume. The successive curves for each temperature are termed *isothermals*.

We may represent on the diagram any change of volume and pressure occurring in a quantity of water substance. For instance, if 1 gramme of water at 0°, and, say, 92 mm., be gradually heated at constant pressure in a closed extensible vessel, it will be represented by a horizontal line starting from the point on *cd* level with c′, and cutting all the water-isothermals till the temperature reaches 50°, when it will have reached *c′*. It now normally begins to turn into steam, and its course is represented by the line *c′b′*, all at 50°. After *b′* is reached the temperature rises again, and successive steam-isothermals are cut along *b′f*. But if suitable precautions as to vessel and freedom from air bubbles are taken, the water may be heated above 50°, still remaining water. This implies that the water-isothermals do not end in their downward course at the points *cc′c″*, but may be prolonged, as represented in Fig. 109 to *ee′*. Or, again, we have seen that water may easily be heated in a very clean glass vessel at the atmospheric pressure to 106° without boiling. This would be represented by the point *g*, where the 106° water-isothermal produced cuts the line of 760 mm. pressure. Dufour's experiment shows that even the 178° isothermal may be prolonged down thus far.

Another case is afforded by a phenomenon sometimes observed in barometer-tubes. If the tube is very clean and the mercury free from air, it is possible after filling and inversion, to raise the top of the tube far above 30 inches without detachment of the mercury from it. This may be done even when some water is above the mercury. The water is then at a negative pressure, showing that its isothermal may be prolonged even below the line OV. Worthington has succeeded in subjecting alcohol to a negative pressure or pull of seventeen atmospheres, and has measured its expansion under this pull (*Properties of Matter*, p. 123).

But not only may the water-lines be prolonged below the normal turning point, but, as described on page 168, if air containing water vapour is filtered through cotton-wool into a clean flask, so that it is quite free from dust, then it may be either compressed beyond, or cooled below,

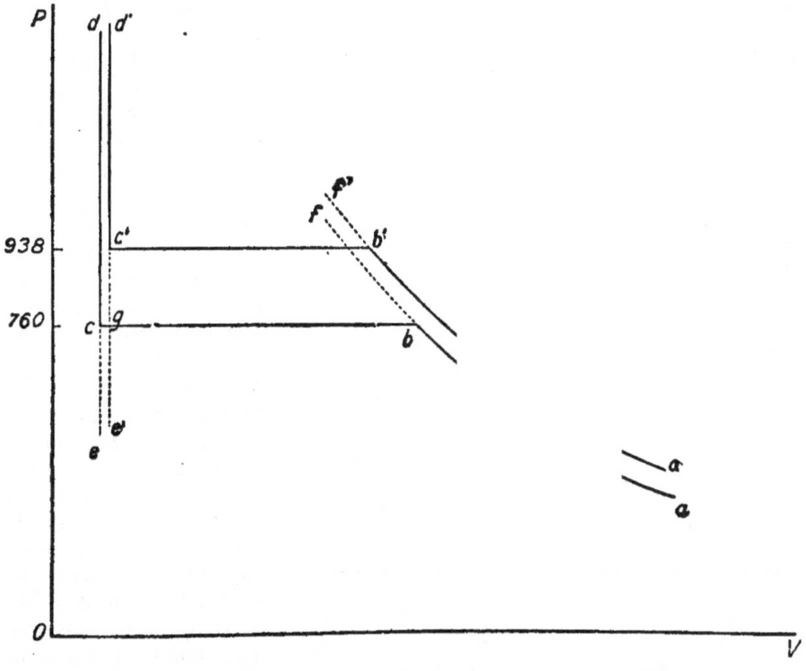

Fig. 109.—Indicator Diagram showing Prolongations of Water and of Steam beyond Normal Turning Points.

the normal condensing-point of the vapour present. The curves *ab*, *a'b'*, may, therefore, be prolonged as represented in Fig. 109 by *bf*, *b'f'*. Just as the change from water to steam appears to require some gas nucleus, so does the change from gas to water seem to require some liquid or solid nucleus. Professor James Thomson suggested that the curves *ce*, *bf* ultimately turn round and join, as represented in Fig. 110.

If this is the case we can hardly hope to experiment on the state represented by *gh*, as it would represent that of a substance increasing in volume with increase of pressure, evidently an unstable condition. But if we could imagine it possible to take the substance round the cycle *ckhgbhc* in a "reversible" manner (chap. xvii.) then the total work done is zero since the temperature is uniform and the area *ckh* must therefore equal the area *hgb*.

Going back to Fig. 108 it will be seen that the length of the horizontal portion of the isothermal decreases as the temperature rises, so that we should expect from this diagram alone that above some limiting isothermal the flat parts of the curve will cease to exist—that is, that the two states, liquid and vapour, will gradually approach each other, as the temperature rises until all distinction ceases.

Critical Point.—This merging of the two states into one, above a certain temperature, was first suggested by some experiments carried out by Cagniard de la Tour in 1822. He showed that if a proper quantity of alcohol was heated in a confined space, in the presence of its own vapour only, up to a certain temperature the contents of the vessel were partly liquid, partly gaseous, but that at this temperature—about 225° C. — the surface of separation between the two disappeared, and above it the contents of the vessel were entirely homogeneous, the two states appearing to meet. Before the disappearance of the surface of separation the surface-tension gradually diminished, tending to show that the distinction was disappearing. He observed the same phenomenon with other liquids, and Faraday continued the researches, making the important suggestion that for such gases as oxygen, hydrogen, and nitrogen, this limiting temperature is far below ordinary temperatures, so that they cannot be liquefied by pressure alone (*Researches in Chemistry and Physics*, p. 99.)

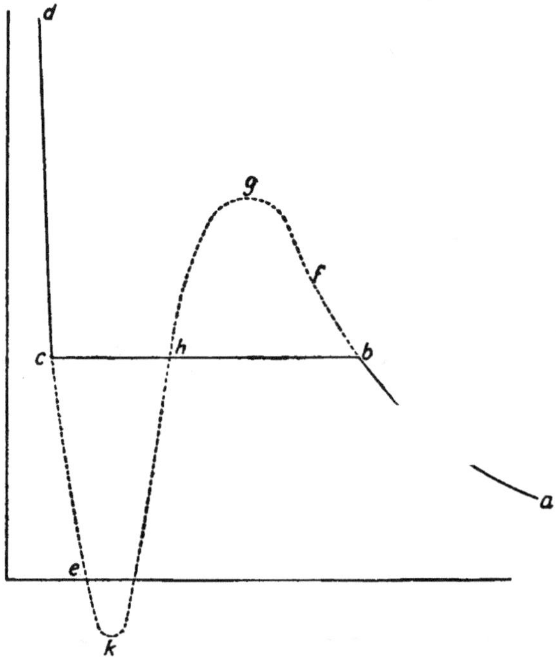

FIG. 110.—Continuous Form of Isothermal suggested by James Thomson.

But our definite knowledge on the subject was greatly increased by the researches of Andrews, who made a careful map of the isothermals of carbon dioxide from actual experiment, verifying the existence of the limiting or critical temperature, and showing the shape of the isothermals both above and below this critical temperature.

His apparatus consisted of a long glass tube (Fig. 111), about 2·5 mm. in diameter from *c* to *b*, and with a capillary bore from *b* to *a*, the whole being carefully calibrated. A current of dry carbonic acid gas was passed through the tube for some hours, until the proportion of air in the gas issuing from the tube was small and constant. It was found always to be not less than $\frac{1}{1000}$ of the whole, and the proportion had to be determined and allowed for. The end *a* was then sealed, the other end closed and afterwards opened under mercury. By suitable manipulation,

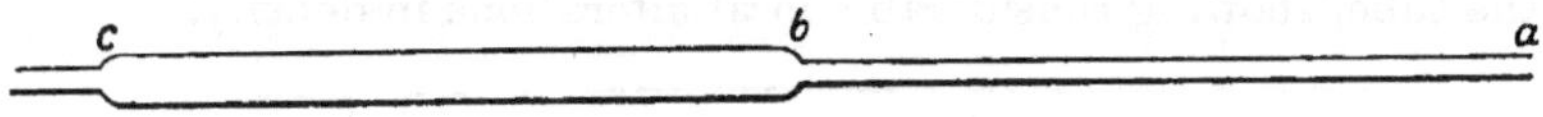

FIG. 111.—Andrews' Tube for Experiments on Carbon Dioxide.

partly by heating, partly by reduction of external pressure, about one-fourth of the gas was expelled, and a short column of mercury introduced into the tube *cb* to act as stopper and index. The tube was then placed in position with the open end *c* in the compression chamber (Fig. 112), which was filled with water. By means of the screw S, any desired pressure could be applied, the amount being measured by a second exactly similar tube and compression-chamber, containing air instead of carbonic acid gas, and placed by the side, a cross tube connecting the two chambers. Each tube was surrounded by a water-bath with plate glass sides, not shown in the figure, that round the air tube being kept at a constant temperature, while that round the carbonic acid gas was raised to any desired temperature.

The diagram (Fig. 113) taken from Maxwell's *Theory of Heat* (5th ed., p. 120) shows Andrews' results. The 13·1° and 21·5° isothermals exhibit the normal change from the all-gas curve on the right to a horizontal line, indicating condensation at constant pressure and a coexistence of the two states, followed by a steep rise of the all-liquid curve. In the actual results, through a small trace of air in the tube, the corners were rounded off and the flat part of the isothermals sloped slightly upwards, showing that the last part of the condensation required a greater pressure than the first. The higher isothermals slope upwards throughout, showing no condensation at all. During the compression the substance remains homogeneous, and the only relic of the coexistence of two states is in the diminished slope at one part of the isothermal. Even this has disappeared at 48·1°. By several experiments the lowest temperature at which condensation does not occur was fixed at 30·92°. This therefore is the *critical temperature* for carbon dioxide.

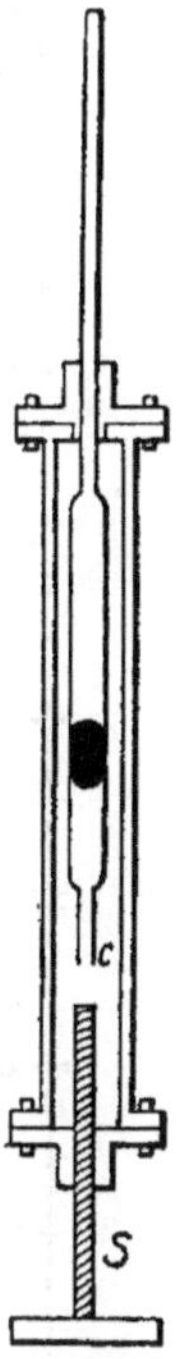

FIG. 112.—Diagrammatic Representation of Andrews' Compression Apparatus.

If we draw a curve (dotted in Fig. 113), the "border curve," through the points where the isothermals below this temperature change into and from the horizontal straight line, its area includes all the conditions of coexistence of gas and liquid at one temperature. To the right the substance is a gas, but one for which the volume decreases more than in accordance with Boyle's law, that is pv decreases as p increases. To the left, the substance is a liquid, and pv increases as p increases.

The critical isothermal just touches the vertex of this dotted curve, and, being there nearly horizontal, a slight change in pressure corresponds to a great change in volume. Andrews noticed that about this point any change in pressure was accompanied by flickering movements in the tube, showing considerable local alterations in density.

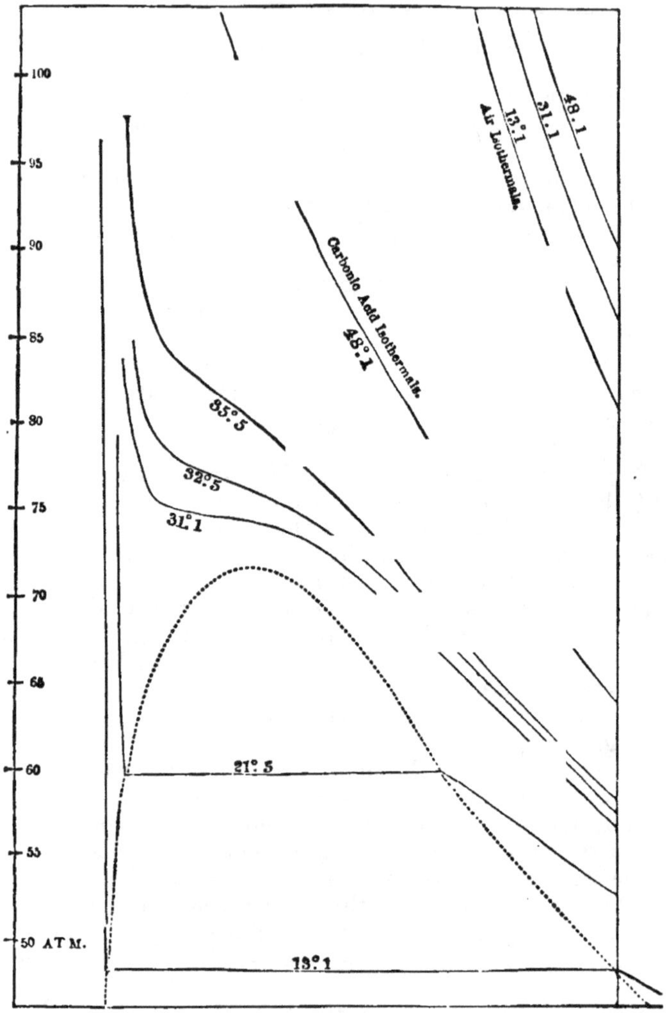

FIG. 113.

The Critical Constants.—The vertex of this dotted curve may be termed the *critical point.* It therefore represents a temperature of 30·92°, and a pressure, termed the *critical pressure*, of about 73 or 75 atmospheres. The pressure could not be determined quite accurately, as the air in the attached manometer did not change volume exactly in accordance with Boyle's law. The volume of unit mass, termed the *critical volume*, was about $\frac{1}{169}$ of the volume at the same temperature and 1 atmosphere.

Above the critical isothermal, the only survival of the vaporous condition is found in the fact that on the right-hand part of each curve, pv decreases as p increases to a certain point. After this it increases as p increases, and this may be regarded as a survival of the liquid condition. The point of minimum value of pv may be considered to represent what in the lower isothermals is a change of state.

Regnault (whose work has been verified by Amagat) found that with pressures up to 20 atmospheres, pv for air and nitrogen decreased as p increased, while it increased for hydrogen. (See *Properties of Matter*, p. 124.)

Andrews found that it was quite possible to take a quantity of carbon dioxide gas round from the undoubtedly gaseous to the undoubtedly liquid condition without any condensation. The mode in which this was effected may be understood from Fig 114.

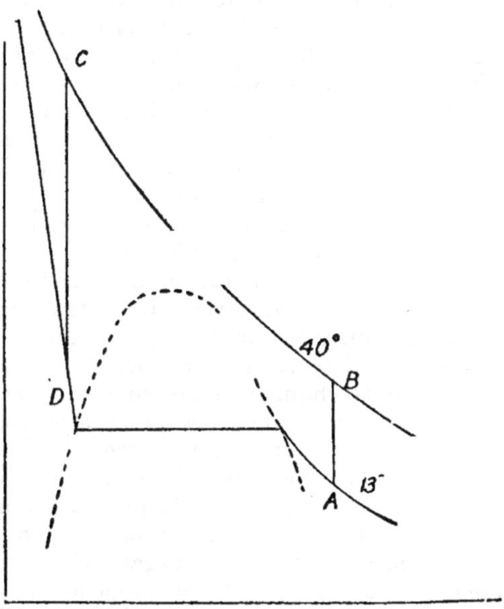

FIG. 114.—A, B, C, D, path from gas to liquid without discontinuous change of state.

Starting with a quantity of gas, say at 13°, represented by A, let it be heated at constant volume till it reaches, say, the 40° isothermal at B. Let it then be compressed at 40° until the volume is less than that occupied by the liquid when just entirely condensed at 13°, this being represented by C. Let it then be cooled at constant volume to D on the 13° isothermal. It is certainly gaseous at A, certainly liquid at D, for, on increasing the pressure at the former point, condensation occurs, and on removing the pressure at the latter point, ebullition occurs. But throughout the passage from A to D the substance remains homogeneous.

Andrews made the suggestion that the term gas should be restricted to the condition of a substance above the critical point, that the substance below that point and to the right of the border curve should be termed vapour, while to the left of the border curve it is liquid.

Determinations of the critical constants of temperature and pressure and, in some cases, volume have been made for a large number of substances by Andrews and others experimenters. Some have used Andrews' method more or less modified, but various other methods have been employed. For example, Cailletet and Collardeau (*Ann. de Chimie et de Physique*, 6, xxv. p. 519), experimenting on water, heated the substance in a steel tube, and observed the point at which the vapour pressure ceased to be one-valued and became dependent on the amount of water present. Up to the critical point the pressure of vapour in contact with the water was definite at a definite temperature, but as soon as that point was passed the contents of the vessel were homogeneous, and the pressure depended on the amount of water dealt with. They found a critical temperature of 365° C., with a critical pressure of 200·5 atmospheres. Traube and Teichner (*Ann. der Physik*, xiii., 1904, p. 620), found the critical temperature to be 374° C. They used a quartz tube filled to $\frac{1}{3}$ with distilled air-free water and heated by mercury boiling under pressure. The meniscus disappeared at 374° and reappeared when the temperature was again brought down to that point.

The refractive index has also been used to find the critical point. In the researches of Prinz Galitzine and Wiliss (*Congrès International de Physique*, 1900, i. p. 668), the refractive index of a substance near the critical point was measured either by using the substance as a cylindrical lens and observing the refraction, or by inserting a small angled prism in a tube containing the substance and finding the refraction due to the prism. The refractive indices of the liquid and vapour approach each other as the critical point is approached, and at that point are equal. For ether the meeting point of the refractive indices was found to be $\mu = 1{\cdot}12$ at 193·7° C.

In these experiments as in others it was found that near the critical point very great variations in density occurred in the successive layers, variations only eliminated by stirring.

The isothermal is very nearly horizontal at the critical point and very minute changes of pressure may bring about large volume changes, and with the volume changes temperature changes may ensue, so that it might be expected that there would be difficulty in obtaining homogeneity of the contents of the vessel even just above the critical point. But the variations are so large that many observers doubt whether the liquid and the vapour do actually meet in all their physical qualities at the critical point. They regard the critical point as that at which the densities coincide. They suppose that slightly above the point the liquid is still liquid, and may have a different density from the gas and a different molecular grouping; but the evidence is hardly conclusive.

Ramsay and Young, and later Young (*Phil. Mag.*, xxxiii., 1892, p. 153, and in other papers) have measured a number of critical constants. The pressures and temperatures were obtained by the direct method of raising the temperature and observing the vapour pressure at the critical point, when the liquid meniscus disappeared. The critical volume was found by heating the substance above the critical temperature, and then cooling it by sudden expansion till there was a temporary separation between liquid and vapour. The volume at which this just took place on slight expansion was taken as the critical volume. MM. Cailletet and Mathias (*C.R.*, cii. p. 1202; civ. p. 1563) have found that in many cases the critical volume can be calculated on the assumption that the mean of the liquid density and saturated vapour density changes uniformly with the temperature, so that on a density temperature curve it would be a straight line. If these densities for a lower range of temperature and the critical temperature are known, the critical density is easily determined.

We give below a few critical constants, but it must be remembered that the results obtained by different observers are not coincident:—

	Temperature.	Pressure in mm.	Volume of 1 gramme.	Molecular Volume.
Benzene . .	288·5	36395	3·293	256·3
Ether . . .	194·4	27060	3·801	280·7
Ethyl alcohol .	243·1	47850	3·636	166·9
Water . .	365	152380	...	...

Van der Waals' Equation—The equation of Van der Waals' (p. 152)

$$\left(p+\frac{a}{v^2}\right)(v-b)=\mathrm{R}\theta$$

is of great interest in connection with the critical constants of pressure volume and temperature, in that above a certain value of θ it is represented by curves having the general form found by Andrews, while below that value the curves are of the form suggested by J. Thomson, and represented in Fig. 110.

The equation may be arranged as a cubic in v, viz.,

$$v^3-\left(b+\frac{\mathrm{R}\theta}{p}\right)v^2+\frac{av}{p}-\frac{ab}{p}=0$$

and this has either one or three real roots for given values of p and θ.

Taking a given isothermal, say the lower curve in Fig. 115, it will be cut only once by an equal pressure line if the pressure is less than that at B, and only once again if the pressure is greater than that at D. Between these values it will be cut three times as in A, C, E.

In the first case the substance is a vapour; in the second, a liquid; while in the third case it may be either a liquid, at A, or a vapour at E. The volume C is only realisable experimentally if ACE represents the vapour-pressure at the temperature chosen. It is then a definite mixture of vapour and liquid. But the equation hardly means this. It rather indicates that it is conceivable that the substance might be brought quite continuously all in one state round the curve EDCBA. We shall suppose that if ACE is at the height of the vapour-pressure then the areas ABC, CDE are equal (p. 187). Now plotting all the isothermals: the equation agrees with observation in showing that ABCDE gets less and less, and that ACE draw nearer and nearer together as the critical temperature is reached. At that temperature they coincide. Hence, if we insert in the equation the critical values p_c and θ_c the equation has three equal roots, each equal to v_c, the critical volume.

Comparing then

$$v^3-\left(b+\frac{\mathrm{R}\theta_c}{p_c}\right)v^2+\frac{av}{p_c}-\frac{ab}{p_c}=0$$

with

$$v^3-3v_cv^2+3v_c^2v-v_c^3=0$$

we get at once

$$3v_c = b + \frac{R\theta_c}{p_c}$$

$$3v_c^2 = \frac{a}{p_c}$$

$$v_c^3 = \frac{ab}{p_c}$$

whence

$$p_c = \frac{a}{27b^2}$$

$$v_c = 3b$$

$$\theta_c = \frac{8}{27}\frac{a}{bR}.$$

Now, if the equation is a correct representation of the behaviour of

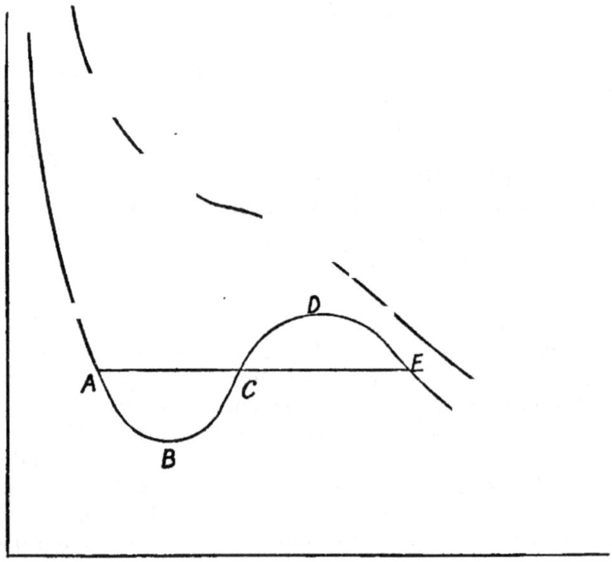

FIG. 115.

gases and vapours, a can be determined from the change of pressure with temperature of the gas when at constant volume, while b can be determined from the change in pv with change of pressure at constant temperature. R may then be found by inserting any three values of p, v, and θ in the equation. For details of the calculations we refer to Van der Waals' paper (*Physical Memoirs*, vol. i., p. 390). He finds for carbonic acid the equation

$$\left(p + \frac{0{\cdot}00874}{v^2}\right)(v - 0{\cdot}0023) = 1{\cdot}00646\ (1 + \alpha t)$$

where the unit of pressure is 1 atmosphere, the unit of volume the volume of 1 gramme at 0° and 1 atmosphere, and t is the temperature centigrade.

From this the critical pressure is $\frac{a}{27b^2} = 61$ atmospheres,

while the critical temperature is $\frac{8a}{27b\text{R}} = 32{\cdot}5°$,

which are values not very far distant from those found by Andrews.

The values of a and b, however, are not determined very exactly, and the best test of the equation is probably to be found in the comparison with experiment of certain deductions from it which we shall now discuss.

Reduced Isothermals—Corresponding Temperatures.—If we express the pressure, volume, and temperature for a given condition of a substance as fractions of the critical values, viz. :—

$$p = ep_c = \frac{ea}{27b^2}$$

$$v = nv_c = 3nb$$

$$\theta = m\theta_c = \frac{8}{27}\frac{ma}{b\text{R}}$$

and substitute these values in Van der Waals' equation, it at once reduces to

$$\left(e + \frac{3}{n^2}\right)(3n - 1) = 8m \qquad (1)$$

from which a and b, which refer to a particular gas, have entirely disappeared.

If the same value of m is taken for different gases which have critical points θ_c, θ_c', &c., the temperatures indicated by $m\theta_c$, $m\theta_c'$, &c., are "corresponding temperatures." Thus the absolute critical temperatures for ether and water are 470° and 638° about. If we put $m = \frac{1}{2}$, then a temperature of 235° absolute, or $-38°$ C. for ether corresponds to 319° absolute, or $+46°$ C. for water. The above equation shows that if m is constant, *i.e.* if different substances are taken at corresponding temperatures, then the relation between e and n is definite, and the same for all substances. We have in fact a cubic in n, a Van der Waals' equation, in which $a = 3$, $b = \frac{1}{3}$, $\text{R} = \frac{8}{3}$, and giving to m a succession of values we obtain a family of curves representing the relation between e and n, of the same general form as the isothermal curves given by the original equation. These general isothermals are called Reduced Isothermals. To apply any reduced isothermal to a particular gas we must multiply the m for it by θ_c to get the temperature, and must multiply the values of e and n at any point by p_c and v_c respectively to get p and v.

The line on a reduced isothermal corresponding to the horizontal vapour-pressure line on a pv diagram is the same for all substances. We may prove this as follows :—The horizontal vapour-pressure line cuts the continuous pv curve in three points, and, as we have seen, makes equal areas above and below, as *bgh* and *hkc* in Fig. 110. This is expressed by

$$P(V_3 - V_1) = \int_{V_1}^{V_3} pdv$$

where pv relate to the continuous curve, P is the vapour pressure, V_1 and V_3 the liquid and vapour volumes.

If we now substitute in terms of the critical values, putting $p = ep_c$, $v = nv_c$, $P = Ep_c$, $V_1 = n_1 v_c$, $V_3 = n_3 v_c$, $p_c v_c$ divides out and we have

$$E(n_3 - n_1) = \int_{n_1}^{n_3} edn$$

an expression which shows that the areas cut off by the E line on the *en* diagram are equal above and below. Substituting from equation (1) put in the form

$$e = \frac{8m}{3n-1} - \frac{3}{n^2}$$

we see at once that we can integrate. Then using the equations which state that E_1, n_1, E_1, n_3 are on the curve, we get two more equations which might enable us to find E, n_1 and n_3 if this were needed. But it is enough to note that for the same value of m, E, n_1 and n_3 are the same for all substances. In other words, at corresponding temperatures the saturated vapour pressures are the same fraction of their critical pressures for all substances, and the liquid volumes and the vapour volumes are the same fractions of the critical volume for all substances.

Corresponding Pressures and Corresponding Volumes.—If we take corresponding temperatures for different substances, the vapour pressures at those temperatures are corresponding pressures. The liquid volumes just before evaporation begins and the vapour volumes just when it is completed are termed corresponding liquid and corresponding vapour volumes.

This constancy of E, n_1 and n_2 for given m gives us a result which can be more easily compared with observation than the original equation between p, v, and θ. Thus Van der Waals used it to calculate the boiling point of carbon dioxide from its known constants and those of ether. The critical pressure of carbon dioxide is 72 atmos. and its critical temperature is 303·9° absolute. For 1 atmosphere, then, $e = \frac{1}{72}$. Now the critical pressure of ether is 36·9 atmos. and the critical temperature is 463°. The corresponding pressure is $36{\cdot}9 \times 760 \times e = 384$ mm., which is the vapour-pressure of ether as obtained by direct experiment at 16·9° C. or 289·9° absolute. Then $m = \frac{289{\cdot}9}{463} = 0{\cdot}625$, and the corresponding temperature for carbonic acid is $303{\cdot}9 \times 0{\cdot}625 = 190°$ absolute, which is very nearly in agreement with observation of the boiling-point. Young (*Phil. Mag.*, xxxiii., 1892, p. 153, and in later papers) has examined the behaviour of a number of liquids and vapours with regard to their corresponding quantities. He finds that Van der Waals' result is not far from true for the halogen derivatives of benzene, but that for many other substances it does not hold. The equation and its deductions are to be regarded not so much as expressing the actual relations for substances in general as an approximation for some substances sufficiently near the truth to suggest lines of experiment, while its divergence from the truth in other cases found by these experiments may ultimately lead to some more correct and more general equation. The modifications hitherto suggested, while more difficult to deal with, are hardly more in accordance with experiment.*

Liquefaction of Gases.—When Andrews made his celebrated researches, several gases, as oxygen, nitrogen, carbon monoxide, and hydrogen, had never been liquefied, and such gases were often termed permanent gases. But the results which Andrews obtained with carbon dioxide gave weight to the suggestion of Faraday that these gases were far above their critical points at ordinary temperatures, and that they could only be liquefied by cold as well as pressure. Acting on this idea, Cailletet and Pictet, working independently, both succeeded at nearly the same time in 1877 in liquefying oxygen and other gases. The general principle of their experiments, though carried out with very different details, consisted in compressing a quantity of gas under 300 or 500 atmospheres, cooling it by surrounding the containing vessel with liquid sulphurous acid or liquid carbonic acid evaporating freely, and then making a sudden expansion of the gas thus compressed and cooled. The work done by the sudden expansion implied a further transfer of energy from the gas, which fell below its boiling point and became liquid. In Cailletet's early experiments the volume of the containing vessel was suddenly increased, and the liquid appeared as a mist. In Pictet's experiments the gas was allowed to issue through a nozzle, the gas behind doing work on that in front, and cooling so far that ultimately it issued as a liquid jet, and sometimes even gave signs of solidification. There were doubtful indications of the liquefaction of hydrogen.†

Much valuable work on the liquefaction of gases was done by Wroblewski and Olzewski, at first working together and later independently. They used the evaporation of liquid ethylene as a cooler, and to find the temperature reached they used either a hydrogen thermometer or a thermo-electric couple. Olzewski ultimately used a platinum resistance thermometer. In some of his experiments he placed a graduated test tube in the tube containing the compressed gas, but separated from it by a narrow space. On the expansion, liquid collected in and round the test tube. That inside the test tube was protected by an evaporating liquid jacket, and could be preserved for some time. The boiling point could then be found. Collecting the vapour or gas given off from a measured volume of liquid, the liquid density could be determined.

Dewar was also one of the early workers in the field, and has made very valuable researches, on a much larger scale than any other worker.

* Further information as to Van der Waals' equation will be found in Nernst's *Theoretical Chemistry*.

† A full account of the various methods of liquefying gases, with a valuable history of the subject, will be found in Harden's *Liquefaction of Gases* and Travers' *Study of Gases*. We only mention a few workers to illustrate methods.

He greatly aided the study of the liquids and their use as cooling agents by the introduction of a special containing vessel, to catch the condensed gas issuing from the expansion nozzle and to keep it for a long time liquid at the atmospheric pressure. The vessel is a double-walled glass globe with a vacuum between the walls, as in Fig. 116, the surfaces in the evacuated space being silvered or, later, covered with a film of mercury, secured merely by condensation of mercury vapour rising from a few drops of mercury left in the space. The evaporation from liquid air or oxygen in such a globe is only a small fraction of what it is in a single-walled globe.

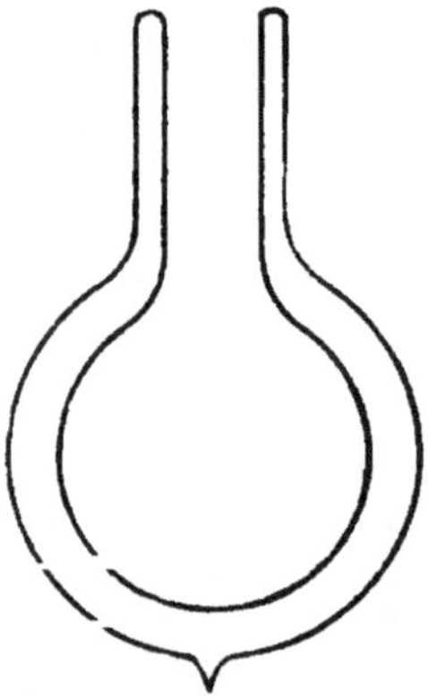

FIG. 116. — Dewar's Vacuum Vessel (Section).

About 1894–95 a new step was taken in the introduction of the "regenerative" method. There has been much discussion as to priority of discovery, but it is enough here to say that it was used very nearly at the same time by Dewar, Hampson, Kammerlingh-Onnes, Linde, and Tripler. The principle of the method may be understood from Fig. 117, which is merely a diagrammatic sketch. The gas is forced by a pump P at very high pressure through a cooler C, which removes the heat developed by the compression, and the gas flows on at the ordinary temperature to the regenerator R, where it passes down a spiral pipe to a nozzle N. From this it issues with a sudden fall of pressure from that of the pump to that of the atmosphere. There is considerable cooling of the issuing gas, which in turn cools the spiral pipe by which succeeding portions are coming to the nozzle. Thus the next issuing gas falls still lower in temperature, cools the spiral still more, and so on, till ultimately the gas issues as liquid and falls into the collecting vessel V. No cooler below the ordinary temperature is required, the cooler C merely reducing the gas to the temperature of the room.

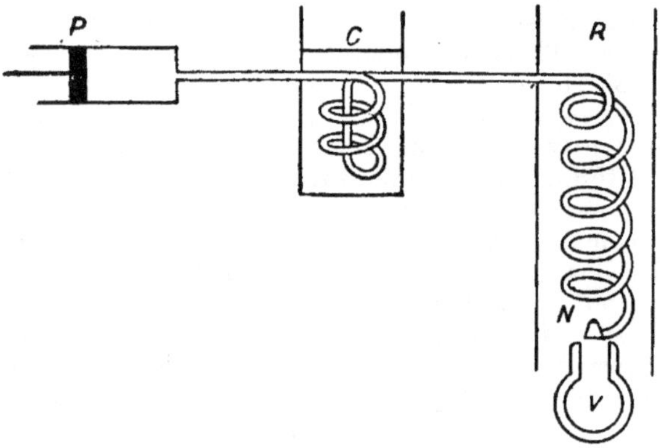

FIG. 117.—Diagram of Regenerator Process for Liquefying Gases.

It is to be noted that the expansion at the nozzle is not adiabatic, *i.e.* not one in which the gas has to use up its own energy in pushing out its surroundings. The pump behind is continually doing work and supplying very nearly all the energy which the gas requires to push its way onwards, when it comes out of the nozzle. The conditions are much more nearly those of the porous plug experiment of Thomson and Joule (chap. xix.), and the cooling is chiefly due to the conversion of molecular kinetic energy into potential energy in the work done against the internal forces. At the ordinary density of a gas so few of the molecules are at any one instant near each other, that we may consider that all the molecules have the potential energy of complete separation from their fellows. But when the gas is at several hundred atmospheres, and still more when it is cooled far on its way to the absolute zero, a very large proportion of the molecules are close together and are exerting cohesive forces. If now expansion occurs, work has to be done in separating them, or at any rate in separating a very large proportion of them, and cooling results.

In Joule and Thomson's researches hydrogen alone showed no certain evidence of such cooling, but even hydrogen at the pressures used in the regenerative method appears to cool with the rest when it is at a sufficiently low temperature initially.

Though hydrogen was not liquefied in the earlier experiments, Olzewski was able to reduce it to a very low temperature, and to calculate its boiling-point from its behaviour, and in 1898 Dewar succeeded in obtaining and keeping it liquid by cooling it first to $-205°$ C. at a pressure of 180 atmospheres, and then allowing it to issue from a nozzle into a vacuum vessel.* In 1899 he succeeded in obtaining solid hydrogen. A small vacuum vessel containing liquid hydrogen was suspended in a larger one also containing the liquid, and on allowing evaporation to take place below 60 mm., a slight air leak caused the hydrogen to solidify as a white froth-like foam in the inner vessel (*Nature*, lx., 1899, p. 514).

The temperatures of the boiling points and melting points of these ordinary gaseous substances are not found with perfect consistency with different instruments, but the various experimenters obtain results not very far apart.

The following are some of these (Travers, *Study of Gases*, p. 247) :—

	Critical Temperature. Absolute.	Critical Pressure in atmos.	Boiling-Point. Absolute.	Melting-Point. Absolute.
Argon . . .	155·6	53	86·9	—
Carbon Dioxide . .	304	72·4	193	208†
Carbon Monoxide .	137	33·4	83	66
Hydrogen . . .	35	15·3	20	16
Nitrogen . . .	124	27·6	75	60
Oxygen . . .	154	58	90·5	below 50

* *Science Abstracts*, i. 1898, p. 562. An account of the liquefaction of hydrogen is given by Travers in the *Phil. Mag.*, 1901, i. p. 411.

† Andrews.

CHAPTER XII.

CHANGE OF STATE. SOLID—LIQUID.

Melting of Ice and Melting of Wax—Melting of Ice at a Definite Point and on the Surface only—Latent Heat—Supercooling—Regelation—Effect of Pressure on Melting Point—Melting Points of Solids—Explanation of Melting on the Kinetic Theory—Resemblance of Solution to Fusion—Evaporation from Solids.

Melting of Ice and Melting of Wax.—The change from the solid to the liquid state may be illustrated by the familiar examples of the melting of ice and the melting of fatty or waxy substances. The latter are probably mixtures, and the process may be complicated by the melting of one constituent before another. But, however this may be, we at once observe a difference between the two cases. A piece of sealing-wax, as its temperature rises, gradually softens from a solid into a viscous liquid without any discontinuity, and on cooling it gradually hardens again into a solid.

Melting of Ice at a Definite Point and on the Surface only.—A piece of ice, however, remains solid to the moment of melting—it is never really soft, and in the converse process of freezing, the water does not gradually get thick, but crystals of quite solid ice form within it. If we take a quantity of ice below the melting point, and gradually supply heat, keeping the ice at a uniform temperature throughout, the ice rises up to 0°, still continuing solid. The surface of the ice then begins to melt, and if care is taken to keep the temperature still uniform throughout, no further rise takes place till the whole of the ice is melted, the temperature remaining at a perfectly definite point. Indeed, on the constancy of this melting point depends our system of graduating thermometers. The ice melts at its surface alone, and there is no softening into a plastic condition, as in the case of wax.

Latent Heat.—The heat taken up by each gramme of ice, in turning from ice at 0° to water at 0°, is constant in amount, and is termed the "latent heat" of water.

Freezing.—Taking the converse process, if a quantity of water be cooled down gradually, the temperature being kept uniform meanwhile by stirring, the water remains liquid without any sudden or remarkable increase in its viscosity. At 0°, crystals of solid ice begin to appear, and grow at the expense of the water. The temperature is arrested at 0° throughout the mixture, if the stirring is kept on while the ice is growing. The same amount of heat per gramme is given up to the surroundings as was absorbed in the change from ice to water.

Supercooling.—If we begin again and, instead of stirring the water, we keep it as quiet as possible during the cooling, then, in general, the temperature falls continuously past 0° to perhaps $-3°$ or $-4°$. Next, freezing sets in, and the temperature immediately leaps up to 0°, and remains there, during the rest of the process of solidification, the latent heat yielded by the ice first formed being sufficient to raise the temperature of the mixture to 0°. This frequently occurs in jugs of water left standing quietly in cold weather. There is no ice in the water while it is quiet, but the moment the jug is lifted up and its contents poured out, a number of minute crystals of ice form in the water, showing that the latter must have been below 0°, for the latent heat of the water given up on freezing can only be accounted for by supposing that it has raised the temperature of the water. If the water be covered with a layer of oil, it is still easier to produce this phenomenon, and without any trouble the temperature may be lowered to $-6°$ or $-7°$ without freezing. In this state the water is said to be superfused. If a crystal of ice be dropped into the liquid, freezing at once commences on this as a nucleus, and the superfusion ceases. Or a sudden shock will often put an end to it.

There is a sudden change of volume on the change from water to ice. As we have already seen, water is most dense at $+4°$, and it expands slowly as the temperature falls. If the water be kept superfused, the expansion is continuous. But if the water changes to ice at 0°, there is a sudden increase of volume, 1 c.c. of water becoming 1·09 c.c. of ice. The ice contracts on cooling below 0°

Similar phenomena are noticed in the melting and freezing of other substances. For instance, phosphorus melts definitely at 44·2; if kept under water it may be cooled in the superfused condition far below this; but a sudden shock, or the introduction of a glass rod previously rubbed with phosphorus, is followed by solidification. With sulphur, melting at 115°, the superfusion is still more marked. These substances also have definite latent heats on melting, and show a sudden increase of volume—1 c.c. of solid phosphorus melting to 1·034 of liquid and 1 c.c. of solid sulphur melting to 1·05 of liquid.

Regelation.—Faraday first directed attention to a remarkable phenomenon, now known as "Regelation." If a number of blocks of ice at 0° be pressed together they adhere at the points of contact. This effect, as remarked by Faraday, is very familiar in another form. Snow near the melting-point will easily bind into snowballs; but if below 0° it is powdery, and will not form a cohesive mass. Similarly the blocks of ice will not adhere if below 0°.

Tyndall showed that ice at 0° could be moulded by great pressure, the ice breaking up under the pressure, and then freezing together at the points of contact. By continued fracture and regelation, the crevices are filled up, and the ice forms a transparent continuous mass which has the shape of the mould.

The motion of glaciers has been ascribed to regelation. It has been observed that the whole mass of a glacier flows like a stream of water down the valley in which it lies, but with exceedingly small velocity, the centre of the glacier flowing more rapidly than the edges. Forbes found, for example, that the Aletsch glacier flowed in "mid-stream" at one point about 14 inches in 24 hours, while near the edge it flowed only 3 inches in the same time. In winter, the speed is considerably less than in summer. Now the ice of a glacier is pretty nearly at 0° through the penetration into all the crevices of the water melted at its surface. The enormous strains to which the ice is subjected continually fracture it, and the fracture allows slight yielding to the strain. Owing to the pressure, regelation immediately occurs and the pieces join together again. This, no doubt, is a partial explanation, but from experiments made by M'Connel and Kidd (*Proc. R.S.*, xliv., 1888, p. 331) it appears that ice when consisting of an irregular aggregation of crystals behaves like a viscous solid. They found that bars of ice of regular crystalline structure showed only exceedingly minute changes in extension and compression under continued pulls and pressures, while bars of irregular structure, formed of aggregations of crystals with their axes in all directions, went on extending or lessening in length under continued forces. The changes may be due to slipping along the cleavage planes. The possession of a sensible vapour-pressure by ice shows a certain mobility or power of escape of the molecules near the surface, and if we suppose this mobility to exist within the mass we may suppose that there will be considerable variations of strain from point to point in the irregular aggregation. The mobile molecules will tend to move from points of greater to points of less strain, and will perhaps fill up the gaps which the slipping tends to form. Thus the mass keeps continuous. The crystalline structure of glacier ice is of this irregular type.

The viscosity of metals is probably due to a similar irregular crystalline structure, and though we have not such clear indication of internal molecular mobility it doubtless exists.*

Regelation has been explained by supposing that the interior of the blocks in contact is slightly below 0°. The points in contact at 0° are by the contact surrounded by portions of ice below 0°, and are thus cooled, and the water on the surface is frozen so as to form a solid bridge. Doubtless there is sometimes an effect of this kind. But it would take place with very great slowness, as the latent heat of the newly formed ice would have to be transmitted through the badly conducting surrounding ice to the colder ice inside the blocks.

In order to remove this difficulty, Forbes and others have supposed that ice passes gradually into water, so that on the surface of a melting block there is a layer which may be regarded as either "plastic ice" or "viscid water." When two such blocks are brought together, any water between them need not give up its whole latent heat, but just enough to become either "plastic" or "viscid," and the connecting-bridge is formed. But there is no independent evidence of this plastic condition. In freezing water, the ice always forms at definite points, crystallising out from the water—the water showing no tendency to become viscid. Indeed, as we have seen, water may be cooled easily below 0°. A mixture of ice and water kept at 0°, and neither gaining nor losing heat from its surroundings, remains ice and water, whereas if the plastic condition existed the whole of it should, as pointed out by Helmholtz, gradually assume the viscid condition at a uniform temperature. The use of ice calorimeters, too, is founded on the supposition that there is no such gradual transition but that the heat given up is used to convert true ice to true water. If there is any intermediate condition, we must make the improbable supposition that the contraction in passing through any part of that condition is proportional to the heat

* See *Properties of Matter*, p. 59.

received. We must, therefore, reject the supposition of a continuous change from liquid to solid, and regard the change from ice to water as discontinuous, like the change from water to steam.

Effect of Pressure on the Melting Point.—The true explanation of regelation is probably given by a fact first predicted from theory by James Thomson, and since verified by his brother, Lord Kelvin, viz.:—that the melting point of ice is lowered by pressure. The following general explanation of this effect is not to be regarded as at all strict, but is given merely as enabling us to think of the main features of the process. A more complete proof will be found in chapter xix.

Let us take a mixture of ice and water at 0°. Any further melting will decrease the volume; any further freezing will increase it. Now, subject the mixture to great pressure, and let us assume that the mixture will yield to this pressure in the most effective way. This will take place if some of the ice melts. In melting, latent heat will be taken up, and the mixture will consequently cool until there is once more equilibrium between the ice and the water at a lower temperature—in other words, the melting point is lowered. If the pressure is still further increased more ice melts to yield to it, more latent heat is taken up, and the mixture will cool still further till a new melting point is reached.

Similarly, with substances in which there is an expansion on melting, pressure will raise the melting point. For, on subjecting a mixture of solid and liquid to pressure, the mixture will yield most effectively by solidification, latent heat is given up and the temperature is raised.

To explain regelation by the lowering of melting point by pressure, let us suppose that two blocks of ice at 0° are pressed together. The pressure lowers the melting point below 0°. Some of the ice at and near the surface is subjected to this pressure, and is, therefore, above the new melting point. It melts, and taking up latent heat it cools the liquid and the surroundings to the new melting point. The cooled water is squeezed out to the space between the two blocks where there is no pressure, so that the water is now below its freezing point. Therefore, being in contact with ice, so that there is no superfusion, it freezes again and forms a bridge connecting the two blocks. In fact, the effect of the pressure is to transfer the ice from points of greater to points of less pressure, the ice, however, liquefying in order to effect the passage. On the viscous theory of ice, it also passes from points of greater to points of less pressure, but without liquefying.

The lowering of the melting point of ice by pressure was verified by Lord Kelvin * as follows: He enclosed a mixture of ice and water in an Oersted's piezometer, consisting of a stout glass cylinder provided with a screw. A thermometer—with ether in place of mercury, to secure greater sensitiveness—was enclosed in a protecting glass tube, and placed in the mixture, the part of the scale to be used being kept free from ice by a lead ring, so that it could be seen. The pressure was indicated by a glass tube closed at one end, and just full of air at the atmospheric pressure. This was put into the piezometer with the open end downwards.

When a pressure of several atmospheres was put on, the temperature fell and remained steady, the fall being about 0·0075° per atmosphere,

* *Mathematical and Physical Papers*, vol. i. p. 165.

this fall agreeing very closely with that predicted by Prof. James Thomson. Dewar (*Proc. R.S.*, xxx., 1880), using apparatus like that of Cailletet for the liquefaction of gases, found that up to 700 atmospheres the reduction was proportional to the pressure and at the rate of 0·0072° per atmosphere.

FIG. 118.—Bunsen's Apparatus for Determining the Raising of the Melting Point by Pressure.

Experiments have also been made by Bunsen and Hopkins. Bunsen used a bent tube (Fig. 118), containing a small quantity of the substance to be tested at the end of the shorter limb CD, which was closed after its introduction. Mercury extended from D round the bend, filling the wider part E of the tube, and rising some distance in the capillary tube AB. The end A was closed when the quantity of mercury was suitably adjusted. The mercury, on heating, expanded and compressed the air which served both as a manometer and as a spring to exert pressure. This pressure and the melting point when it was exerted could be observed. Bunsen found that 100 atmospheres raised the melting point of spermaceti about 3·2°, and that of paraffin about 3·6°.

Hopkins, using an iron cylinder in which the melting was indicated by the fall of a ball of iron within the cylinder, was able to put on greater pressures, raising the melting point of spermaceti, for instance, from 51° to 80·2° by a pressure of 792 atmospheres—a result nearly agreeing with Bunsen's.

Melting Points of Solids.—Various methods have been used to find the melting points of solids. The simplest is to employ an arrangement similar to that used for fixing the zero point of thermometers. Heat is very gradually supplied to a mixture of the liquid and solid, and the steady point observed. The melting point of a substance like paraffin may be determined by filling the capillary part of a bent tube (Fig. 119), with the substance when liquid, freezing it, and then immersing the tube in a bath, and noting the temperature at which melting occurs. Carnelly determined the melting points of many salts which melt at a high temperature, by heating a small quantity in a platinum crucible of mass large compared with that of the salt, and then, at the moment of melting, dropping the crucible into a calorimeter. The quantity of heat given up allowed the temperature at melting to be calculated from the known specific heat of platinum—the heat given up by the salt being negligible.

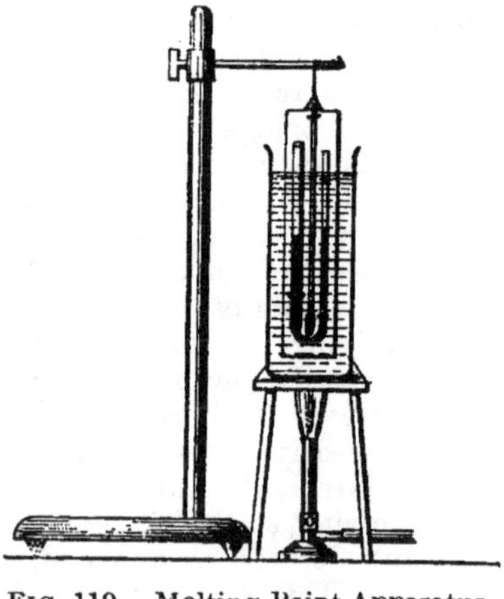

FIG. 119.—Melting-Point Apparatus.

Latent Heat.—The latent heat of many solids may be found by an experiment resembling the determination of the specific heat by the method of mixtures. If the specific heat of a substance in the liquid state is σ, and in the solid σ', while its latent heat is L, then on cooling w grammes of it from the liquid state, θ above the melting point, to the solid state, θ' below the melting point, the total heat given up is

$$w\,(\sigma\,\theta + L + \sigma'\,\theta'),$$

and knowing the capacity of the calorimeter in which the cooling takes place, and the rise of temperature of the calorimeter, this heat may be measured. σ and σ' may be found beforehand and then one observation will give L. If not, separate experiments in which θ and θ' are varied will enable us to determine all three quantities.

This method was employed by Person, who warmed the substance in a small copper vessel, and then immersed the vessel and its contents in the calorimeter.

For the latent heat of ice, Bunsen made use of his calorimeter, determining the quantity of ice melted by pouring into the calorimeter a known quantity of water at a known temperature.

The following are a few values of latent heat:—

	Person.	Bunsen.	Smith.
Water, . . .	79·2	80·2	79·818
Phosphorus, . .	5·0	—	—
Lead, . . .	5·36	—	—

The Explanation of Melting on the Kinetic Theory.

We have already pointed out that, in cases of fusion resembling that of ice, there is no gradual change from solid to liquid—no softening throughout the mass—and that the melting always occurs at the surface.

Again, in the converse process, ice is not spontaneously formed within the body of the water till at least many degrees below zero, even if then, and, in general, solidification requires the presence of ice or some disturbance to start it.

We cannot, therefore, regard the two states as in any way continuous one with the other. At melting, there is an abrupt change. If we have a piece of ice in water at 0°, on supplying heat the water grows at the expense of the surface ice, and on withdrawing heat the ice grows at its surface at the expense of the water. This closely resembles the change from liquid to gas by surface evaporation, which we explained by supposing interchanges of molecules between the gas and liquid, and the resemblance suggests an explanation of a similar kind for the change between solid and liquid, in which this also is regarded as a matter of exchange of molecules.

Perhaps we may make this explanation clearer by considering an ideal experiment.

Let A and B (Fig. 120) be two bulbs containing ice and water respectively, and connected by a tube C, the space above the ice and water in A and B and in the tube C containing water vapour only. Let the bulbs be surrounded by a constant temperature bath, so that both are at exactly the same temperature.

If this temperature is 0° C., the vapour-pressures of ice and water are each equal to 4·6 mm., and each will be in equilibrium with the vapour, neither growing nor diminishing. If the temperature is $-1°$ C., the pressure of water vapour exceeds that of ice vapour by nearly $\frac{1}{25}$ mm. Hence, when the water vapour has saturated the space, the pressure in A will exceed the maximum for the ice, and the ice will grow. Distillation will go on into the bulb A till all the water has gone from B.

If the temperature of the bath rises ever so little above 0°, the vapour-pressure of ice exceeds that of water, and the vapour from the ice tends to supersaturate the space above the water, and condensation goes on in B. But the condensation is not now confined to B; it occurs on the surface of the ice, and melting takes place there.

Suppose now that we have a mixture of ice and water at 0°; imagine each block of ice to be separated from the surrounding water by an indefinitely thin vacuous layer; evaporation will take place into this till the common vapour-pressure is reached, and then a steady state of equal interchanges is arrived at. If the temperature falls below 0°, the ice

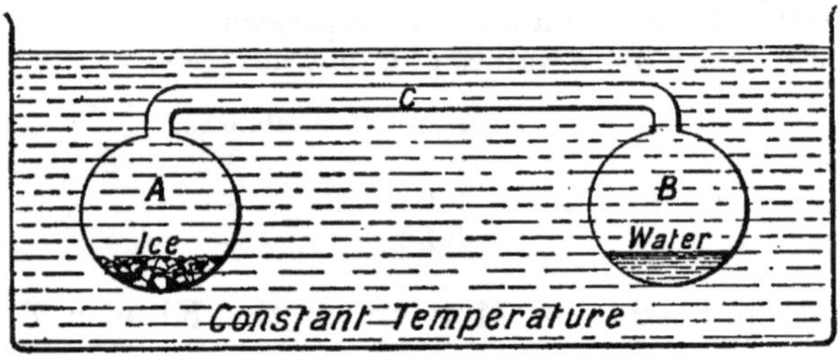

FIG. 120.

receives more than it gives up; if the temperature rises above 0°, the ice gives up more than it receives.

Now let us do away with this vacuous space, and replace the evaporation by the "mobility" of the molecules, using this term to describe the tendency of the molecules to escape from their position—in this case from the surface. This mobility is to some extent proved to exist by the evaporation. Then, at 0° the mobilities are equal. Above 0° the mobility of the ice molecules is the greater. Below 0° the mobility of the water molecules is greater. We may remark that probably the passage of molecules across the bounding surface of ice and water is much greater than the passage across the surface of ice or of water into a space only containing gas. For in the one case, we have the pull of the solid or liquid molecules in front assisting the escape, while in the other we have only the feeble pull of the gas in the space.

There is one fact which remains to be explained on this theory, that is, the melting of ice at 0°. No attempt to raise ice above 0° has as yet succeeded—melting always takes place at the surface.

The effect of pressure in lowering the melting point is explained by the greater mobility due to the pressure. We know that in general pressure increases mobility. Thus, in viscous solids there is a transfer from points of greater to points of less pressure, or there is more mobility at points of greater pressure; and we know, too, that the vapour-pressure of a liquid or of a solid is slightly greater at greater pressures (see chap. xix), which again implies that pressure increases the mobility. It can be shown that the effect on the vapour-pressure of a given increase of pressure is greater for ice than for water. Hence, we may suppose the mobility of ice increased more than that of water by a given pressure, and by putting a sufficient pressure on to a mixture of ice and water below 0°, the mobilities can again be made equal.

If we seek to represent the behaviour of ice and water on an indicator diagram, we get curves whose general course is shown in Fig. 121. There is no attempt here to draw to scale, the diagram merely representing the characteristic features. It is probable that the horizontal portions would diminish gradually after a time as the expansion of water for fall of temperature increases, and ultimately there may be a critical point at which the volumes of ice and water become the same. Below this temperature there would, perhaps, be a gradual passage from ice to water.*

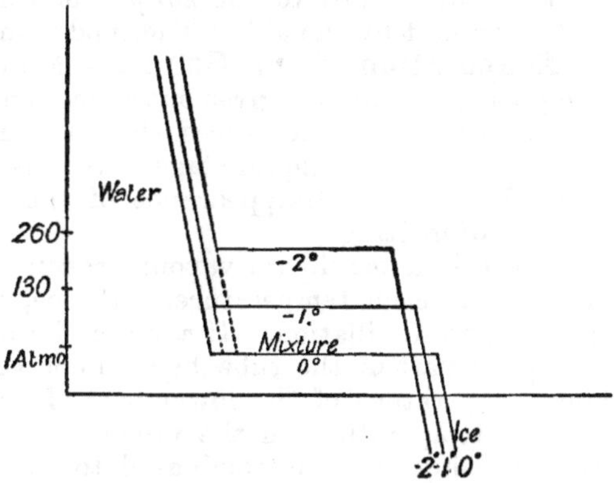

FIG. 121.—Ice and Water Isothermals.

In mixtures, softening may take place gradually with rise of temperature through the melting of one constituent, which then, perhaps, dissolves the others. If this occurs on reversing the process, supersaturation and superfusion may come in to lower the solidifying temperature. This is observed with fats. If the melting is a true continuous change from solid to liquid beyond the critical point, then, on reversing the order of change of temperature, the change of state should be exactly reversed. Perhaps this kind of melting may be found to occur in metals which soften as the temperature rises.

Resemblance of Solution to Fusion.—In several respects solution and precipitation from solution resemble fusion and solidification. If a salt is placed in excess in a solvent, a state of equilibrium is reached, when the liquid is said to be saturated. If the mixture of salt and solution is subjected to pressure, more salt is dissolved if the solution occupies a volume less than that previously occupied by the constituents. If the volume of the solution is greater than that of the constituents, the pressure is accompanied by deposition from solution.

Again, in most cases, rise of temperature increases the amount of salt taken into solution. If the warm saturated solution is poured off the salt into a clean vessel, it is easy to cool the liquid without deposition of the salt, though the amount dissolved is now greater than that corresponding to the saturated condition. The liquid is then said to

* Tammann (*Ann. der Physik*, 1900, II. 1, 424) finds that at −22° and under 2130 atmos. ice tends to pass into two other crystalline forms. Possibly ordinary ice could not be subjected to much greater pressure than 2130 atmos.

be supersaturated. On adding to the liquid a small quantity of the salt, deposition at once occurs.

This may be illustrated by a solution of sodium sulphate in water. At 33°, water dissolves its maximum amount of this salt. Filtering the saturated solution into a clean vessel at this temperature, the liquid may be cooled down without crystallisation. But on the introduction of a crystal of sodium sulphate, the excess of salt at once crystallises out, forming a mass of crystals throughout the liquid.

Solution resembles fusion also in requiring, in general, a supply of heat to effect the change of state, though in some cases this "heat of solution" is negative.

From these general resemblances, it appears probable that solution is an exchange phenomenon, a saturated solution in contact with its salt taking up and depositing equal quantities of the salt.

We shall return to the subject of solution in chap. xix., where we shall discuss it by the aid of Thermodynamics.

Evaporation from Solids.—Many solids evaporate sensibly. Camphor, for example, gives out a characteristic odour through evaporation at ordinary temperatures, the solid turning at once into gas without passing through the liquid stage. Ice also evaporates slowly, as may be shown by the slow disappearance of snow and ice in dry winds too cold to allow of melting.

There is a maximum vapour-pressure for solids, just as for liquids, definite for each temperature. Hence, we may have a process corresponding to distillation. If a piece of camphor is heated in a test-tube, the upper part of the tube being kept cool, the vapour-pressure at the high temperature of the lower part of the tube is above the maximum for the temperature in the upper part of the tube, and some of the vapour rising there is condensed to the solid form on the side of the tube.

This process is termed "Sublimation." It is used extensively in the preparation of sulphur, the product of sublimation being known as flowers of sulphur.

If a solid is supplied with heat in such a manner that evaporation can take place freely, it appears that there is a definite subliming-point, at which there is an arrest of temperature corresponding to the boiling-point, the temperature remaining at the point at which the pressure of the vapour equals the external pressure (Ramsay and Young, *Phil. Trans.*, Part I., 1884, p. 37).

CHAPTER XIII.

WATER IN THE ATMOSPHERE.

Hygrometry—Relative Humidity—Dew-Point and its Determination—Regnault's Researches on the Density of Water-Vapour — Cloud — Convective Equilibrium—Halos and Parhelia—Coronas—Rate of Fall of Cloud Drops—Hail —Fog—Dew.

Hygrometry, *or the Measurement of the Amount of Water-Vapour present in the Atmosphere.*—The atmosphere always consists in part of water-vapour, which is present in varying quantity, the mass in any given mass of air depending on its preceding history. If it has been moving over the sea or over damp ground, meanwhile being at a high temperature, it will contain a large quantity. If it has lately been cooled down, so that much of its vapour has been condensed, and has fallen as snow or rain, and if it has since been moving in dry regions or been maintained at a low temperature, it will contain a small quantity.

Relative Humidity.—It is of great importance in meteorology to measure the quantity present at any given time, for should this approach the quantity required to saturate the air, a slight lowering of temperature may result in condensation. It is usual to determine the amount present in terms of the pressure which it exerts, and the ratio which this bears to the maximum vapour-pressure at the temperature is frequently termed the "relative humidity."

Dew-Point and its Determination.—The simplest mode of determining the pressure of the water-vapour present in the air is by means of some form of "dew-point" apparatus in which a surface is gradually cooled down until dew is deposited on it. At the temperature at which deposition begins, the "dew-point," the air is just saturated, and from the tables of the pressure of water-vapour, the pressure for the dew-point may be obtained. Now in merely cooling the air the pressure of the water-vapour in it does not change, since both dry air and vapour contract equally. Hence the pressure of the vapour in the air before cooling is the maximum pressure at the dew-point.

A very good and easily used form of dew-point apparatus was devised by Regnault. It consists essentially of a tube AB (Fig. 122) like a test-tube, the lower part AB being of very thin highly polished silver. The upper end is closed by a cork through which passes a thermometer T and a narrow glass tube G reaching nearly to the bottom of the silver tube. A side tube DM is connected through the stand MN and by the tube NO with an aspirator. Some ether is placed in the tube up to PQ, and when the aspirator is set working it draws air down through the tube G and up through the ether. The free evaporation into the air-bubbles lowers the temperature of the ether and of the silver surface, and when the dew-point is passed the bright surface clouds over. The instant this clouding is seen the thermometer is read and the aspirator is closed. But the observed temperature will be a little too low, as the very first deposition of dew will be too small to be visible. Communication with the aspirator being cut off, the ether gradually rises in temperature, and when the dew-point is exceeded the dew will evaporate again. When the surface is once more clear the thermometer is again read, but the temperature now observed will be slightly above the dew-point, as the evaporation will not be instantaneous. After a few trials the two temperatures, one above and the other below the true dew-point, are made to close in upon it until the mean of the two can only differ from it by a very small quantity. The tube A'B', similar to AB, is not connected with the apparatus. It only contains a thermometer T', which gives the temperature of the air, while the silver surface always remains bright and serves as a standard with which to compare the surface on which the dew is deposited.

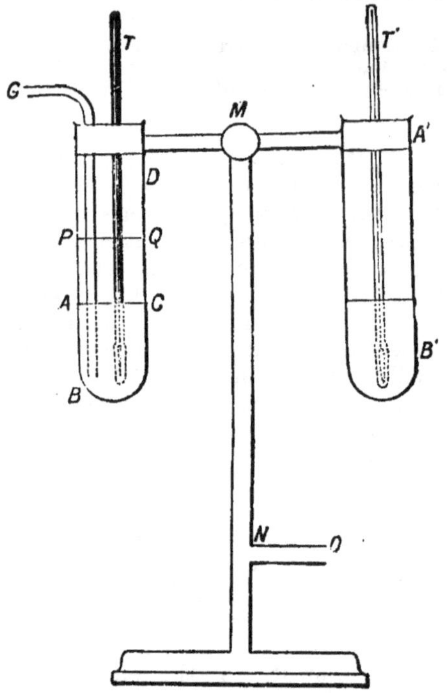

FIG. 122.—Regnault's Dew-point Hygrometer.

The form of the Regnault hygrometer has been modified in various ways so as to make the deposition of dew on the silver surface as evident as possible, and with practice an observer determines the dew-point speedily and accurately.

Dines's Hygrometer.—A somewhat simple instrument is that devised

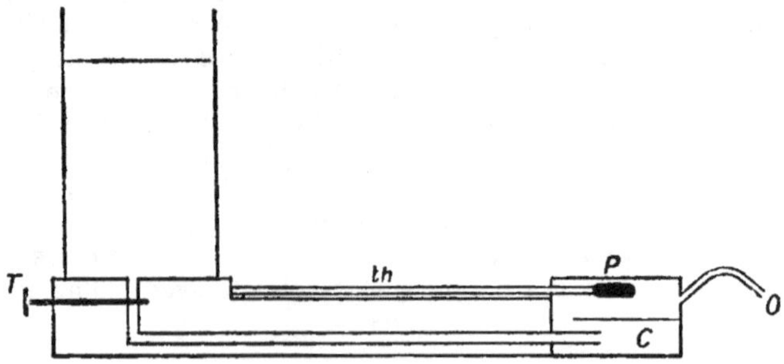

FIG. 123.—Dines's Hygrometer.

by Dines. V is a vessel containing water in which ice is placed to cool it. T is a tap which allows the cold water to trickle out of the vessel into a chamber C of which the top is closed by a black glass or silver plate P. Immediately under the plate is the bulb of a thermometer *th* round which the water flows. The water trickles out at O. As it passes the plate P it cools the bulb and plate approximately at the same rate, and at the dew-point P is clouded over. The tap T is then turned off till the dew disappears. The deposition and disappearance are observed several times, just as with the Regnault hygrometer, until the two temperatures are sufficiently near together, when the mean is taken as the dew-point.

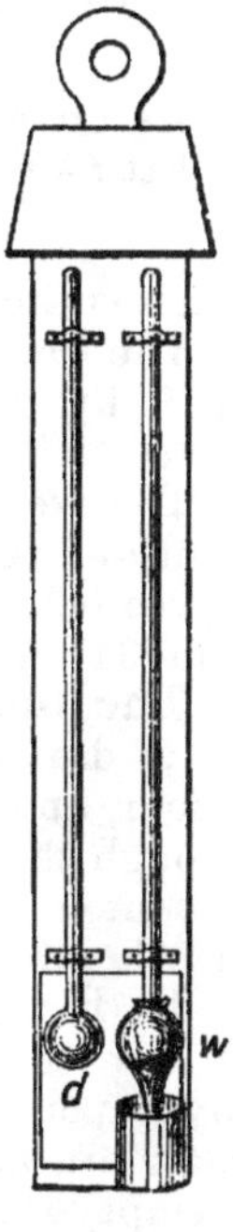

FIG. 124.—Wet and Dry Bulb Hygrometer.

Wet and Dry Bulb Hygrometer.—This consists of two vertical thermometers arranged side by side on a frame with the bulbs projecting below the frame (Fig. 124). One bulb *d* is freely exposed to the air, while the other *w* is covered with muslin or wick which dips down into a cistern of water just below the bulb. The water rising up the threads keeps the bulb always moist. The hygrometer is hung up so that the air can always flow past it. If the air is saturated it takes up no water from the wet bulb, and the wet and dry bulbs show the same temperature. If the air is not saturated then evaporation goes on from *w*, and the latent heat of evaporation is abstracted from the air, which cools and so keeps the temperature of the wet bulb below that of the dry bulb. We can see, as follows, without attempting an exact investigation, how the instrument will work. Let t be the temperature of the dry bulb thermometer, *i.e.* the temperature of the air arriving, let t' be the temperature of the wet bulb, *i.e.* the temperature to which the air is cooled by the latent heat taken up. Let F be the pressure of the water-vapour present in the air, f the saturation pressure at the temperature of the wet bulb. If the air passing the wet bulb becomes saturated, the pressure of the vapour is raised from F to f and the latent heat taken is proportional to this rise, or is equal say to $A(f-F)$. But the air is cooled from t to t' giving up heat proportional to the fall or equal say to $B(t-t')$

Equating the two quantities of heat

$$A(f-F)=B(t-t'),$$

or

$$F=f-\frac{B}{A}(t-t').$$

Apjohn gave for $\frac{B}{A}$ the value $\cdot01147\frac{p-f}{30}$ when the wet bulb is above 32°F. and the value $\cdot01042\frac{p-f}{30}$ when it is below 32°F., p being the barometric height in inches and F and f being in inches. Various modifications of the formula have been proposed with the aim of making it more in accordance with facts. But at the best the instrument is not very exact, its indications varying with its situation and its consequent exposure to wind, and it is probably best to use a simple formula, and be content with an approximation to the truth. In Hazen's Tables * the formula given is

$$F = f - \cdot 011(t - t')$$

using a constant barometer reading of 29·4 inches. A variation of two or three inches in the barometer does not seriously affect the result.

If the pressures are in millimetres and the temperatures are centigrade the formula may be replaced by

$$F = f - \cdot 00068(t - t')750$$

As an example of the first formula suppose $t = 65°$F. and $t' = 50°$F., so that $t - t' = 15$. At 50°F. $f = \cdot 3598$ in., whence

$$F = \cdot 3598 - \cdot 165 = \cdot 1948 \text{ inch.}$$

which is the vapour-pressure at 34°, and this is the dew-point.

The relative humidity is obtained by dividing the actual pressure ·1948 by the vapour-pressure at 65° which is ·6163. This gives the value ·31.

To save time and arithmetic, tables are constructed in which the dew-point and relative humidity are given for every depression read to half a degree below the temperature of the dry bulb. Thus opposite 65° in the table 34° and ·31 are entered under $t - t' = 15$.

The Chemical Method.—In this method a measured quantity of air is drawn through drying tubes which take from it all the water vapour, and weighing the tubes before and after the passage of the air through them, the gain in weight gives the amount of vapour which was present in the air. For details of the method we refer the reader to Glazebrook and Shaw's *Practical Physics*, p. 233.

In Fig. 125 A is the aspirator from which the water sipnons out; CC are the drying tubes; B an intercepting drying bottle to prevent any vapour passing back from the aspirator to CC. The volume of water drawn through from the beginning to the time when the aspirator is empty is measured. Suppose it to be V. Let the mean temperature of the air entering at a be t and let the final temperature of the air in the aspirator be t'. Let the barometric height be H. Then we have a volume V of air at t' saturated with vapour at t'. If the saturation pressure at t' is P′ the pressure of the dry air is H – P′ and the weight of the dry air drawn through the tubes

$$W = V\frac{H - P'}{760} \times \frac{\cdot 001293}{1 + at'}$$

If the gain in the weight of the tubes is w, the weights of water-vapour and dry air present in any volume of air are in the ratio $w : W$.

Since, according to the results given below, a given volume of water-vapour weighs 0·622 as much as an equal volume of dry air at the same temperature and pressure, the pressures exercised by these are in the ratio $\frac{w}{\cdot 622} : W$, their sum making up the barometric pressure H.

If then P is the actual pressure of water-vapour

$$P : H = \frac{w}{\cdot 622} : \frac{w}{\cdot 622} + W$$

* *Handbook of Meteorological Tables* (Washington).

or
$$P = \frac{\frac{w}{\cdot 622}H}{\frac{w}{\cdot 622} + W} = \frac{wH}{w + V\frac{H - P'}{760} \times \frac{\cdot 001293 \times \cdot 622}{1 + at'}}$$

This method has the disadvantage that it takes a long time to execute.

Regnault's Researches on the Density of Water-Vapour.—Regnault investigated the density of water-vapour at pressures below the maximum, both at ordinary temperatures, and near the normal boiling-point. He also determined the density of the vapour present in saturated air at ordinary temperatures, expressing the result in all cases as the ratio of density of water-vapour to that of dry air at the same

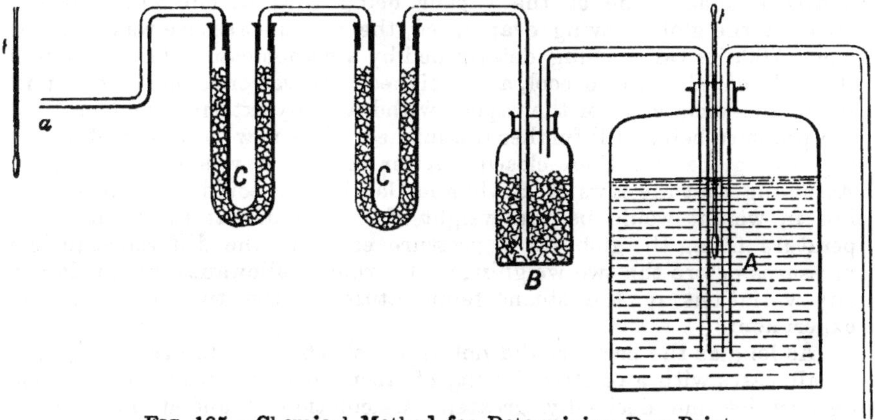

FIG. 125.—Chemical Method for Determining Dew-Point.

temperature and pressure, in order to test the applicability of the laws of Boyle and Gay-Lussac.

If we calculate the density of water-vapour from the densities of its constituents, assuming that two volumes of hydrogen combine with one of oxygen to form two of water-vapour at the same temperature and pressure, the density compared with that of air is

$$\frac{\cdot 0896 \times 2 + 1\cdot 43}{2 \times 1\cdot 293} = \cdot 622$$

Regnault's results, as will be seen below, show that under the various circumstances of the experiments, the density is very nearly this, or we may, to a close approximation, apply the ordinary rules connecting temperature, pressure and volume.

At Ordinary Temperatures.—Regnault introduced a small glass bulb containing a known weight of water into a large glass globe, which was then exhausted, the small quantity of residual air being carefully dried. A manometer attached to the globe gave the pressure within it. The bulb was then broken and the temperature of the globe was raised to a point above that at which all the water evaporated. The vapour-pressure was, therefore, below its maximum. The pressure indicated by the manometer was then observed, and subtracting that due to the residual air, the difference gave that of the water-vapour.

He found that if the pressure did not exceed ·8 of the vapour-pressure for the temperature of the experiment, the relative density of the water-vapour was about ·621.

When near saturation the density appeared much greater, but this must have been due to some error in the method, inasmuch as the density of saturated vapour in air agrees very nearly with the value ·621. Probably the vapour condensed on the glass before saturation.

At Temperatures near the Boiling Point, and at Pressures far below 760 mm.—A large globe containing some water was kept for a long time in steam until all the air was expelled, its place being taken by the water-vapour. Then the globe was connected with a condenser outside, some of the vapour being thus withdrawn. All the water in the globe having evaporated, the vapour-pressure was reduced below 760, its value being determined by a manometer. The globe was then closed, allowed to cool, and weighed. It was once more heated in steam and some more of the vapour withdrawn by external condensation, the pressure being still further diminished. The new value was observed and the globe was then closed. After cooling it was weighed again. Assuming that the laws of Boyle and Gay-Lussac could be applied, the loss in weight would be the weight of the vapour at the steam temperature filling the globe at a pressure equal to the difference in the pressures before the two weighings. Of course allowance was made for any variations in the steam temperature in the two parts of the experiment.

As long as the pressure did not approach 760 mm. the results agreed in all cases with a relative density of about ·620, but near 760 mm. the density became decidedly greater. Calculations founded on thermodynamic principles also show that the relative density of saturated steam is decidedly above ·623.

Using the method of Dumas, Cabours has shown that at the atmospheric pressure the relative density falls again to about the normal value when the temperature is raised considerably above 100°.

At Ordinary Temperatures in Saturated Air.—Air was drawn by an aspirator through a balloon filled with wet sponge into a box in which hung wet linen cloths, so that it was thoroughly saturated at the temperature of the box. It was then drawn on through drying tubes, where all the water-vapour was left behind, into the aspirator, where it became thoroughly saturated once more. Let us suppose the volume of the aspirator to be V, and for simplicity let its temperature be taken as equal to that of the box, both being $t°$; let the gain in weight of the drying tubes while the aspirator is once filled with air, be w. Let P be the maximum pressure of water-vapour at $t°$, and let us suppose that the relative density of the water-vapour is ·622, then

$$w = \cdot 622\mathrm{V} \times \frac{\mathrm{P}}{760} \times \frac{\cdot 001293}{1 + at}.$$

The observed values very nearly agreed with this calculated value though in all cases they were slightly lower—in general by about 1 per cent.—a difference which may be due either to a real difference in density or some error in experiment. Still the agreement is sufficiently close to justify the statement that the relative density of the water-vapour present in the atmosphere is practically the same as if the vapour alone were present at the same pressure.

Cloud.—When a mass of air is cooled below the saturation point—the dew-point—the excess of water-vapour is deposited as cloud, which consists of minute drops of water forming in general on "dust-particles" as nuclei (see p. 168). The cooling may sometimes result from a mixture of warm and cold air, and it is quite possible that a mixture of two quantities of air at different temperatures, neither saturated, may form a supersaturated mass. This may easily be seen from Fig. 126. Let PRQ represent the vapour-pressure curve, PM, QN being the saturation pressures at t_1 and t_2. Let equal masses at t_1 and t_2 contain vapour at pressures AM and BN respectively. When mixed, the temperature will be $\frac{t_1 + t_2}{2}$, and the vapour will be sufficient to produce pressure CL. But this is in excess of the saturation pressure RL, and CR represents the amount which can be deposited as cloud. Clouds appear to be formed in this way on changes of the wind. When a south-west wind comes on in the upper regions over a north-east wind in the lower regions of the air, clouds are formed where the two streams mingle.

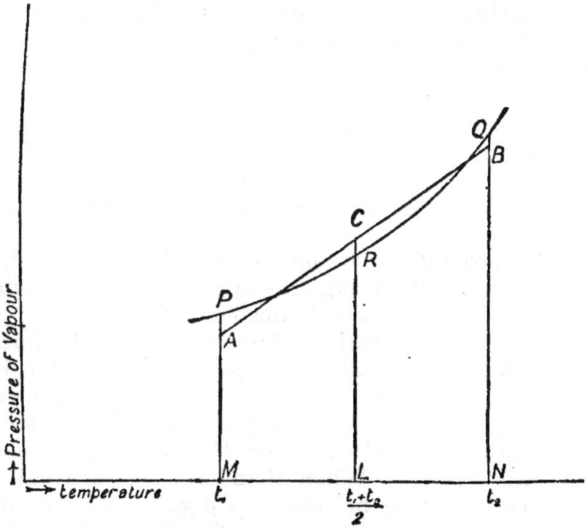

FIG. 126.—Mixture of Equal Masses of Unsaturated Air at Different Temperatures.

But a much more obvious mode of cloud formation is that in which a large mass of air expands and, in doing work, cools below its saturation point, depositing the excess of vapour as cloud. Such formation may be often watched on mountain sides and summits. A stream of nearly saturated air may be blowing against a mountain side, and being deflected upwards it comes into regions of less pressure, where it expands and cools, and the hitherto transparent vapour is converted into visible cloud. Sometimes beyond the mountain top the air descends again, and the clouds trailing off are gradually dissolved in the air which is now becoming warmer.

Expansion may also result from ascent of masses of air due to destruction of vertical equilibrium of the atmosphere. The limiting condition of equilibrium of the atmosphere, if it is uniform in composition, is that known as a condition of convective equilibrium.

Convective Equilibrium.—In this condition the temperature falls as we rise at such a rate that any mass carried upwards will cool by expansion so that it will always be at the same temperature, and of the same density, as its new surroundings. This means that the fall of temperature with decreasing pressure is that given by the adiabatic relation, and near the earth's surface it is easily shown that the rate is about 1° F. for each rise of 183 feet, or 1° C. for about 100 metres. If the lower layers of air are cooled so that the rate of fall of temperature upwards is less than this, or if the temperature rises as we ascend, the equilibrium is stable. But if the lower layers are heated so that the fall is more rapid, equilibrium is destroyed and circulation occurs. The lower air rises, and the upper air falls to take its place. This circulation frequently occurs on a hot summer day, when the surface of the earth, heated by the sun, in turn heats the air in contact with it, and there are small streams of ascending and descending air. These streams are rendered evident by the quivering of objects seen through the surface layers, the light being refracted by passage through a medium of density varying from point to point. The quivering is often seen with the naked eye, but is still more evident when a distant outline is looked at through a telescope. Some of the haziness of outline is no doubt due to this quivering.

If a large mass becomes simultaneously heated, it may rise as a whole and be thrust up through the upper layers. In rising it expands and cools, and if it rises far enough and cools down enough, it may become supersaturated at a certain height and begin to form cloud. Often such formation may be watched in progress, a big piled-up mass of "cumulus cloud" somewhat as in Fig. 127 being formed, the outline of the cumulus marking the upper boundary of the rising column, the horizontal base marking the cloud level or the point at which the air is cooled to saturation point. When thunderstorms are gathering in hot weather, the formation may be seen on a gigantic scale, the columns pushing far up into the higher regions. The "strato-cumulus" clouds, which are those with rather irregular globular or fleecy outlines, seen, say, in a south-west wind with a broken sky, were probably formed originally as cumulus clouds at the top of ascending columns, but they have drifted far from their birthplace, as their form shows. The upper part is always in advance of the lower, the upper wind travelling faster than the lower.

The so-called "Ripple Clouds," almost like ripple marks in sand, have been explained by Helmholtz as due to alternate expansion and contraction of the air. When one stream of air moves over another, there will be undulations in the surface of separation, and a mass of air at this surface will alternately rise and fall as it travels on. If it is just at the saturation point, when it rises to the top of an undulation it expands, cools, and deposits cloud, which it dissolves again in falling into the succeeding trough, where it contracts and heats. The parallel lines of cloud then mark the summits of the undulations, and the air may be travelling on at quite a different rate from the clouds.

Halos and Parhelia.—If the temperature of the air is below the freezing point, any excess of water-vapour is deposited as ice in the form of minute crystals. The presence of these crystals, when forming only a thin veil of cloud, is shown by halos round the sun and moon and by parhelia to the right and left of the sun when it is near the horizon. The halo at 22° from the sun or moon is the most frequent and conspicuous effect, and has value as a weather sign. Very often after clear weather, the arrival of a cyclone and rain is preceded by a damp wind in the upper air, so cold that ice crystals are formed in it. These are in the form of regular hexagonal prisms with angles of 120°. Alternate faces therefore make angles of 60° with each other, and through prisms of 60° the minimum deviation for ice with refractive index 1·31 is 22°. At this minimum deviation there is a concentration of the rays, and so the rays from the sun striking a crowd of such 60° prisms will be somewhat concentrated in a direction making 22° with the original direction. When an observer looks through the crowd towards the sun and then outwards from it, he will receive no light refracted through the 60° prisms till he has reached a distance of 22°, when there will be a concentration of refracted light. Of course the radius of the bright halo will differ slightly for different colours, but partly through the appreciable diameter of the sun and the consequent overlapping of halos due to different parts of his disc, and partly due to the scattering of other light in all directions, the colours are hardly evident, the inner edge of the halo at the best only showing a red tinge.

The ends of the ice prisms are planes perpendicular to the faces, and so form a second series of prisms with angle 90° and minimum deviation 46°. These form a second halo, which, though rarely observed in our latitude, is sometimes seen farther north.

FIG. 127.—Cumulus on Horizontal Base of Cloud.

When the sun is near the horizon, and the air is in such a condition that a large proportion of the hexagonal prisms have their axes in a vertical position, the refraction is much greater in the horizontal direction through the sun to right and left, and sometimes quite brilliant patches of light are seen with well-developed colour, at 22° on each side of the sun just above the horizon. If at the same time a large proportion of the prisms have their axes horizontal, a patch is also seen above the sun at the same distance, though this is much rarer. These patches are termed parhelia. Sometimes the parhelia lie on a circular halo. The phenomena are more frequent and much more marked in northern latitudes. Other circles, not all yet explained, have been occasionally observed passing through the parhelia.*

Coronas.—The coronas of colour seen within a few degrees of the moon when shining through the edge of fleecy clouds, and round the sun (seen best when reflected in a pool so that the excessive light is diminished), are due to diffraction by equal sized water drops. The explanation is given in the volume on Light.

Rate of Fall of Cloud Drops.—The drops of water in a cloud are subject, of course, to gravity, and they therefore begin to fall directly they are formed. But the viscosity of the air through which they pass introduces a resisting force which increases with their velocity, and is

* Tait, *Light*, p. 132.

very soon equal to their weight, and then the velocity of fall is uniform. Stokes (*Camb. Phil. Trans.*, vol. ix. p. 94) showed that this limiting velocity is proportional to the square of the radius of the drop, and his formula, with the now accepted coefficient of viscosity of air, gives the velocity for drops ·001 inch radius as about 0·8 inch per second. From the observed radii of coronas round the sun and moon, we know that the drops in clouds are often much less than this, so that the velocity of fall is very much less than 1 inch per second.

The change of the cloud-drops and crystals into rain-drops and snow-flakes has not been made out in full detail. It appears probable that rain-drops are formed by coalescence of cloud-drops. Defant (*Sitz d. Wiener Akad.*, May 1905; *Sci. Abst.*, 1905, p. 634) found drops of various sizes in the same rainfall, but those most frequent had weights in the ratios 1 : 2 : 4 : 8, apparently formed by the coalescence of drops of equal size.

In thunderstorms the clouds are usually of very great depth. A drop therefore tends to grow, perhaps by coalescence, through a very considerable range of fall, and this may account for the largeness of the drops as they reach the ground.

Hail.—The genesis of hail is much more complicated than that of the ordinary raindrop. The fall of hail accompanies very great atmospheric disturbances, in which there is ascent of large masses of damp air to very great heights. The hailstones usually show a structure of several coatings of ice and snow round a core of snow, and it is supposed that they may begin as snow in the upper regions, then descend some distance where the air is above the freezing-point. Here they condense on themselves a coat of ice. Then they may be carried up again into the snow region and acquire a coat of snow. Descending again they will receive another coat of ice, and the number of coatings is supposed to show the number of alternations.*

Fog is cloud formed on the surface of the earth. The condition for its formation appears to be that the upper air is warmer than the surface layer and nearly saturated. The lower layers cooling below the saturation point, the excess is deposited as fog. As the number of dust particles, especially in town air, is very great near the ground, the drops may be very minute and very numerous, and the fog may be much more opaque than ordinary cloud. When the upper air is warmer than the lower air, the smoke rising from chimneys does not tend to rise so high, as its excess of temperature above its surroundings is less. Hence in a town fog the smoke hangs near the ground and mixes with the fog instead of rising away from the surface as when the upper air is colder. In confirmation of this account of fog, it may be noticed that often a fog comes on just before the break up of a frost. A warm damp wind has already come on overhead, and the vapour diffusing from this into the still, cold surface layers, supersaturates them and produces deposition of the excess of vapour as fog.

Dew.—The formation of dew was first successfully explained by Wells in a celebrated "Essay on Dew," in which he pointed out that when the surface of the earth radiates out its heat at night so far as to

* An account of the atmosphere and of the mode of determining the condition of the upper regions by means of kites is given in Rotch's *Sounding the Ocean of Air.*

cool the surface layers of air below the saturation, or dew-point as it is now termed, the excess of water vapour is deposited as dew, and he made out the chief conditions for a plentiful deposit. When the sky is clear the radiation is greater, and when the surface layers are at rest they are more rapidly cooled below the dew-point. Hence the conditions favourable are, air with considerable relative humidity, at rest and under a clear sky to allow free radiation. The nature of the ground has, of course, some effect. If it is a good radiator and a bad conductor, its surface will become the more rapidly chilled.

Aitken (*Proc. R.S. Edin.*, 1886) has supplemented Wells' theory of dew by showing that there are really two kinds, the dew which comes from the air and the dew which is the extruded water or the condensed vapour exhaling from the surface. The surfaces of leaves are always tending to give off vapour, and if the air is not saturated it takes this vapour up. But if the air is saturated, then the exhalation forms drops of dew on the surface.

CHAPTER XIV.

GENERAL ACCOUNT OF RADIATION.

Radiant Energy—Radiometers—Radiant Energy and Light resemble each other—Light is Radiant Energy to which the eye is sensitive—Radiant Energy has a much greater range of Wave-Length than Light—Radiometers only measure Energy Streams and do not indicate Quality—Comparison of Emissive Powers—Radiation of different Wave Lengths—Comparison of Absorptive Powers—Comparison of Reflecting Powers—Diffusion—General results—Radiating and Absorbing Powers vary together—Illustrations—Transparency and Opacity—Radiation and Absorption by Gases and Vapours.

Radiant Energy.—When heat is received from a hot fire through an intervening space of air, the air is not permanently affected by the passage of the energy through it; there is no continuous downward slope of temperature from the fire to the receiver, and frequently, as on a cold winter's day, the air may be far colder than the receiver. The energy of the hot body spreads out on all sides from the body as a centre if unimpeded by obstacles, and it is therefore said to be radiated. The process is quite distinct from conduction, a good conductor like copper being as efficient an obstacle to the propagation of the energy as a bad conductor like wood, while air (as we have seen, an exceedingly poor conductor), allows the energy to pass freely. Before leaving the radiating body the energy radiated is evident as heat, and on reaching the receiver it is again evident as heat; but since in transit it does not make itself evident by warming the medium, we must ascribe to it a form distinct from heat which we term *radiant* energy.

In this chapter we shall investigate the nature of radiant energy, and discuss its emission from sources and its absorption and reflection at surfaces upon which it falls.

Radiometers.—It is necessary, of course, to have some instrument to receive, and indicate the rate of reception of, the energy arriving at a given point. We may term such an instrument a radiometer, though that name is frequently applied to one particular instrument, Crookes's Radiometer, described on p. 150.

When the rate at which the energy is arriving is large, we can detect it directly by the temperature sense, that sense which tells us when the skin is warmed or cooled. But obviously the temperature sense can only give us qualitative indications, and is no more suited for radiation measurements than for measurements of temperature.

Differential Thermometer.—Leslie, a pioneer in research on radiation, used the differential thermometer, which consists of a U tube with a bulb at each end of the U. A small quantity of coloured liquid is introduced before the bulbs are sealed off, and the air in the two bulbs is so adjusted that the liquid extends round the bend of the U and stands at the same level on each side when the bulbs are at the same temperature. One of the bulbs is exposed to the radiation to be measured, while the other is protected from it, and the liquid is depressed by the expansion of the air in the exposed bulb.

Thermopile. — Melloni, another pioneer, used a thermopile, which is much more sensitive than the differential thermometer. It consists of a number of bars of bismuth and antimony connected at the ends as represented in Fig. 128, the bars being insulated from each other along their sides. The beginning of the first bar and the end of the last bar are connected to binding screws, which in turn are connected to a galvanometer. If one set of alternate junctions is heated by exposure to radiation, a current flows from bismuth to antimony across the hot junctions, proportional to the difference of temperature of the hot and cold junctions, and the current is indicated by the galvanometer.

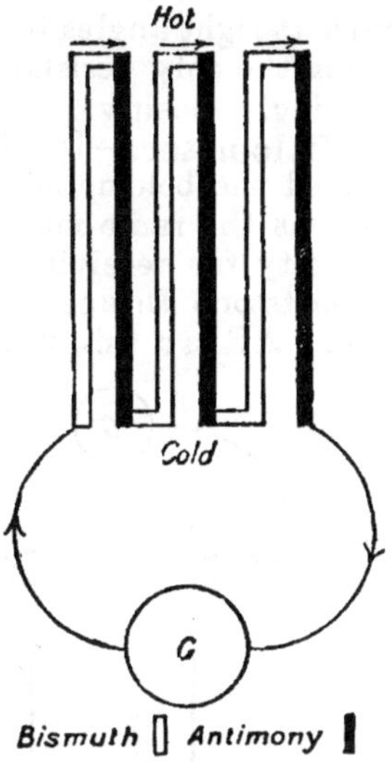

FIG. 128.—Thermopile.

Thermopiles have been made containing many hundred pairs of bars, the E.M.F. being proportional to the number of junctions. But the resistance increases with the number of bars and with a properly designed low-resistance galvanometer very little is gained by this multiplication of bars.

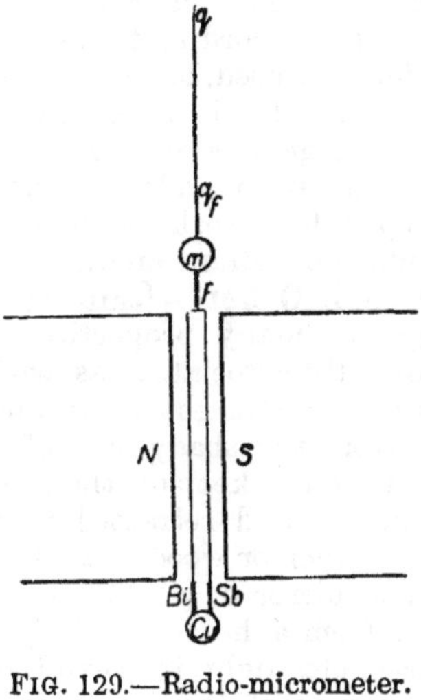

FIG. 129.—Radio-micrometer.

Radio-Micrometer.—This point is well brought out in the radio-micrometer, an instrument first invented by D'Arsonval and subsequently reinvented and made into an instrument of extraordinary delicacy by Boys.* It may be regarded as a combination of a thermopile of two bars and a galvanometer, all in one. Fig. 129 shows the principle of the instrument. The lower ends of two short bars, respectively of bismuth and antimony, are soldered to a small copper disc Cu in the figure. To the upper ends are soldered the ends of a copper wire, completing the circuit and making it a very narrow rectangle. This circuit depends from a glass fibre *ff*, on which is a mirror *m*, and this in turn from a quartz fibre *qq*. The rectangle hangs between the two poles of a strong permanent magnet. The whole is enclosed in a protecting case of wood, but there is a window for a beam of light to fall on *m*, and a tube is directed on to the disc Cu. If radiation passing through this tube warms the disc, it in turn warms the junctions with the

* *Phil. Trans.*, 1889, A., p. 158.

bismuth and antimony bars, and the same E.M.F. is produced as if the bars were directly joined and raised to the same temperature. A current flows counter-clockwise round the circuit as represented, and it tends to turn at right angles to the plane of the paper, and the torsion of the fibre, being the only resistance to turning, the instrument may be made exceedingly sensitive.

Bolometer.—In 1881 Langley * described an instrument which he named the bolometer, which is probably of the same order of sensitiveness as the radio-micrometer, but has the advantage of portability and capacity for receiving radiation in any direction. The bolometer is a Wheatstone Bridge and may be represented diagrammatically by Fig. 130. AB is a thin flat strip of metal—in the original instrument a strip of iron 7 mm. long, ·177 mm. broad, and ·004 mm. thick. It had a resistance of 0·9 ohm. CD is a wire of equal resistance. EF and GH are the ratio arms, here of course nearly equal, and capable of adjustment to give a balance in the galvanometer. The bridge is enclosed in a protecting case, but opposite AB is a shutter which can be opened to admit radiation. When AB receives radiation which warms it, its resistance rises and the balance in the bridge is destroyed. The current through the galvanometer is proportional to the rise in temperature of AB. The instrument is variously modified according to the purpose for which it is used, but it still retains its simplicity of principle. By its aid Langley has greatly extended our knowledge of radiation, and it is generally used by other workers, having largely displaced the thermopile in investigations on radiant energy.

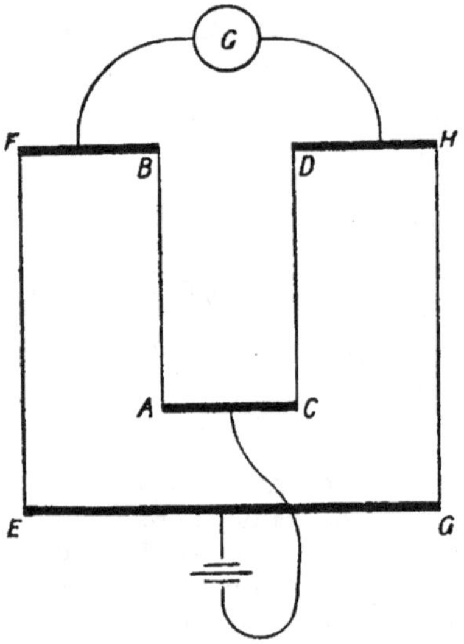

FIG. 130.—Bolometer.

Radiant Energy and Light resemble each Other.—Quite common experience tells us that radiant energy has many properties in common with light, with which it is so frequently associated, as with sunlight, firelight, and so on. It is propagated in straight lines, and forms shadows—as we know when we cross from the shady side of a street on a winter's day, or when we use a fire-screen to keep off the heat as well as the light of a fire. It is also reflected and refracted with light, as we know by the charring of a piece of paper or wood placed in the focus of the sun's light formed by either a mirror or a "burning" lens. We may also reflect the dark radiation from a heated ball by a mirror, just as we reflect light. It further resembles light in travelling with great rapidity, for the obstruction of the light of the sun by the most distant cloud is accompanied by a diminution of the warmth received from him ; and in eclipses by the moon there is a marked fall of temperature as well as a loss of light.

* *Nature*, vol. xxv., 1881, p. 14. For the construction of bolometers we may refer to Kurlbaum, *Wied. Ann.*, xlvi. p. 204, 1892 ; or Langley, *Am. Journ. Sci.*, v. p. 241, 1898.

Then again, radiant energy and light have the same laws of propagation. The amount radiated varies with the slope of surface, just as in the case of light, and the quantity of radiant energy falling on a given area, from a given source in a given time, varies inversely as the square of the distance—the law of light-propagation.

This may be proved very simply by placing a thermopile in front of a blackened tin vessel full of hot water, and between the two a screen with a hole in it (Fig. 131). So long as the hot vessel entirely fills up the field of view of the thermopile through the hole in the screen, the deflection of the galvanometer remains the same, whatever the distance between the screen and the hot vessel, provided that the distance of the thermopile from the hole in the screen remains unchanged. That this proves the law may be seen from a simple case. Let the hot vessel be placed successively at two distances 1 and 2 respectively from the pile, the screen remaining fixed. Then the areas seen from the pile, and from which it is receiving radiation, are as 1 : 4. But the total energy received is the same in each case. Then the energy received from the single area at the double distance is $\frac{1}{4}$ of that received from it at the single distance. Further, we may alter the slope of the hot vessel without affecting the heat received by the thermopile, just as we may alter the slope of a light-giving surface. Common observations and simple experiments, therefore, show us that radiant energy resembles light in its most evident properties. We may supplement these observations and experiments by various other experiments, and they all confirm the conclusion.

FIG. 131.—Proof of the Law of Inverse Squares.

Light is Radiant Energy to which the Eye is Sensitive.—Formerly, it was thought that this resemblance was all that we could assert and that corresponding to our two distinct sense-perceptions, one by the retina and the other by the warmth-sense in the skin, there were two distinct external agents travelling together. This was quite natural before physicists had begun to think of phenomena in terms of energy and its transformations. But now we clearly realise that light itself is a form of energy sent out by a heated source, which, in parting with the light, loses an equivalent of heat, and that this light energy, falling on an absorbing surface, is transformed into heat again. In this respect it is undistinguishable from radiant energy. We have now no reason whatever to suppose that there is the double agency. When the radiant energy falls on the retina—if it is of suitable quality—it gives rise to the sensation of light. The same energy falling on an absorbing substance, say, a blackened thermometer, will be transformed into heat. The difference, then, between light as received by the eye and radiant energy as studied by the thermometer, thermopile, or other radiometer is not in the external disturbance, but in the effect upon which we concentrate our attention.

Radiant Energy has a much greater Range of Wave-length than Light.—But though we now believe that all light is radiant energy, the converse is obviously not true, for we may receive heat by radiation from an absolutely dark body in a dark room. To illustrate the distinction let us consider the formation of a spectrum. If a beam of sunlight, or the light from an electric lamp falls on a slit S (Fig. 132), and passing thence is received on a lens L, it will be brought to a focus at a point F, the focus conjugate to S, and an image of the slit is formed on a screen placed at F. If a prism P be interposed at the angle of minimum deviation, then the light is bent round, each colour at a different angle, and forming its own image, so that a spectrum consisting of an infinite number of overlapping slits of different colours is formed on a screen *rb*, placed at the same distance from the prism as F, the red being the least, the violet the most refracted. In various experiments on interference of light we obtain a similar spectrum, arranged in the same order from red to violet. Now, whatever view we hold as to the nature of the disturbance constituting light, there can no longer be any doubt that it is a wave-form of energy, this supposition affording the only conceivable explanation of interference phenomena. And it can be shown that difference in position in the interference-spectrum must correspond to difference in the length of wave constituting the light. We learn thence that red light has the longest wave, the length diminishing through the spectrum to the violet. The total range of wave-length of the visible spectrum appears to lie between $\frac{1}{30.000}$ and $\frac{1}{60.000}$ of an inch, or between ·000075 cm. and ·00004 cm.

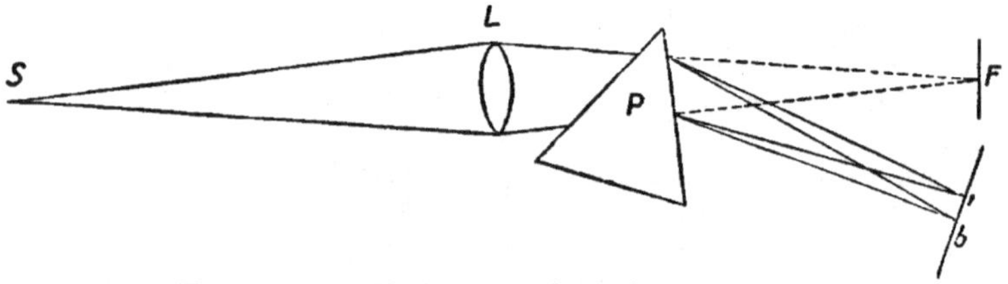

FIG. 132.—Radiation spread out into Spectrum.

We may turn the light into heat by placing a thermopile at various parts of the spectrum. At the red end it is sensibly affected, but less and less as it moves from that end, the heating being quite insensible to ordinary experimental arrangements in the blue. It was formerly supposed that the visible spectrum was accompanied by a heat spectrum especially strong in the red.

Now, were there any difference in kind between heat radiation and light, we might expect by some process to filter out the one from the other. But we cannot separate red radiation into two parts, one warming and the other lighting. When we diminish the light, we equally diminish the heat, as Jamin has shown.* We conclude, then, that there is a radiation of one kind only for each wave-length. The eye translates the radiation of the different wave-lengths into different colours;

* *Cours de Physique : Étude des Radiation*, p. 62.

but when they are absorbed by lampblack, say, they all become merely heat, of the same quality after absorption, whether absorbed red or absorbed yellow. The absence of heating effect at the blue end is merely due to want of sensitiveness of the heat detector, the eye being far more easily excited than the thermopile.

But this does not tell us anything about the radiation received from dark bodies. We may easily, however, explain the nature of this dark radiation in answering a question which naturally arises. Does this narrow range in the wave-length of the visible spectrum correspond to limits existing in the external disturbance, or merely to limits in our light-sense perception?

Suppose that we replace the eye, as the recording instrument, by a thermopile deflecting a galvanometer-needle. Placing this in various parts of the visible spectrum, we get, as we have seen, deflections hardly visible in the blue and green, but rising towards the red, showing that the light is converted into heat in the thermopile. Now, taking the thermopile out beyond the red, it still gives indications which show clearly that longer wave-radiations are falling on the screen there, and these are termed "infra"-red radiations But just as a coloured glass absorbs light of a certain wave-length in the visible spectrum, while transmitting that of other lengths, so what we call transparent glass absorbs some of the radiation of long wave-lengths, though transmitting that of the visible spectrum. The effect beyond the red is, therefore, diminished by absorption in the glass lens and prism. If we use rock-salt instead of glass for prism and lenses the absorption is much more limited, and the effect on the thermopile is greater On moving the thermopile to the ultra-violet, we find no sensible effect, and we must use a more delicate test for the presence of the energy. If we paint a sheet of paper with a solution of sulphate of quinine in dilute acid, and expose it in the ultra-violet region, there is an action on the solution, and it gives out visible radiation. Or, there are ultra-violet short wave-length radiations, which are absorbed by sulphate of quinine, and given out again by it as radiations of wave-length long enough to excite the retina. Or still more marked is the effect on an interposed screen of the kind used for the detection of X-rays, brilliant bands of green light appearing in the region beyond the violet. These ultra-violet radiations are especially active in exciting the chemical changes employed in ordinary photography. That these extreme radiations are also of wave form, has been shown from the fact that they exhibit the phenomena of interference and of polarisation, which we can explain only by the wave theory.

We conclude then, that radiant energy is, in form, a wave disturbance, the lengths of the waves extending over a wide range, but to one particular part of the range the retina is sensitive, and we call the energy within this range *light*. This is probably the part of the energy which is present in greatest quantity in sunlight. As we see most things by reflected sunlight, we see them more easily by being especially sensitive to that kind of radiation which is present in greatest quantity.

Radiometers only Measure Energy Streams and do not Indicate Quality.—The various radiometers, unlike the eye, take no account of the wave-length of the stream of radiation which they measure. Their indications are proportional to the rate at which they are absorbing energy. For when they are in a steady state they are giving out as much as they receive. Now, with the small excess of temperature produced by the radiation, the amount given out is proportional to the excess, and therefore the excess is proportional to the amount received. It is just the same whether the energy was blue light, yellow light, or ultra-red radiation before it was absorbed and converted into heat. If the total energy absorbed is the same the indication of the instrument is the same. When we wish to study wave-length as well as quantity we must first sort the radiation out by a prism or grating. In much of the earlier work this was not attempted, and it was really by the invention of the bolometer by Professor Langley that the detailed study of quality as well as quantity of radiation was rendered practicable. But a great deal of valuable knowledge was obtained by the earlier workers who merely investigated radiation totals, and we shall give here a brief account of their results.

Comparison of Emissive Powers.—Leslie* compared emissive powers by an arrangement represented in Fig. 133. C was a tin cube

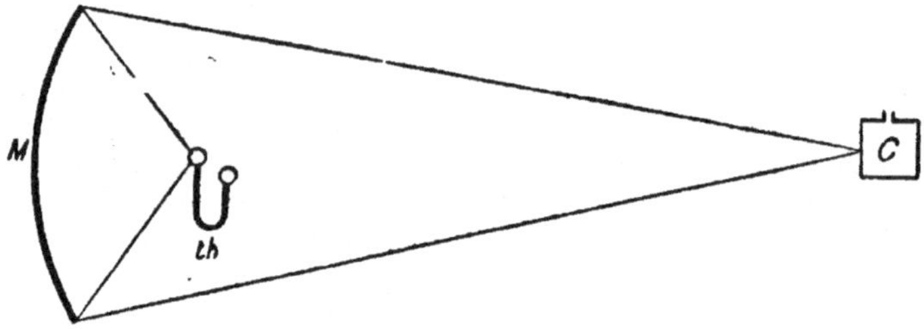

FIG. 133.—Leslie's Comparison of Emissive Powers.

filled with boiling water. M was a mirror of tinplate beaten into parabolic form and about 12 inches aperture. *th* was the thermometer bulb placed at the focus conjugate to C. The cube had its vertical face either plain or coated with the substance to be tested. He obtained the following results:—

Emissive Power of Surfaces at 100°. Lampblack = 100.

Lampblack	100
Writing paper	98
Indian ink	88
Red lead	80
Tarnished lead	45
Clean lead	19
Polished iron	15
Tinplate, gold, silver, copper	12

Melloni obtained results confirming those of Leslie, but De la Provostaye and Desains showed that through faulty arrangements both the earlier observers overestimated the emissive power of metals. They obtained

* *An Experimental Inquiry into the Nature and Propagation of Heat.* 1804.

the following results, the temperature of the radiating body being 120°—

Lampblack	100
Indian ink	88
Platinum	11
Copper	5
Silver	2 to 3

They also found that the relative powers changed with the temperature. For this purpose, they used a platinum plate coated on one side with lampblack, and on the other side with the substance to be tested, and heated it to various temperatures with an electric current. The radiations on the two sides were compared by means of two thermopiles, one placed on each side of the plate. Now, had the relative emissive powers been independent of the temperature, the indications of the two thermopiles would have always been in the same ratio. But the ratio was found to change. Thus the radiation of borate of lead was equal to that of lampblack up to 100°, but after that it fell in comparison, being only $\frac{3}{4}$ at 550°. This is exactly what we should expect, since the radiation is a mixture of different wave-lengths or "colours," and we have no reason to expect that the rise of temperature would produce the same increase in emission of all the different wave-lengths.

Radiation of Different Wave-Lengths.—The difference of wave-length in invisible radiation was shown conclusively by Forbes, who succeeded in showing that as the temperature of a source rose, the refractive index of the most energetic radiations emitted, as determined by a rock-salt prism, rose also.* Langley,† using his bolometer, showed that, while the refractive index of rock-salt for light ranges from about 1·58 to 1·53, he was able to obtain radiation from a Leslie cube at 178° C., which passed through a rock-salt prism with deviation corresponding to a refractive index not greater than 1·45, and a wave-length probably many times that of the D line. For the details of the work the reader should consult the original papers, but a sketch of Langley's method may serve to show how the quality of radiation may be investigated.

When a beam of light from a slit falls on a diffraction grating, and the rays are brought to a focus, a central bright band is formed, and a series of spectra is arranged on each side, the distance of a given colour in any spectrum from the centre being proportional to its wave-length. The blue rays are therefore the nearest, and the red furthest, from the centre in each.

Considering the first spectrum on one side of the centre C, Fig. 134, the various rays appear at points distant from the central bright band C proportional to their own wave-length. If then S_1 (Fig. 134) is the position of the D line, and we erect a height S_1D_1 equal to the wave-length of the D line, $589\mu\mu$, and join CD_1 producing it onwards, the height of this line above any point of CS_4 will represent the wave-length of the ray of the first spectrum diffracted to that point. But even in the solar spectrum there is no radiation perceptible below about $290\mu\mu$. Hence, the spectrum only begins at A_1 where A_1B_1 equals 290. But each

* For an account of Forbes's researches on radiation, see Balfour Stewart's *Heat*.
† *Phil. Mag.*, xxi. 1886, p. 394; xxii. 1886, p. 149.

ray is also diffracted into the other spectra; in the second, at double the distance from C; in the third at treble the distance from C, and so on. Hence if we make $CS_2 = 2\ CS_1$; $CS_3 = 3\ CS_1$, &c., the D line of the second spectrum is at S_2, of the third at S_3, &c. Making $S_2D_2 = S_3D_3 = S_1D_1$, and joining CD_2, CD_3, &c., the heights of these lines above any point give us the wave-lengths of the various spectra diffracted to that point. Drawing through B_1 a parallel to C_1S_1, we need not consider any line below that, as the corresponding radiation is insensible.

It is evident that, except near the centre, the spectra overlap. At the point S_4, for instance, we have

$$\begin{aligned} \lambda &= 589 \text{ of the 4th} \\ &= \tfrac{4}{3} \times 589 \text{ of the 3rd} \\ &= 2 \times 589 \text{ of the 2nd} \\ &= 4 \times 589 \text{ of the 1st.} \end{aligned}$$

The first of these is, of course, the yellow line easily seen, the second is in the far red, the others dark rays.

If, now, the slit of a spectroscope be placed at S_4, and a rock-salt train

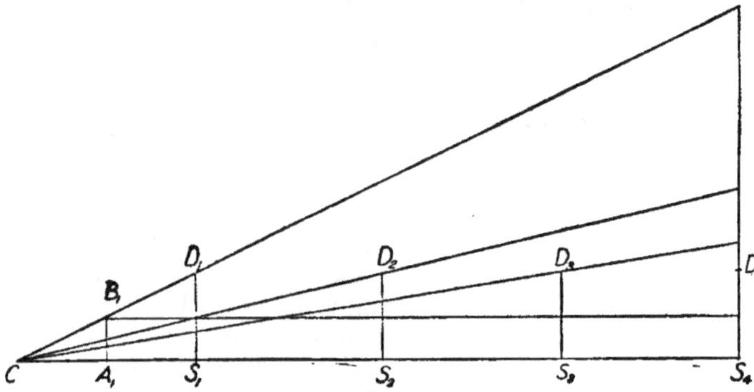

FIG. 134.

of lenses and prisms be used to minimise absorption, since the different rays have different refractive indices, they are sorted out into different positions in the field of the spectroscope. The bolometer is made to travel along, first being placed in the yellow line, which is visible. As it passes each of the others in succession, the effect is apparent, and though the ultra-red rays are invisible the bolometer feels them. By this method, Langley was able to detect radiation received from bodies at quite low temperatures, and in positions in the spectrum corresponding to very great wave-lengths.

Comparison of Absorptive Powers.—The absorptive powers of various substances were first compared by Leslie, who covered his thermometer bulb with a layer of lampblack, and placed it in the focus of the mirror, noting the rise, and then compared this with the rise when the bulb was covered with other substances. This method is not satisfactory, as the thermometer is steady only when the absorption is equal to the emission, and the emission as well as the absorption varies with the substance. Melloni put plates of thin copper between the source and the thermopile, the side towards the source being covered with the substance and that towards the pile with lampblack.*

If the quantity of heat emitted in one second by one side, per degree-rise of temperature of the plate above the surroundings, is E for lampblack, and E_1 for the substance, and if the excess of temperature is t_1, the emission is

$$(E+E_1)\,t_1.$$

If the quantity absorbed by the surface towards the source is A_1, then in the steady state

$$A_1=(E+E_1)t_1.$$

If both surfaces are lampblacked, and the excess of temperature is t, then if A is the quantity absorbed by the lampblacked surface towards the source,

$$A=2Et$$

Then

$$\frac{A_1}{A}=\frac{E+E_1}{2E}\cdot\frac{t_1}{t}$$

$$=\frac{1+\frac{E_1}{E}}{2}\cdot\frac{t_1}{t}$$

Now the indications of the galvanometer in circuit with the pile are proportional to t_1 and t, so that we know $\frac{t_1}{t}$. Knowing also $\frac{E_1}{E}$ from the previous determinations of emissive power, we can find $\frac{A_1}{A}$,

Just as the light absorbed by various surfaces depends on the nature of the incident light, red paper, for instance, absorbing more of blue light than it does of red, so the absorptive power for radiation in general depends on the source. For example, among other results Melloni obtained the following absorptions from the sources given at the head of the columns, and taking in each case the absorption of lampblack as 100 :—

Absorbing Surface.	Locatelli Lamp.	Incandescent Platinum.	Copper at 400°.	Copper at 100°.
Lampblack . . .	100	100	100	100
Indian ink . . .	96	97	87	85
White lead . . .	53	56	89	100
Metal . . .	14	13·5	13	13

De la Provostaye and Desains worked in another way, coating a

* A simple method of illustrating the difference of absorptive power of metal and lampblack consists in placing two equal metal vessels in front of some source of heat—equidistant, say, from a gas-flame, the surface of one being lampblacked on the side turned towards the source. The rise in temperature of the lampblacked vessel is much more rapid than that of the other. A good way of arranging the experiment is to make each of the vessels into an air thermometer on the principle of Fig. 32, chap. iv.

thermometer with the substance to be tested, and finding its rate of cooling for different excesses of temperature above the surroundings. It was then exposed to radiation from a source, and the steady state was attained. The heat lost per second was proportional to the rate of cooling previously found for the temperature now observed, and the loss was equal to the heat gained by absorption. The results obtained agreed in their general character with those of Melloni.

Comparison of Reflecting Powers.—The earliest work on this subject was done by Leslie. He employed an arrangement represented in Fig. 135. C was a Leslie cube filled with boiling water and with a blackened face turned towards a tin spherical mirror M. The focus conjugate to C was *f*, but the rays were intercepted on their way thither by a plate of the substance to be tested placed at *ab*. The rays came to a focus at *f'*, and at that point was one bulb of a differential thermometer. The rise in temperature given by the thermometer was

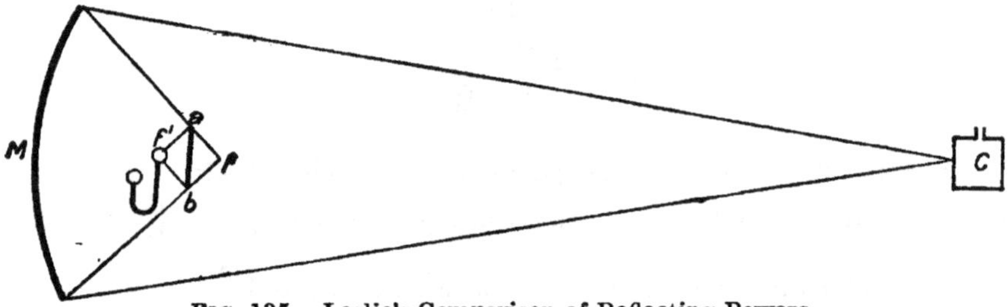

FIG. 135.—Leslie's Comparison of Reflecting Powers.

proportional to the reflecting power. The following are some of his results:—

Reflecting power of surfaces for radiation from a black cube at 100° C. in terms of brass taken as 100—

Brass	100
Silver	90
Tin	80
Steel	70
Glass	10
Lampblack	0

Lampblack practically reflected none of the radiation falling on it.

Melloni obtained results generally confirming those of Leslie. But we know that the reflecting power for light varies with the quality of the incident light and also with the particular specimen of the substance tested. And there is no doubt that there is a like variation in the case of invisible radiation. We should expect the results of different observers to differ, and they do differ.

As illustrating this we give the following results of De la Provostaye and Desains obtained from a Locatelli lamp as source. It will be seen that the order is slightly different from that found by Leslie:—

Reflection of radiation from a Locatelli lamp—

Silver . .	97	per cent. of incident light.
Gold . .	95	" "
Brass . .	93	
Speculum metal	86	
Tin . .	85	
Steel . .	83	"

Diffusion.—When radiation falls on a surface it may be in part diffused or irregularly reflected. The diffusion depends on the polish of the surface.

A method of studying diffusion will be understood by the aid of Fig. 136, which represents the apparatus used by Melloni to determine the variation of diffusion of radiation received from different sources. The incident beam falls on a thin plate of copper *a b*, coated with lampblack. Two thermopiles *c*, *d* are arranged symmetrically in front and

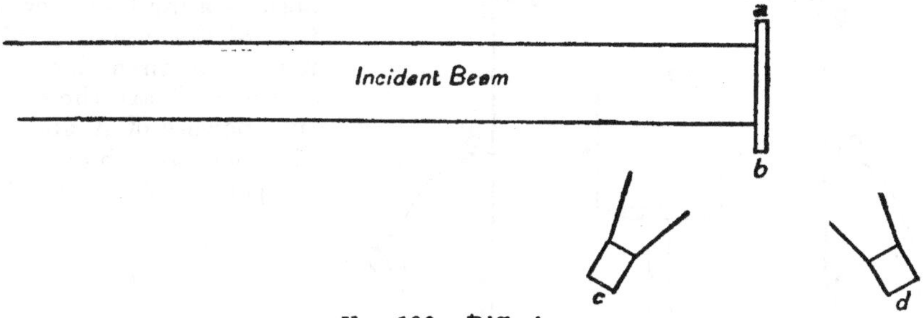

FIG. 136.—Diffusion.

behind the plate. The front one *c* registers emission plus diffusion, while *d* registers emission alone. The heat received by the former was about 1·18 of that received by the latter. On coating the front of *a b* with white lead the difference was very much greater, the diffusion by the white lead being much larger.

Knoblauch found that lampblack diffused about 10 per cent. of the incident radiation whatever the source from which it was received.

General Results.—Even though no reliance is to be placed on the exactness of the numbers obtained for the different qualities of absorption, emission, and reflection, a comparison of their magnitudes at once leads to an important conclusion. For instance, lampblack has no regular reflection, low diffusive power, and the highest emissive and absorptive power. Indian ink has emissive power about 88 per cent. of that of lampblack, and absorptive power 85 per cent. Metals reflect over 75 per cent. of the incident radiation. Their emissive and absorptive powers are less than 20 per cent. of those of lampblack. These may serve as typical cases, and we conclude that high reflectors, which must therefore have low absorbing powers, have also low emitting powers, while low reflectors, which, if opaque, have high absorbing powers, have also high emitting powers. In other words, *Radiating and absorbing powers vary together. High radiators are high absorbers. Low radiators are low absorbers.*

This statement must not be taken too generally. It requires limitation on account of the variation of the absorbing power with the nature of the radiation falling on it, and we shall see later that the limitation is that the law applies to each kind, or wave-length, of radiation emitted. So that a high or low absorber of radiation of a given wave-length is a high or low emitter of that same radiation. If, as in the case of lampblack, nearly all wave-lengths are absorbed to a very great degree, the surface is a high radiator of nearly every kind of radiation.

The way in which absorbing and radiating powers go together may be illustrated in a very striking way by the experiment represented by Fig. 137. A Leslie cube AB is placed midway between two tin plates A′ B′. These have small thermoelectric junctions at the back, and the circuit is completed, as shown in the figure, through a galvanometer G. If the plate B′ is at a different temperature from the plate A′ a current goes through the galvanometer. If A and A′ are lampblacked surfaces, while B and B′ are bright tin, no current is generated. A radiates far more than B, but B′ absorbs a far less proportion of the radiation falling on it than does A′, and if E, E′ are the emitting powers of A and B, while $a\, a'$ are the absorbing powers of A′ and B′,

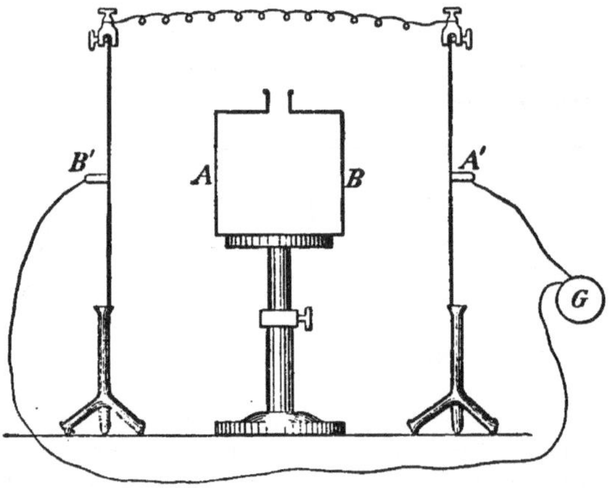

FIG. 137.

$$a'E = aE'$$

or

$$\frac{E}{a} = \frac{E'}{a'}.$$

A further illustration of this equality of emitting and absorbing powers is given by a thermometer placed in an enclosure at a different temperature. If the thermometer is $\theta°$ above the enclosure, it will fall towards the temperature of the enclosure at the same rate as it will rise towards it if it is $\theta°$ below the enclosure.

Illustrations of Emission and Absorption of Radiation.—The low radiating and absorbing power of metals is frequently used in physical experiments. Thus the external surface of a calorimeter is brightly polished, and is usually surrounded in turn by a second metal vessel brightly polished. Most of the small quantity of heat lost by radiation from the calorimeter is reflected by the outer vessel, and the low radiating quality of the outer vessel protects the inner vessel from external changes. It is probably advantageous to make the two vessels of different metals, since, if of the same metal, the outer vessel will absorb more of the rays emitted by the inner one than if it is of a different metal. It should as far as possible reflect them back

A familiar example of the use of low radiating power of bright metal is afforded by a metal teapot.

If we wish to keep ice from melting, it is advisable to enclose it in a box of low conducting power, lined outside with bright metal. The outer surface will ultimately take a temperature below that of the room, and such that the temperature slope from the outside to the inside of the box will just conduct to the ice all the heat supplied by the surroundings. The lower the absorbing power of the metal surface, the less will be the slope in the non-conducting case needed to convey the heat to the ice, and the less the heat so conveyed. This may be seen more exactly as follows:—

If A is the absorption of heat per second by the metal surface for each degree that it is below the temperature of the room, and C the quantity of heat conducted in through the box per second per 1° difference of temperature between inside and outside, and if t° is the temperature of the room, θ° that of the outer surface of the box, 0° that of the inside, the steady state is given by

$$A(t-\theta)=C\theta$$

whence

$$\theta=\frac{At}{A+C}.$$

The quantity of heat received by the ice is, therefore,

$$C\theta=\frac{ACt}{A+C},$$

which diminishes as A and C diminish.

Deposition of Dew on Different Surfaces.—A very interesting illustration of the effect of variation in radiating power is afforded by the phenomena of dew formation. We owe the explanation of the formation of dew, now generally accepted, chiefly to the experiments and observations of Dr. Wells, made about 1812. He found that dew is most freely deposited on calm, clear nights, on substances close to the surface of the earth and not shielded from the sky. These observations led him to suppose that the deposition is due to the cooling of the surface by radiation, and the consequent cooling of the air in contact with the surface below the saturation-point of the vapour contained in it. He verified this cooling by radiation by showing that the surface cooled down far below the air a little above it, and he showed also that the cooling was greater for high radiators than for low. He also found that any protection from the sky, by hindering the radiation outwards, lessened the cooling. The lesser deposit on surfaces raised above the surface of the earth he attributed to the falling down from them of the cooled and therefore heavier air. There was, therefore, a constant renewal of the air in contact with the elevated surface, and this air had not time to cool sufficiently to deposit much of its vapour. Wells's theory has since been supplemented by Aitken, who has shown that much of the dew found on the surface of vegetation arises from the vegetation. When the air is not saturated, the water evaporates from the surface; but when the air is cooled to saturation-point, it cannot take up this water, which therefore remains as dew.

Formation of Ice in India.—Ice has been formed in India on nights when the air does not fall to freezing-point by forming square pits (according to Dr. Wells) about 2 feet deep and 30 inches wide, and filling them about 8 inches or 1 foot deep with straw. On this rows of small unglazed earthen pans were placed, about $1\frac{1}{4}$ inch deep, filled with boiled soft water. Evaporation and radiation, acting together, cooled the water below freezing, and ice was formed. Dr. Wells pointed out that as the formation was most successful on calm clear nights, the effect was chiefly due to radiation. Probably evaporation would aid the first cooling until the dew-point was reached, any subsequent cooling being due to radiation.

Transparency and Opacity.—Substances differ as markedly in their transparency for dark radiations as for light, and all the phenomena of selective absorption are repeated with these longer waves.

Many bodies which are transparent to light are opaque to long wave radiations, but the opacity differs with the source, that is, with the quality, or wave-length, of the radiation received. Thus, Langley finds that glass is more or less transparent to all the radiations reaching us from the sun. But Melloni found that a plate of glass about 2 mm.

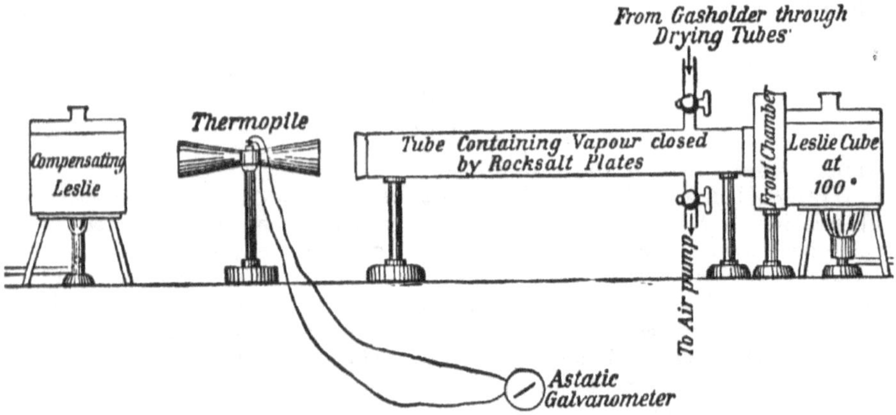

FIG. 138.—Tyndall's Experiment on Emission and Absorption of Gases.

thick would absorb half the radiation from an argand burner, while in front of a Leslie cube at 100° it was as efficient a screen as a plate of metal.

Again, it was found by Melloni that rock-salt transmits a very large proportion of dark radiations, and it was formerly supposed to be transparent to all kinds. But Balfour Stewart showed that a plate of cold rock-salt was exceedingly opaque to the radiations from hot rock-salt.

Lampblack, which is opaque to nearly all radiations, has been found by Langley (*Phil. Mag.*, 5th series, xxi. p. 403) to transmit some of very great wave-length. Tyndall has found that a solution of iodine in bisulphide of carbon, though quite opaque to light, is transparent to many long wave radiations. A very thin plate of ebonite is also transparent to long wave radiations, and if placed in front of the condenser of a lantern, allows enough radiation to pass through to heat the hand if held at the focus.

A thin ebonite prism spreads the invisible long-wave radiation into a spectrum, which is quite sensibly detected by the radio-micrometer.

Radiation and Absorption by Gases and Vapours.—Tyndall

made a great number of researches on the transparency and opacity of gases and vapours, his principal mode of experiment consisting in passing a beam of radiation through a tube about 4 feet long and 2 or 3 inches in diameter, with its two ends closed by rock-salt plates, from a source which, in many of the experiments, was a Leslie cube filled with hot water, as represented in Fig. 138. The beam fell on one face of a thermopile, and the heating effect was compensated for by the radiation from a second cube falling on the other face of the pile. A chamber which could be exhausted intervened between the cube and the nearer rock-salt plate, so that no absorption could occur by the air before the radiation reached the gas. The air could be pumped out of the tube, and the dry gas to be experimented on could be introduced from a gasholder. When vapours were experimented on, a flask containing some of the liquid to be vaporised replaced the gasholder. By these experiments Tyndall showed that air, oxygen, nitrogen, and hydrogen exercised only a very slight effect, but that compound gases showed decided absorption. When the source was a copper plate heated by a Bunsen burner, Tyndall found the comparative results given below. It will be seen that at a low pressure the differences are still more striking than at the pressure of the atmosphere. In the latter case, probably, the absorption was in some instances so complete that practically all the radiation which could be absorbed was sifted out of the beam before it had traversed the entire thickness, so that the whole length of the tube was not effective for the gases lower on the list.

	Amount absorbed at Atmospheric Pressure. Absorption by Air = 1.	Amount absorbed at 1 inch Mercury Pressure. Absorption by Air = 1.
Air	1	1
Oxygen	1	1
Nitrogen	1	1
Hydrogen	1	1
Chlorine	39	60
Hydrochloric acid	62	—
Carbon monoxide	90	750
,, dioxide	90	972
Nitrous oxide	355	1860
Sulphuretted hydrogen	390	2100
Marsh gas	403	—
Sulphurous acid	710	6480
Olefiant gas	970	6030
Ammonia	1195	5460

The effect of vapours was, in general, very great—in some cases enormously greater than that of the simple gases, and he found also that they followed the same general order as the liquids from which they evaporated.

Tyndall also investigated the emissive powers of gases and vapours, either by directly heating the gas and measuring the radiation from it, or by a very ingenious method suggested by an apparent anomaly noticed in the course of his work. On admitting air into a tube previously containing vapour, he noticed that, apparently, the vapour became much more transparent, as the thermopile indicated a larger amount of heat falling on to it. But he found that this was really due to the heating of the vapour by compression. This suggested that he might do away with the external source of heat, and allow the gas in his tube to be heated by compression. He was able to measure its radiation when so heated. He thus found that radiating and absorbing powers were in the same order. One result obtained by Tyndall is that water-vapour has a very considerable absorption for dark radiations, a result which was contradicted by other experimenters. The various methods, however, by which Tyndall obtained evidence of the absorption, and the latter experiments of Paschen and others, leave little doubt that his experiments warranted his conclusion. It is confirmed by the observations of meteorologists on the greater or slower rate of cooling of the earth at night, when the air contains less or greater amounts of water-vapour. If we accept this conclusion, we can also account for the great variation in the absorption of the solar spectrum noticed by Langley * and others as occurring with changes of weather.

We may notice one very interesting result obtained by Tyndall, that water-vapour was especially opaque to the radiation from a hydrogen flame, while a thickness of about a quarter of an inch of liquid water quite cut off this radiation.

The foregoing account of the results obtained by measurement of quantity without regard to exact quality of radiation, amply confirms the general conclusion that radiating and absorbing powers go together, both being either great or small together. But they do not tell us anything as to the relation to each other of the wave-lengths of radiations emitted and absorbed by the same body; and for the long wave-length radiations the experimental difficulties in the way of such an investigation have only recently been overcome. But certain general conclusions have been arrived at, chiefly by the aid of the Theory of Exchanges, of which we shall now give some account. These conditions have been directly tested and found to hold good with regard to light radiations and, in some few cases of dark radiations, and we cannot doubt that they are true with regard to all radiations.

* *Researches on Solar Heat*, p. 184.

CHAPTER XV.

THEORY OF EXCHANGES.

Theory of Exchanges—Uniform Temperature Enclosures—Full Radiation—Propositions regarding Uniform Temperature Enclosures—Bodies exchanging Radiation at different Temperatures—Bodies in the same Physical State continue to absorb the same kind of Rays independently of Change of Temperature—Radiation of every kind emitted by a Body increases as the Temperature rises—Application to Special Cases.

Theory of Exchanges.—If an enclosure is maintained at a constant temperature, and we put another body—say, a thermometer—inside it, the thermometer, if at a lower temperature, will receive energy from the surrounding walls by radiation, while, if at a higher temperature, it will radiate out energy, till, in either case, it arrives at the temperature of the enclosure, after which everything remains steady. But according to the kinetic theory of matter, the molecules of the enclosure and the body are still in vibration, still in a condition to send out radiation. We are led, therefore, to suppose that the radiation is going on just as before, and that the apparent cessation of transfer, when equilibrium of temperature is reached, is really due to a balance of exchange, the thermometer radiating out just as much as it absorbs. The space between the two bodies is not to be regarded as no longer affected, but as the medium of two equal and opposite streams of radiation. Before the balance is arrived at, the thermometer, if hotter, is sending out more than it receives; if colder, it is receiving more than it sends out. We may suppose, in fact, that there is, in a given transparent medium, (we say "given" since, as we shall see later, the medium has an effect) a total radiation from a body depending on its own temperature alone, and not on the surrounding bodies. When a body ceases to lose or gain energy, it is receiving an amount equal to its own radiation. We imagine a process more complex than that for which we have direct experimental evidence.

Perhaps we may illustrate the nature of the theory by considering the money payments between two commercial firms, each supplying the other with goods. Leaving out of consideration altogether the goods passing, at each settling-day a certain balance would generally have to be paid over by one of the firms to the other. But, examining the books, we should find each transaction entered separately, and the effect is, therefore, the same as if each purchase were paid for separately—as if each firm paid for the whole of the goods it purchased, and as if it received the whole sum due on the goods it supplied. The book-keeping takes account of the separate items, but the money passing only concerns the balance. Now, in the case of radiation, experimentally we only detect the balance paid over, but the theory of exchanges supposes the payments all gone through on each side; it is, as it were, a complete system of book-keeping. Our justification for its use is, that it tells us how the balance to be paid over is made up, and under what conditions the balance will exist. How far the double process is gone through in nature is, perhaps, open to question. In the closely related theory of secondary waves in light, we have a similar "book-keeping" account of the disturbance at any point to which a wave travels. We consider that this is made up of disturbances sent by all the points in the wave-front in some previous position, and we find the actual disturbance to be the resultant or balance of these. But we do not suppose that each secondary wave exists, in the sense that each is sending energy to the point. The supposition of their existence is to keep an account of the process of wave-propagation, and whenever on reckoning up we find a balance of uncompensated disturbance at a point, observation shows us that the disturbance actually exists, and that the calculated amount of energy has been supplied for it.

The theory of exchanges consists of two parts:—

In the first we consider the transactions when bodies are entirely surrounded by bodies at their own temperature, that is, are in uniform temperature enclosures.

In the second we consider the transactions when bodies are exchanging radiations with bodies at a temperature different from their own.

1. Uniform Temperature Enclosures.

We shall not at present consider the effect of the medium on the radiation, but suppose that we have the same medium in all cases and that its refractive index for all radiations is unity.

By experiment we find that a body placed in a uniform temperature enclosure finally comes to the temperature of the enclosure, whatever may be the shape or nature of the walls and whatever the position of the body with respect to them. On the theory of exchanges, therefore, the body is receiving everywhere just as much as it radiates, that is, it is receiving the same total quantity at all points. Not only is it receiving the same quantity, but also the same quality. For, suppose the body able to absorb more of one kind of radiation than of another—to fix our ideas, say more of red than of yellow—and suppose that at A the proportion of red to yellow is greater than at B. Placing the body at A, it will be able to absorb more energy than when placed at B, but it radiates equal amounts at the two points. Hence, if the temperature remains constant when the body is at B, it must rise when the body is at A, which is contrary to experience. The proportion of the two kinds must, therefore, be the same throughout. The radiation must also be the same in quantity and quality in all enclosures of the same temperature, for, removing the body from one enclosure to another, the preceding argument applies.

It is, therefore, independent of the nature of the surface of the enclosure. Suppose part of the surface lampblacked, part covered with polished silver, and part with a plate of rock-salt. The lampblack absorbs nearly all that falls on it, but makes up for this by its great radiation; the silver radiates very little, but then it reflects nearly all that falls on it. The rock-salt neither reflects nor radiates much, but transmits freely most of the radiation from the surface against which it is placed.

There is, therefore, a stream of radiation going in all directions in a constant-temperature enclosure, the amount and quality of the stream depending only on the temperature. Since a lampblacked surface absorbs nearly all the radiation falling on it, it must send out a stream nearly equal to this total radiation, and hence the radiation in a constant-temperature enclosure is sometimes termed the lampblack radiation. But we know that even lampblack does not entirely absorb the radiation falling on it, diffusing and transmitting small fractions. Hence it does not radiate out quite as much radiation as falls on it in the enclosure. It is better, therefore, to term the stream in a constant, uniform temperature enclosure, *the full radiation for the given temperature.*

From the foregoing we may immediately deduce the following results:—

(*a*) *If a body absorbs any kind of radiation it must also emit the same kind of radiation at the same temperature, and if it is placed in a uniform-temperature enclosure the emission equals the absorption.*

This follows at once from the fact that, in order to make the stream of issuing radiation full, it must restore to the incident stream just what it took from it.

(*b*) *A body at a given temperature cannot emit more of a given kind of radiation than exists in the full radiation for that temperature.*

For, if it could, and if it were placed in a constant-temperature enclosure, the issuing stream would contain more of the radiation than the incident stream—or, the stream would not be uniform in character.*

(*c*) *A body opaque to any kind of radiation must emit that kind of radiation.*

For, construct a uniform-temperature enclosure entirely lined with the substance, and let the temperature be such that the given radiation is a constituent of the full radiation at that temperature. The given radiation now existing in the enclosure must have come from the lining substance, for it could not get through from behind. Hence, opacity is always accompanied by the power of emission, and, therefore, of absorption.

To enable us to think a little more definitely of such an enclosure, let us suppose a hollow iron ball, coated inside partly with lampblack, partly with silver, partly with rock-salt. Let it then be heated red hot and kept at a constant temperature. If we can imagine ourselves placed within, but protected in some way from the high temperature without affecting the full radiation, then, on looking round, everything will appear of the same colour. The lampblack radiates out most but it reflects little, the silver radiates little but reflects much. The rock-salt is nearly transparent to the radiation from the iron behind it, and the uncovered iron partly radiates and partly reflects enough to make up the full radiation. We shall, therefore, have one uniform glare, and shall be quite unable to see the boundary lines between the different substances.

* This does not necessarily hold if the body is phosphorescent, or if it is undergoing chemical or molecular change. In these cases molecular energy may be transformed into radiant energy without a corresponding rise of temperature.

If any bodies are suspended in the enclosure, they too will be indistinguishable. A piece of coloured glass will absorb special rays, but will also emit an equal amount of those rays. A plate of tourmaline will polarise the light transmitted through it, but it will emit light polarised in a perpendicular plane, making up the issuing stream to the full radiation without polarisation.

A difficulty in accepting these statements arises, perhaps, from our want of experience of true constant-temperature enclosures. A room without a fire in daylight may possibly be at a nearly constant and uniform temperature, yet our experience shows that it is far from answering to the above description, since we can see the various objects in it quite plainly. But it is to be marked that we see them *not* by their own radiation, but by reflected daylight which is originally sunlight. In fact, the room is not an enclosure. The window being transparent, the sun forms part of the boundary from which radiations are received, and it is at an enormously higher temperature than the rest of the room, while the sky or clouds probably form another part of the boundary far colder than the room. In order, then, to have a uniform-temperature *enclosure*, the radiations received at any point within it must all proceed from bodies at the same temperature. Hence it would be impossible to make glass or rock-salt enclosures, unless, indeed, the thickness were so great that the impurities and the natural absorption rendered the walls opaque to external radiation.

A room darkened by thick shutters, or a cellar without windows, gives a very near approach to a uniform-temperature enclosure, but our eyes are not sensitive to the radiations given out at ordinary temperatures. A very hot coal-fire, however, frequently has cavities forming nearly complete enclosures, into which we may look, and so verify some of the properties stated above. The boundaries between the different pieces of coal are very indistinct. Pieces of glass of different colours put into the cavity soon become nearly indistinguishable, each restoring to the stream of radiation what it takes from it. Similar observations may be made on heating an ordinary clay crucible, or even a clay tobacco-pipe, to red heat and putting small pieces of glass or of any other non-combustible material at the bottom. On looking in, the same indistinctness of outline will be marked.

Effect of the Medium on Radiation.—We have seen that in all enclosures at the same temperature, containing the same medium, the stream of radiation is the same, our proof depending on the assumption that the radiation from an enclosed body is independent of the nature of the radiating enclosure.

But we have no ground for supposing it to be independent of the surrounding medium. A body surrounded by rock-salt may be able to send out more or less energy in a given time than the same body surrounded by air.

Let us now investigate the effect of a change of refractive index on the stream of radiation. The following proof is due to Balfour Stewart (*Treatise on Heat*):—

Let ACA′ (Fig. 139) represent a spherical constant-temperature enclosure, the lower half being filled with a substance of refractive index μ, for some given radiation. Let B be a very small area on the surface of the substance at the centre of the sphere. Let CBD represent a cone of rays of very small angle, with ABA′, the normal to B, as axis, and draw the cone BC′D′ such that

$$\sin ABD = \mu \sin A'BC'$$

or, since the angles are small, $ABD = \mu\, A'BC'$

If B is sufficiently small the refracted part of all the rays reaching B through the cone CBD must pass through the cone C′BD′.

We have an exchange, then, of radiations through the area B, and the total radiation, measured by the energy received per second, sent downwards must equal the total radiation returned upwards, and this will be true for the two corresponding elementary cones CBD, C′BD′.

Now, the quantities of radiation incident on B through the cones will be proportional to the solid angles of the cones, which, since they are small, are proportional to the squares of the angles ABC, A′BC′, that is, are as $\mu^2 : 1$.

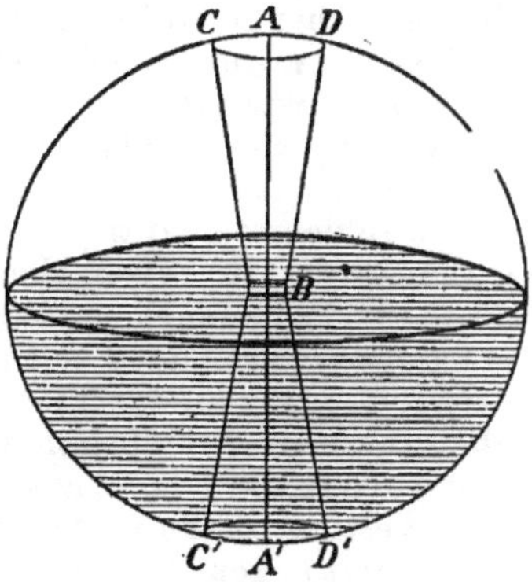

FIG. 139.—Effect of the Medium on Radiation.

If, then, R, R′ represent the amount of the given radiation in the full radiation in each medium, then the radiation incident on B through CBD is to the radiation incident on B through C′BD′ as $\mu^2 R$ is to R′.

But by optical theory and experiment the fraction of the incident light transmitted is the same, whether the radiation falls on the surface from above or below.

Hence, the radiations transmitted into the two cones are also in the ratio $\mu^2 R : R'$.

But these are equal, or

$$R' = \mu^2 R.$$

To illustrate the meaning of this result, suppose that a body is suspended in an enclosure colder than itself, the space between the body and walls being a vacuum—it will radiate out more than it receives, and tend to fall in temperature. If, now, the body is surrounded by a layer of rock-salt of which the refractive index is about 1·5, the stream of radiation from the body and the stream to it will each be multiplied by μ^2. The difference is, therefore, increased in the same proportion, and, therefore, the body will cool more than twice as rapidly, even neglecting the effect of conduction.

We may summarise the foregoing by the statement that in a uniform constant-temperature enclosure there is a definite stream of radiation in all directions depending only on the temperature and the medium, and not on the nature of the radiating surfaces. Further, that any body placed within the enclosure and at the same temperature absorbs and radiates the same kind or kinds of radiation and to the same amount, leaving the radiation issuing from it the same as the full radiation incident upon it.

The principle that bodies absorb radiations of the kind which they emit is perhaps a special case of the general principle of resonance, of which we have examples in sound and mechanics, according to which principle bodies naturally vibrating in a given period are set vibrating by receiving waves of that period from different sources.

2. Bodies exchanging Radiations at different Temperatures.

We can apply what we have learnt from uniform-temperature enclosures to cases of unequal temperature by the aid of two general principles:—

(*a*) *Bodies in the same physical state continue to absorb the same kind of rays independently of change of temperature.*

The experimental evidence in favour of this is that bodies keep the same colour through wide ranges of temperature, that is, they absorb the same constituents of the radiation falling on them. If they change colour, we usually find some evidence of chemical action or of change of molecular aggregation, as when yellow phosphorus changes to red phosphorus, a change which might almost be considered chemical.

(*b*) *The radiation of every kind emitted from a body increases as the temperature rises, as long as the physical state remains the same.*

We may verify this for light rays by heating an opaque body till it becomes incandescent. When first visible, it only gives out the longest red rays in sufficient quantity to be seen and these only in small quantity; but as the temperature rises, the spectrum gradually extends towards the blue till the body becomes white-hot, when it gives out a full spectrum. But as fresh rays are added to the spectrum, those previously existing become stronger—the red, for instance, being much more intense than it was when the body first began to glow. The same holds good for the dark long wave radiations, as has been shown for blackened copper by Langley. He investigated the wave-lengths emitted at various temperatures from 40° C. up to about 800° C., and found that, as the temperature rose, the spectrum became extended towards the shorter wave-lengths, and that the previously existing radiations also became more intense.

In the case of solids and liquids, the spectrum is always more or less continuous, and we may lay down the general rule that, at low temperatures, only the very long wave-lengths are sensible, but that, as the temperature rises, the spectrum gradually extends to the shorter wave-lengths, and ultimately comes into the visible range.

Gases, however, only emit, at least in the visible spectrum, waves of definite length or in definite groups, and their spectra consist of bands or lines. As the temperature rises, these bands or lines become brighter. We may easily show this by comparing the sodium flame in a spirit-lamp with the hotter sodium flame in a Bunsen burner.

Application of the foregoing Principles to Special Cases.—We know that a black surface such as lampblack absorbs nearly all the radiations which fall upon it. If we raise its temperature, it will still continue to absorb. It must, therefore, also emit rays of nearly all kinds, and at a given temperature it will emit more radiation than other bodies, or when incandescent will shine more brightly. In illustration of this, if a piece of platinum-foil with inkmarks on it, which will withstand heat, is heated to incandescence, the inkmarks shine more brightly than the rest of the foil. If a piece of white porcelain has a dark pattern on it, the white porcelain reflects a great part of the light-radiations falling on it, and the dark pattern absorbs most of them. On heating the porcelain to incandescence, the pattern is still the best absorber, and it is, therefore, also the best radiator, and the pattern stands out as bright on a less bright ground.

Blue glass absorbs red more freely than blue rays. If heated to incandescence, it radiates red more freely than blue, and therefore it appears a very bright red. Transparent glass absorbs little, if any, of the light rays, and therefore it radiates poorly, and a piece of colourless glass-tubing heated white-hot only gives out a faint white light as compared with a wire heated to the same temperature.

We have a converse case with sodium vapour in a Bunsen flame, which emits yellow light, its spectrum consisting of the two well-known D lines. It, therefore, also absorbs yellow light of the same wave-lengths. If, then, we place a cooler sodium flame—that of a spirit lamp—in front of a hotter sodium flame given by a Bunsen burner, the cooler flame will absorb from the hotter flame behind it, but will not restore to the outgoing radiation as much as it absorbs. Hence it appears smoky, especially round the edges, where the absorbing vapour is coldest.

This is applied to explain the existence of the dark lines in the spectrum of the sun. The body of the sun is intensely hot, and so dense that it sends out radiations of all kinds, forming a continuous spectrum. But round it is an atmosphere consisting of a mixture of gases and vapours cooler than the interior mass, though they are still exceedingly hot, sufficiently so to send out their own radiations. But they absorb from the stream passing outwards more than they restore, each selecting the kind of radiation it emits. Thus, sodium vapour absorbs the two D wave-lengths, while the light on each side passes on unabsorbed. On examining the solar spectrum, then, these two D lines appear dark, not from absence of sodium light, but because the sodium light is weakened relatively to the neighbouring radiation. We may quote as another example the cases already mentioned of absorption by cold rock-salt of the radiation from hot rock-salt, and of the absorption by water of the radiation from a hydrogen flame.

CHAPTER XVI.

RADIATION AND TEMPERATURE.

Variation of Rate of Radiation with Temperature—Newton's Law of Cooling—Dulong and Petit's Law—Rosetti's Law—Stefan's Law—Constants of Radiation—Radiation from Surfaces which Absorb Selectively—Rate of Solar Radiation—Solar Constant—Pouillet's Pyrheliometer—Violle's Actinometer—Langley's Researches—Crova's Researches—Effective Temperature of the Sun—Source of Solar Energy.

Variation of Rate of Radiation with Temperature.—When a body is in a constant temperature enclosure, at an excess e above the temperature θ of the enclosure, it is radiating out at a rate depending on

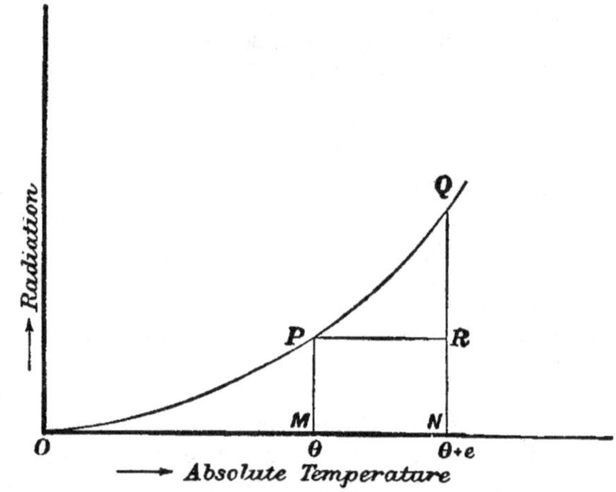

FIG. 140a.

its own temperature $\theta + e$, but if its absorbing power is the same as when it was at θ it is absorbing as much as it would radiate out at θ. For at that temperature it would radiate out as much as it receives. Hence if we represent its total radiation per second at any temperature t, by R_t it is losing energy by radiation at a rate

$$R_{\theta+e} - R_\theta$$

If we could surround the body by an enclosure at the absolute zero the enclosure would probably cease to radiate and we might put $R_\theta = 0$. The rate of cooling would then represent the total radiation per second from the body at the temperature e.

Various attempts have been made to find how the quantity R_e, representing the radiation from a body at absolute temperature e, in an enclosure at the absolute zero, depends upon the temperature. Since we cannot have a zero enclosure it is obvious that we must deduce R from observations of the difference

$$R_{\theta+e} - R_\theta$$

where θ is the temperature of the enclosure or surroundings and e the excess of the radiating body above that temperature. Let us suppose that we have succeeded in determining the form of the curve representing the radiation of a body at different temperatures in a zero enclosure, and that it is OPQ in Fig. 140*a*. Let us draw the ordinate PM at θ and QN at $\theta + e$ and PR parallel to the temperature axis. Then PQ, with axis PR,

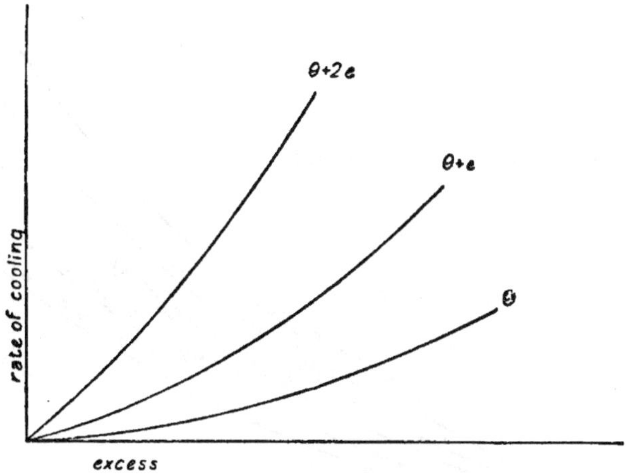

FIG. 140*b*.—Curves of Cooling for Temperatures θ, $\theta + e$, $\theta + 2e$ of the Surroundings.

will represent the cooling curve in an enclosure at temperature θ, for $QR = R_{\theta+e} - R_\theta$. Suppose that we obtain successive cooling curves, say first between θ for the enclosure and $\theta + e$ for the body, then for $\theta + e$ for the enclosure and $\theta + 2e$ for the body, and so on. Let these curves be represented on a single diagram, as on Fig. 140*b*, each curve being marked with the temperature of the enclosure; by putting these curves end to end we can build them up into the radiation curve of Fig. 140*a*.

Newton's Law of Cooling.—Newton * made the first experiments on rate of cooling, and found that the rate was proportional to the excess of temperature above the enclosure. This may easily be verified by raising a thermometer some 20° or 30° above the temperature of the room and observing its fall. Plotting the logarithm of excess of temperature against the time, the result will be found to be a straight line. If e be the excess at time t, and if E be the initial excess, the observations show that

$$\log \frac{e}{E} = -at$$

* *Phil. Trans.*, 1701, p. 828. Newton really investigated the convection effect in a current of air. See Russell, *Phil. Mag.* (6), xx. p. 599, 1910.

Differentiating

$$\frac{1}{e}\frac{de}{dt} = -a$$

or

$$-\frac{de}{dt} = ae$$

that is, the rate of fall is proportional to the excess.

If this law held good for any value of the excess the radiation curve would be a straight line. But it is only true for small excesses, so small

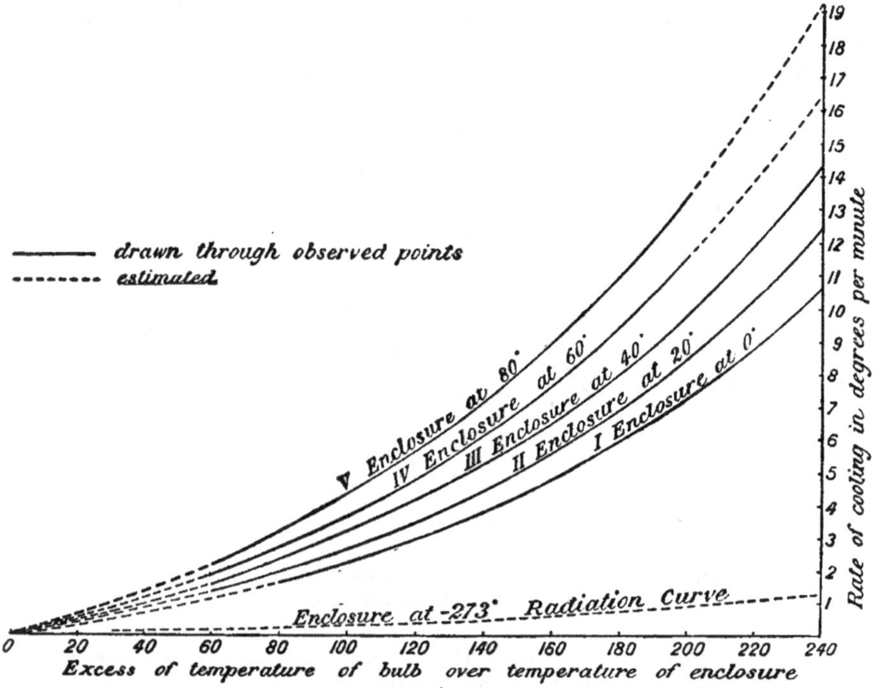

FIG. 141.—Dulong and Petit's Cooling in Vacuo

that we may regard the radiation curve as straight, and consider that we are moving along the tangent. It therefore gives us no information as to the shape of the radiation curve.

Dulong and Petit's Law of Radiation.—Very extensive researches were made on the cooling of thermometers in enclosures by Dulong and Petit about 1817.* Their chief work was carried out by suspending a thermometer raised to any temperature up to 300° C. in a copper globe surrounded by water. The globe was then exhausted to a pressure of 2 or 3 mm. and the rate of cooling of the thermometer was observed. Corrections were applied for the effect of the residual air and for other disturbances, and the results were taken to give the cooling due to radiation. They are represented in Fig. 141 by the curves I. to V., the tem-

* *Ann. de Chemie*, vii., 1817, or Thomson's *Annals of Philosophy*, xiii., 1819.

perature of the enclosure being marked on each curve. Now the heights of these curves are in a constant ratio to each other, each being 1·16 times as high as the preceding. Then if we have obtained the rate of cooling for a given temperature of the enclosure, when we raise that temperature by 20° we must multiply the ordinate of the curve by 1·16, when we raise it 40°, by $1{\cdot}16^2$, when we raise it 60°, by $1{\cdot}16^3$, and so on. Then for a difference of 1° we should multiply by $1{\cdot}16^{\frac{1}{20}}$, or by 1·0077, and for a difference of $n°$ by $1{\cdot}0077^n$. Taking curve I., where the enclosure is at 0° C, if we multiply the ordinates by $1{\cdot}0077^{-273} = {\cdot}123$ we shall obtain the curve of cooling when the enclosure is at 0° absolute, and this is the radiation curve on the assumption that the results obtained between 0° C. and 300° C. allow us to extrapolate.

We have already shown that the cooling curves are successive pieces of the radiation curve, and if we break the radiation curve up into successive 1° steps, beginning at the absolute zero, as in Fig. 142, the height

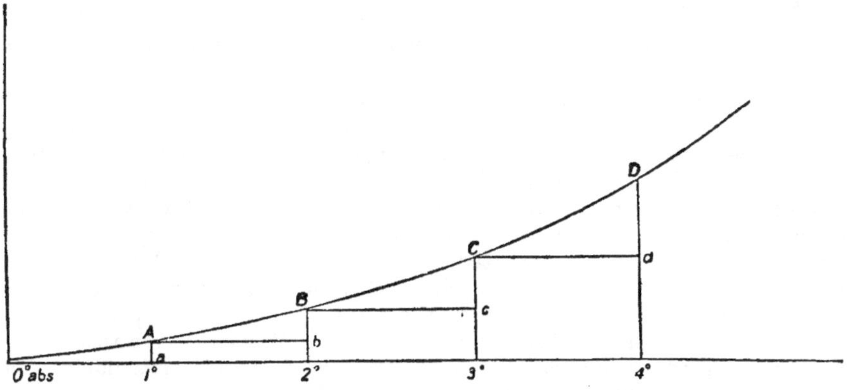

FIG. 142.—Radiation Curve broken up into Cooling Curves.

of each step above the last is, by Dulong and Petit's result, 1·0077 times the preceding height, or $Bb = 1{\cdot}0077\ Aa$; $Cc = 1{\cdot}0077\ Bb$, and so on. But these 1° steps up are each $\frac{dR}{d\theta}$ for the radiation curve so that $\frac{dR}{d\theta}$ is multiplied by the factor 1·0077 for each step of 1°, and at θ we shall have

$$\frac{dR}{d\theta} = 1{\cdot}0077^{\theta} \cdot Aa$$

where Aa is the height of the first step. Integrating we get

$$R = m(1{\cdot}0077^{\theta} - 1)$$

where m is a constant for the radiating body.

For a long time this result was supposed to represent the radiation curve, at least within the range of the experiments, but there is no doubt that the residual gas played some part not eliminated, and that the radiation is not truly represented by the formula.

Rosetti's Law.—Rosetti* found that with a Leslie cube or an iron vessel filled with mercury, but not surrounded by a lampblack enclosure, the cooling between 300° C. and 0° C. was much more nearly represented by

$$aT^2(T-\theta)+b(T-\theta)$$

when T is the absolute temperature of the body, θ that of the enclosure, and a and b are constants for the body.

Stefan's Law.—In 1879 Stefan † gave an entirely new turn to the researches on radiation by the suggestion that the rate of radiation is proportional to the fourth power of the absolute temperature. He was led to this by observing that a result of Tyndall's accorded with it. Tyndall found that the radiation from a platinum wire at 1200° C. was 11·7 times its radiation at 525° C. Now $\left(\frac{1200+273}{525+273}\right)^4$ is very nearly 11·7, or the two radiations were almost exactly in proportion to the fourth powers of the absolute temperatures. Re-examining the work of Dulong and Petit and their successors, Stefan found that on taking conduction by the surrounding gas into account, the radiation was more in accordance with the fourth-power law than with the formulæ previously used. In 1875 Bartoli applied thermodynamic reasoning to radiation, and in 1884 Boltzmann followed up Bartoli's work. Starting with the supposition, now proved,‡ that radiation presses against any surface on which it falls, Boltzmann treated the radiation in an enclosure as the working substance in a Carnot cycle, and showed that the full radiation at any temperature must, if the method is justifiable, be proportional to the fourth power of the absolute temperature. We shall give an account of this thermodynamic treatment in chap. xx Here it is sufficient to say that the subject has been developed by Wien § and others, and that Wien has shown that if a body is emitting full radiation then the wave-length of maximum radiation is given by

$$\lambda_m\theta=\text{constant},$$

where θ is the absolute temperature of the surface, and that the energy radiated through a small range $d\lambda$ at this maximum is

$$E_m d\lambda=\text{constant}\times\theta^5.$$

Wien on certain assumptions deduced a formula to express the energy radiated for every wave-length. But the results of experiment did not confirm Wien's formula so well as a modification obtained by Planck.||

* *Phil. Mag.*, viii., 1879.

† *Wien. Akad. Ber.*, 1879, lxxix, pp. 391–428.

‡ Lebedew, *Congrès International de Physique*, ii. p. 133.

§ Wien, *Congrès International de Physique*, ii. p. 23; Larmor, "Radiation," *Encyclopædia Britannica*, xxxiii., 10th ed.

|| *Deutsch. Phys. Gesell. Verh.*, **2**, xiii. p. 202; **2**, xvii., p. 237, 1900.

Wien's formula for radiation in the small range $d\lambda$ is—

$$E_\lambda d\lambda = \frac{C_1}{\lambda^5} e^{\frac{-C_2}{\lambda\theta}} d\lambda$$

Planck's modification is—

$$E_\lambda d\lambda = \frac{C_1}{\lambda^5} \frac{1}{e^{\frac{C_2}{\lambda\theta}} - 1} d\lambda$$

where C_1 and C_2 are constants. These last formulæ must for the present be regarded as of less weight than those for λ_m and E_m, as they are obtained by the aid of special molecular hypotheses.

Since the enunciation of Stefan's law and the foundation of the thermodynamics of radiation, much work has been done to test the formulæ obtained by the theory.* Lummer and Pringsheim used as radiating source a constant-temperature enclosure with a small hole in the side, the enclosure being of different materials according to the temperature. The radiation emerging from the hole was practically the full radiation for the temperature of the enclosure. For imagine that a stopper at the temperature of the enclosure is put into the hole, the full radiation strikes on the stopper; but it only exceeds that passing out through the hole by the amount emitted by the stopper and reflected by the sides of the enclosure back to the stopper again, and this is negligible if the hole is small enough.

The issuing radiation was measured by a bolometer, and the fourth-power law was very exactly verified. Using a fluorspar prism, the radiation was dispersed, and a bolometer travelling along the spectrum gave the position of maximum energy and the amount of that energy. The formulæ $\lambda_m\theta = \text{constant}$ and $E_m = \text{constant} \times \theta^5$ were verified.

In subsequent work † Planck's formula has been shown to accord very closely with observations on the energy in different parts of the spectrum—more closely than that of Wien. But it should be noted that C_2 is in Wien's formula $5\lambda_m\theta$, where λ_m is the wave-length for which E is a maximum and in Planck's very nearly the same. Hence we have practically $e^{\frac{5\lambda_m}{\lambda}}$ in the denominator, and this is great compared with 1, so long as λ is less than λ_m. The two formulæ nearly agree, therefore, on the shorter wave-length side of the maximum.

The confirmation of these formulæ justifies their use to determine the temperature of bodies emitting full radiation. Thus Lummer and Pringsheim ‡ have determined the temperature of a uniform-temperature enclosure (1) by measuring the total radiation emitted, (2) by measuring the wave-length of maximum energy, (3) by comparing the brightness of a given part of the spectrum with the brightness of the same part of a spectrum emitted by a surface at a lower known temperature and using Planck's formula. All three methods gave the temperature as 2325° absolute within 20°. The constants in the formula had been previously determined.

* An account of the subject will be found in the *Congrès International de Physique*, ii. p. 41, by Lummer, who has been one of the chief workers.

† Paschen, *Ann. d. Physik*, 4–2, Feb. 1901, pp. 277–298.

‡ *Berichte der Deut. Phys. Ges.*, 1903, p. 3.

Constants of Radiation.—These constants may be taken as having nearly the following values* :—

If R_θ is the energy emitted per second per square cm. from a full radiator at temperature θ on the absolute scale,

$$R_\theta = \sigma\theta^4$$

where $\sigma = 5 \cdot 32 \times 10^{-5}$ ergs per second per sq. cm., or $5 \cdot 32 \times 10^{-12}$ watts per sq. cm.

In $$\lambda_m \theta = A$$

where λ_m is the wave-length of maximum energy at θ expressed in terms of $\mu = 10^{-6}$ metre as unit

$$A = 2940.$$

In Wien's formula $$E = \frac{C_1}{\lambda^5} e^{-\frac{C_2}{\lambda\theta}}$$

$$C_2 = 5A = 14700.\dagger$$

C_1 is of the order 1000 for $\lambda = 1\mu$.

The Radiation from Surfaces which Absorb Selectively.—The foregoing results do not apply to surfaces which absorb different proportions of different wave-lengths. Even if a surface always absorbed the same proportion of the same wave-length it need not absorb the same fraction of the full radiation at different temperatures. Suppose, for instance, that I., Fig. 143 represents the full radiation for one temperature, while II. represents it for another. Suppose that at L a certain surface has an absorption band, L being the wave-length of maximum intensity of II., and suppose this the only absorption band, then in II. a certain amount at the maximum is absorbed, while in I. it is not at the maximum and is therefore not such a large fraction of the whole.

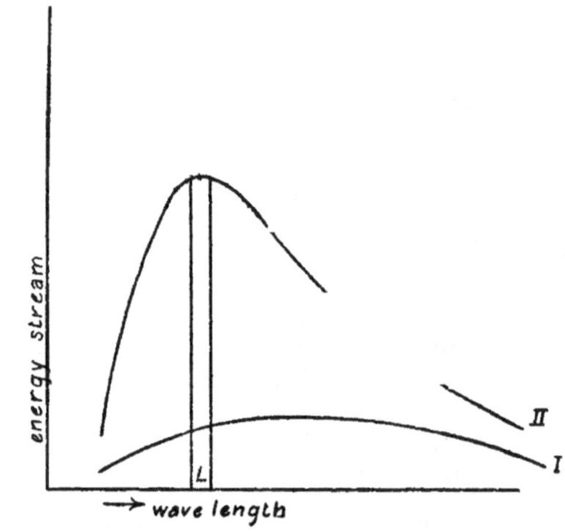

FIG. 143.—Radiation Curves for Spectrum at Two Different Temperatures.

But we cannot assume that a surface does always absorb the same proportion of the same wave-length. It appears possible that the fraction increases in general as the temperature rises, for exceedingly hot surfaces appear to approach fulness in the quality of their radiation even though at low temperatures they are low radiators and only give a small fraction of full radiation.

* *Congrès International de Physique:* Lummer, *loc. cit.* The value of σ is due to Kurlbaum; *Wied. Ann.*, lxv., 1898, p. 748. A simple experiment giving an approximate value of σ is described by Denning: *Phil. Mag.*, x., 1905, p. 270. Uncertainty arises from the fact that the receiving surface, though black, is not a full absorber. See Kurlbaum, *Wied. Ann.*, lxvii. p. 846, 1899.

† Holborn and Valentiner (*Ann. d. Physik*, xxii., 1, 1906, p. 1) give 14200.

So far no general formulæ have been obtained for such surfaces, though special cases, such as bright and black platinum, have been studied, and formulæ have been devised to suit the results obtained.

The Rate of Solar Radiation.

The Solar Constant.—One of the most interesting problems in solar physics consists in the determination of the rate at which the sun is radiating energy. The rate is expressed in terms of the amount falling on 1 sq. cm. at the distance of the earth but outside the earth's atmosphere. When this is stated in calories per minute it is called the solar constant. If we multiply by (radius of earth's orbit/radius of sun's

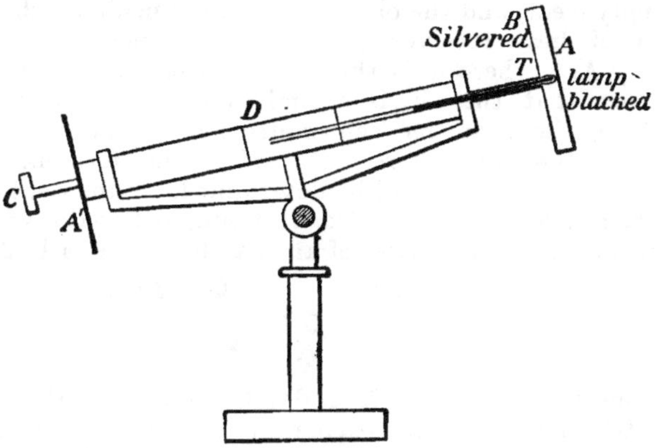

FIG. 144.—Pouillet's Pyrheliometer.

radiating surface)^2, *i.e.* by $(92400000/430000)^2 = 46000$ we get the amount passing out from each square centimetre of the sun.

Pouillet's Pyrheliometer.—The first successful research on the solar constant was made by Pouillet by means of the pyrheliometer. This consists of a flat, thin, cylindrical vessel A, Fig. 144, of which the upper face is lampblacked and the rest is silvered. This contains water and serves as a calorimeter. It is mounted at one end of an axis D round which it can be rotated to secure mixing of the contents, and the thermometer stem lies along this axis, so that the temperature of the calorimeter can be read easily.

The instrument is movable about another axis on the stand, so that it can be pointed in any direction, and a disc A′ of the same diameter as A, is fixed near the other end of the axis D, and this is exactly covered by the shadow of A when D is directed towards the sun.

Pouillet's method of observation was as follows. He first directed A to a part of the sky away from the sun, protected the instrument from sunlight, and noted its fall of temperature, say θ_1, during five minutes. He then directed it towards the sun for five minutes, and noted the rise, θ. Finally he again directed it away from the sun, and noted the fall θ_2 during five minutes. He assumed that the mean cooling during the time of exposure to the sun was $\frac{\theta_1+\theta_2}{2}$, so that, had all the heat received from the sun been retained, the rise would have been $\theta+\frac{\theta_1+\theta_2}{2}$

Knowing the capacity of the calorimeter, the total heat received per unit of surface per unit of time could be measured. This, of course, only gave the radiation reaching the earth after much had been absorbed by the atmosphere. That this absorption is very considerable is sufficiently shown by the fact that we receive sensibly more heat from the sun as he rises higher, and thus diminishes the thickness of air through which his rays must pass.

To determine the absorption by the air, Pouillet made observations when the sun was at different heights, and found that as long as the air was exceedingly clear, and the observations were made on the same day, the quantity of heat received per centimetre per second might be represented by $A\alpha^d$, where d is the thickness of air passed through on the supposition that the air is a limited ocean of uniform density. A is, therefore, the heat per centimetre per second which would be received outside the atmosphere. By aid of this formula and the experimental results, A and α could be determined. Suppose, for instance, that observations are made with a vertical sun, and again when it is 60° from the zenith. The thicknesses of air may be taken as 1 : 2.

Hence, $$Q_1 = A\alpha^d \qquad Q_2 = A\alpha^{2d},$$

Therefore, $$\frac{Q_1^2}{Q_2} = A$$

It was found that α changed from day to day, and with the clearness and hygrometric state of the atmosphere, but the mean absorption of the radiation from a vertical sun was, according to Pouillet, about 20 per cent.

He obtained 0·02939 calories per second per sq. cm., for the value of A, or 1·7634 calories per minute per sq. cm.

Multiplying by 46000, the radiation from a square cm. of the sun would be about 1350 calories per second, or $1350 \times 4{\cdot}2 = 5650$ watts $= 7{\cdot}6$ horse-power. This gives about 7000 horse-power per square foot. Later observers have obtained values somewhat higher than that of Pouillet. His calorimeter was probably somewhat too slow in responding to the heating effect of the sun. The outer surface being at a higher temperature than the water as registered by the thermometer, the radiation outwards was most likely greater than that allowed for.

Again, he assumed that the air absorbed all rays in the same proportion, whereas it absorbs or diffuses some much more than others. For instance, the blue colour of the sky shows that there is greater diffusion of the shorter wave-lengths, and the spectroscope gives us clear evidence of marked "selective absorption." As Langley has shown, Pouillet's supposition that the radiation may be represented by $A\alpha^d$ would always lead to an underestimate of the solar constant.* This may be seen by considering a simple case. Imagine that the radiation outside the atmosphere consists of two kinds, respectively A and B in quantity. Let A

* *American Journal of Science*, September 1884; *Phil. Mag.* xviii., 1884.

pass through unchanged, but let B be reduced by one-half in passing vertically through the atmosphere. Then with the sun in the zenith we observe $A+\frac{B}{2}$. If we observe the radiation when the sun is 60° from the zenith we obtain $A+\frac{B}{4}$. Pouillet's rule would give the outside radiation as

$$\frac{\left(A+\frac{B}{2}\right)^2}{A+\frac{B}{4}}=A+\frac{3B}{4}+\frac{\frac{B}{4}}{1+\frac{4A}{B}}$$

which is necessarily less than $A+B$.

Again, if any ray were totally absorbed by the atmosphere even with vertical sun, Pouillet's method would take no account of it.

Violle's Actinometer.—Violle used an instrument which lessens the fault of Pouillet's Pyrheliometer by diminishing the lag behind the true temperature. A thermometer *t*, Fig. 145, is placed in the middle of a double-walled enclosure AB, the space CC between the walls being filled with melting ice or with water at a constant temperature. An opening provided with a shutter allows the sun's rays to fall on the thermometer bulb. The shutter being closed the thermometer comes to the temperature of the enclosure, which is noted. The shutter is then opened and in about fifteen minutes a stationary temperature is reached when the heat from the sun balances that lost to the enclosure. To find the latter, the shutter is closed and the initial rate of cooling of the thermometer is noted. The water equivalent of the thermometer being known, we have the heat gained from the sun. Violle found for A the value ·04233 calories per second per sq. cm. or 2·5398 calories per minute.*

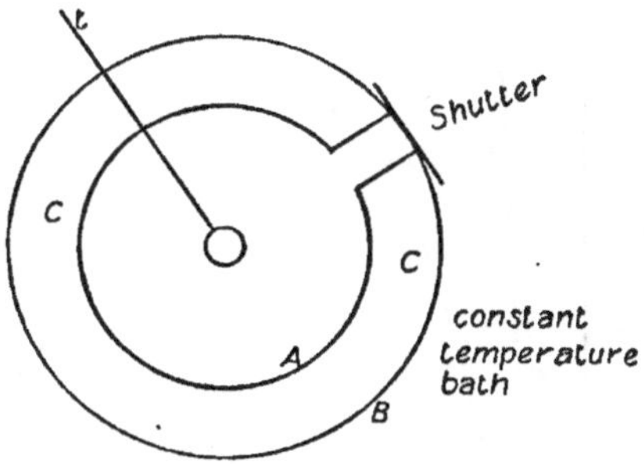

FIG. 145.—Violle's Actinometer.

Langley's Researches with the Bolometer.—Langley was the first to take selective absorption by the atmosphere thoroughly into account. In his researches on Mount Whitney † he sought to measure the absorption of each part of the spectrum separately, using a diffraction grating to separate out the various wave-lengths, and a bolometer to compare the heating effect with different thicknesses of atmosphere passed through. His results for wave-lengths between $\cdot 4\mu$ and $1\cdot 0\mu$ are given in Fig. 146, μ being one-millionth of a metre.

Curve I. is for a given position of the sun at noon, termed "high sun," II. for a position—"low sun"—in which twice the thickness of atmosphere is passed through. Curve III. is obtained by drawing ordinates bearing the same ratio to those of I. that the ordinates of I.

* Violle's researches are described in various papers in the *Comptes Rendus*.
† *Researches on Solar Heat* (Washington 1884): *Phil. Mag.*, xv., 1883, p. 153.

bear to II., and therefore III. is the curve he would have obtained had he been outside the atmosphere altogether. The general absorption steadily increases as the wave-length diminishes, though there is marked absorption of particular rays in the infra-red. This can easily be seen by inspection of the ratios of the ordinates of I. and II. at H with those at A.

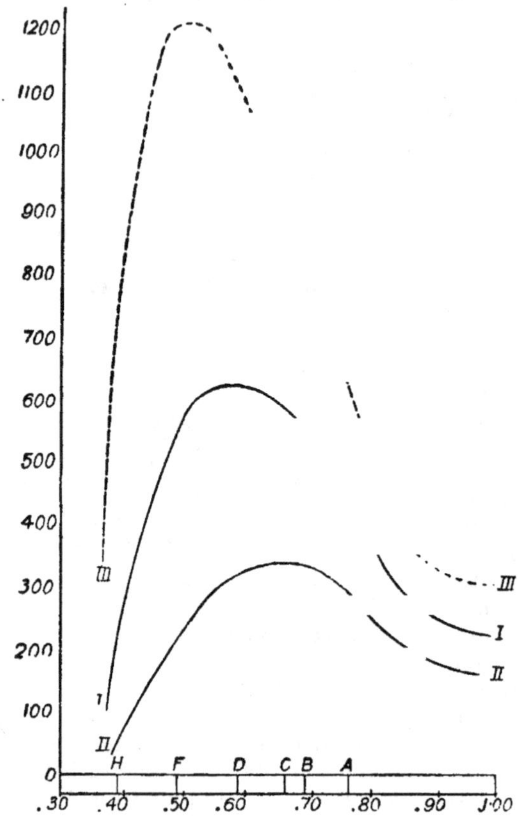

FIG. 146.—Langley's Curves of Solar Radiation.

I. High sun, through one atmosphere.
II. Low sun, *i.e.* through two atmospheres.
III. Estimated curve outside atmosphere.

Curve III., therefore, contains a larger proportion of the blue rays than I. or II., and Langley concludes that outside the atmosphere the sun would appear of a bluish tinge. Continuing the curves beyond 1·0 as far as he was able, and comparing the areas of curves I. and II., he found

$$\frac{\text{Area high sun}}{\text{Area low sun}} = 1{\cdot}57$$

Hence also area of curve III. = 1·57 area of curve I. Now by using Violle's actinometer, he obtained as the total radiation received per square centimetre per minute at high sun on clear days, 1·81 calories.

Then outside the atmosphere this would be increased by the fraction 1·57, giving 2·84 calories.

Langley believed that even 2·84 was an underestimate and put the solar constant as at least 3.

Crova's Researches.—Crova,* who devised and carried on a series of researches with an actinometer, put the value of the constant still higher, believing that it is probably as high as 4 calories per minute per square centimetre.

Recent work tends to show that these values are too large. Abbott and Fowle,† using the Bolometer, found almost the same value at Washington and at Mount Wilson, 1800 metres higher, just below 2·1 calories per minute per square centimetre. Millochau,‡ working with the Féry pyrometer on the summit of Mont Blanc and at different levels below,

* *Cong. Int. de Phys.*, iii. p. 453, contains a discussion of the value of the solar constant by M. Crova.

† *Terrestrial Magnetism*, xiii. p. 79, 1908; *Ast. Phys. Journ.*, xxix. p. 281, 1909.

‡ *Journ. de Physique*, viii. p. 347, 1909.

found a value about 2·4, and Bellia * found from his own and previous work a value 2·1.

There are indications that the so-called constant is not constant, but varies with the condition of the sun. It would be better if some such title as "Solar Radiation Stream" were adopted to replace "Solar Constant."

Taking 2·1 as the value, it is equivalent to $0{\cdot}147 \times 10^7$ ergs per second or 0·147 watts.

Multiplying by 46000, the radiation from a square centimetre at the surface of the sun is 6800 watts, about 9 horse-power.

The Effective Temperature of the Radiating Surface of the Sun.—We receive radiation, no doubt, from different layers of the sun, and these are at different temperatures. We cannot, therefore, accurately say that the radiating surface has one definite temperature. But we may find the one definite temperature of a fully radiating surface which is giving off energy at the same rate as the sun, and this is defined to be the "effective temperature" of the sun.

Before there was any approximation to a correct law of radiation, a temperature was assigned to the sun which we now know to be immensely exaggerated. Thus Watterston, using Newton's law, according to which the temperature is proportional to the radiation, found a value 7,000,000° C.

Violle, on the other hand, found far too low a temperature by using the law of Dulong and Petit. He obtained an effective temperature of 1500°, and considered that the actual temperature might be 3000°.

The first determination on what we may describe as modern methods was made by Rosetti.† He measured the heat received by a thermopile from a lamp-black surface at various temperatures up to 300° C., then from a copper ball suspended in a flame, the temperature being determined by a specific heat experiment. By a subsidiary experiment he determined the ratio of emission of bare copper to that of lampblacked copper, and ultimately arrived at the law of radiation already quoted (p. 248). He then observed the radiation received from the sun and, correcting for the absorption by the atmosphere, he obtained for the sun's effective temperature a value about 10,000°.

Another important determination was made by Wilson and Gray.‡ They compared the radiation from the sun with the radiation received from a platinum strip which they could raise by an electric current to any desired temperature, the radiation being measured by Boys's radio-micrometer. That of the sun was reflected into the instrument by a mirror of known reflecting power. They first found that the radiation from the platinum increased according to Stefan's fourth-power law, up to the highest temperature which they could measure. Then they used a result found by Rosetti, that at high temperatures the emission from bright platinum is 1/2·9 of that from lampblacked platinum, and so they calculated the temperature of lampblacked platinum at which it would give off the same energy per sq. cm. as the sun, and they found for this a value varying from 6200° C. to 7400° C., according to the value assigned to the absorption by the earth's atmosphere.

In a later paper Wilson § described a modification of the experiment,

* *Science Abst.*, xiii., No 660.

† *Phil. Mag.*, viii., 1879.

‡ *Phil. Trans.*, A., 1894, p. 361.

§ *Proc. R.S.*, lxix., 1901–2, p. 312.

in which the radiation from the sun was compared with that from a hole in the wall of a constant-temperature enclosure. The enclosure was an iron and porcelain tube heated in a gas furnace. Its temperature was given by a Callendar platinum thermometer. One end of the tube was open, and in front of the opening was a rectangular area of variable width, and this was arranged so that the radiation from the sun was equal to that from the enclosure.

If p is the fraction of the sun's rays passing through the atmosphere, and if q is the fraction reflected by the mirror, we may put

$$pq\theta^4 \times \text{solid angle subtended by the sun at the receiving surface}$$
$$= (T^4 - T_0^4) \times \text{solid angle subtended by the rectangular aperture.}$$

where θ is the effective temperature of the sun,
T the temperature of the enclosure,
and T_0 that of the receiving surface,

T_0^4 was negligible in comparison with T^4, since T was of the order of 1000°. Then we may put

$$\theta^4 = \frac{T^4}{pq} \; \frac{\text{solid angle of aperture}}{\text{solid angle of sun}}$$

The value of θ depends on the value assigned to p. If, with Rosetti, it is put at 0·71, the effective temperature of the sun comes out as 5768° absolute. If, with Langley, it is put at 0·59, the effective temperature comes out as 6085° absolute.

It is interesting to compare with this the value given by the solar constant and the constant of radiation determined by Kurlbaum (p. 250). With constant 2·1, the radiation from the sun is 6800 watts per sq. cm. The radiation constant, according to Kurlbaum, is 532×10^{-14} watts.

If the effective temperature of the sun is θ, then

$$5 \times 3210^{-14}\theta^4 = 6800$$

whence $\theta = 5980°$ nearly.

If we adopt Langley's value of the solar constant, which is 3, we get

$$\theta = 6500° \text{ about,}$$

slightly above the value obtained by Wilson.

Another value may be obtained from the equation (p. 250)

$$\lambda_m\theta = \text{constant} = 2940$$

in conjunction with Langley's value for λ_m, viz.,

$$\lambda_m = 0{\cdot}5\mu.$$

This gives $\theta = 5880°$ absolute.

But this should hardly be taken as another mode of determining the sun's temperature, since the equation $\lambda_m\theta = 2940$ applies only to a surface emitting full radiation. The close agreement of the result it gives should rather be taken as indicating that the sun is nearly a full radiator, when we consider its radiation as a whole.

Whatever the temperature of the solar surface, the interior on the average is doubtless much hotter. If, as we are entitled to suppose from the high temperature, the interior of the sun is in a mobile condition, the tendency will be towards a condition of "convective equilibrium," the fall of temperature outwards from the centre being such that a mass moving outwards and expanding as it moves will cool, owing to diminished pressure through expansion, at such a rate that it is in equilibrium with its surroundings.

If the temperature falls as we go outwards less rapidly than this, a mass raised up from the inside will be colder than its surroundings, and therefore denser, and will tend to sink, or the equilibrium is stable. But if the temperature falls more rapidly, a mass raised up from the inside will be warmer than its surroundings, and will be less dense, and will, therefore, continue to rise, or if equilibrium exists it will be unstable. Hence, as the cooling goes on from the outside, there is a limit to the slope of temperature according to this convection law. As soon as the slope is greater than this limit, the arrangement is unstable, and circulation takes place, cooler material from the outside sinking down and being replaced by hotter material from the inside, until the slope is equal to or less than that of convective equilibrium. We probably see such disturbances taking place in sunspots, which appear to be due to the increased absorption of the colder layers thickened by the disturbance.

Source of Solar Energy.—The enormous rate of loss of energy by the sun—a radiation of probably 10,000 to 15,000 h.p. per square foot—requires some explanation. We cannot suppose that the supply of energy is merely yielded by the cooling of the body of the sun, for at such a rate the sun would have cooled very appreciably within historic times, and of this there is no evidence. There must be, therefore, some other form of energy continually being converted into heat to keep the temperature up. This cannot be chemical energy, for not only is the sun above the temperature at which we believe chemical combination to be possible, but even had it been composed originally of separate elements giving out the greatest known amount of energy by chemical combination, the energy would last but a few thousand years. One very probable source was pointed out by Helmholtz, who suggested that the sun's heat is the equivalent of the potential energy of the matter of the sun gradually converted into kinetic energy, and then into heat, as the body of the sun contracts.

As the sun cools slightly, the matter comes nearer together. The potential energy thus disappearing is converted into heat, which not only keeps the temperature up nearly to its original value, but yields an enormous surplus which is radiated out.

If the sun were of uniform density throughout, and if as it contracted every part were of the same density, a contraction of the radius of the order of 200 feet or 6000 cm. per annum would yield potential energy mechanically equivalent to that radiated out. This would not amount to more than one second of arc in 6000 years, so that we can hardly expect as yet to obtain direct evidence for the contraction.

It appears possible, too, that the solar energy is in part maintained by the disintegration of radio-active bodies present in the sun. This was suggested by Rutherford and Soddy,* and Rutherford states that 2·5 parts of radium per million by weight in the sun would account for its present rate of emission of energy.

* Rutherford, *Radio-activity*, chap. x.

CHAPTER XVII.

THERMODYNAMICS.

The Second Law of Thermodynamics—The Indicator Diagram—Isothermals—Adiabatics or Isentropics—Heat Engines—Carnot's Reversible Heat Engine—Carnot's Cycle—Conditions for Reversible Working—Examples of Reversible Processes—Of Irreversible Processes—Efficiency of an Engine—Absolute or Work Scale of Temperature—Efficiency expressed on the Absolute Scale—Comparison of the Absolute with the Air Scale—Water-wheel Analogue—Reversible Cycles in general—Entropy—Entropy-Temperature Diagram—Quantities Analogous to Entropy—Entropy tends to increase—Dissipation of Energy—Intrinsic Energy—Available Energy—Possible efficiency of a Steam Engine.

We shall now investigate the conditions under which the transformation from heat into other forms of energy occurs, and the extent of the transformation under given conditions. We shall make use of a general principle, which we may conveniently state here, known as—

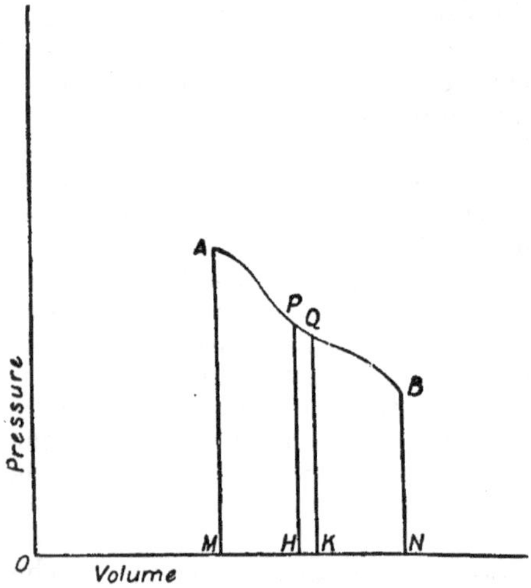

FIG. 147.—Indicator Diagram.

The Second Law of Thermodynamics.—*We cannot transform heat into work merely by cooling a body already at the lowest available temperature.*

The significance of this statement will be appreciated better when we apply the law to reversible heat engines. Being a negative statement, it is hardly founded on direct experiment, but rather on the general observation that when we seek to transform heat to work we usually derive the heat from a hotter body, we transform some of it, and are obliged to give some of it to a colder body, actually heating that colder body. Any apparent exception will, on examination, be found to involve something beyond mere cooling of the already coldest body. For instance, if a quantity of air be under great pressure it may be allowed to do a considerable amount of work in expanding, and its temperature will fall very appreciably, perhaps far below that of the surroundings. But it is not *merely* cooled. It has not given up its heat by conduction. It has changed its volume while doing the work. If the air could be put through any process by which it returned finally to the same volume, but was appreciably colder, and if it had meanwhile produced a balance of work, then the second law would be untrue.*

The application of this principle will enable us to find the maximum fraction of heat which under given conditions can be transformed to work, and will further give us the temperature changes occurring when bodies undergo known strains or are subjected to known stresses. Incidentally, it will give us a scale of temperature which is definite and quite independent of any particular substance—Lord Kelvin's work scale.

As our investigations will be carried out chiefly by the aid of the Indicator Diagram, we shall first give some account of that diagram as a mode of setting forth the conditions of a body and the relations between its temperature, pressure, and volume.

The Indicator Diagram.—In this diagram two axes at right angles are chosen as volume and pressure axes respectively. A point on the diagram represents by its ordinate the pressure of the body, supposed to be uniform throughout, and by its abscissa the volume. Thus in Fig. 147 A represents the condition of a body which has volume OM and pressure AM.

If any change occurs in the body the successive conditions may be represented as regards pressure and volume by a series of points which will together form a line either straight or curved, joining the points representing the initial and final conditions, as APQB in Fig. 147.

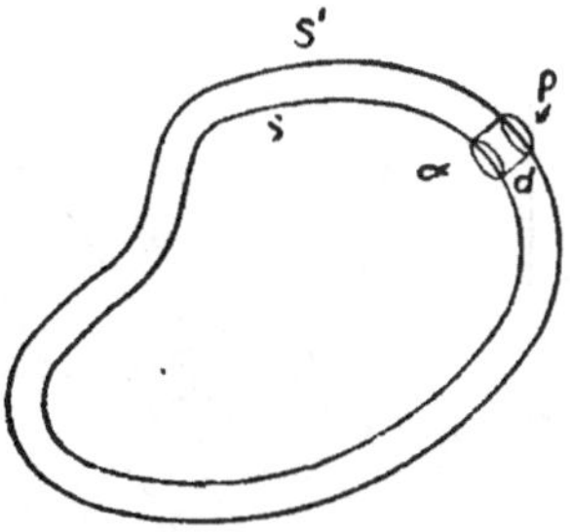

FIG. 148.—Work done by a Body in Expansion.

We may represent the work done by or on a body during a change of volume by an area on the diagram. Let S (Fig. 148) represent the surface of the body when in the condition represented by P on the diagram, and let S′ represent the surface when the body is in the condition Q very near P, so that S′ is only very slightly larger than S. Let a be a small area on S, which has moved against the pressure through a distance d.

If p and p' are the initial and final pressures, the change is so small that we may suppose the pressure to vary uniformly from p to p'. Then the mean pressure is $\frac{1}{2}(p+p')$, and the work done is $\frac{1}{2}(p+p')ad$. Summing up all over the surface, the total work

$$=\frac{p+p'}{2}\Sigma\, ad.$$

$$=\frac{p+p'}{2}\times \text{ volume between S and S}'.$$

$$=\frac{p+p'}{2}\times \text{ increase in volume.}$$

* A statement less general but sufficient for our purpose in Thermodynamics would be: We cannot by a cyclical engine derive work continuously by conducting heat into the engine from the coldest part of the surroundings.

But $$\frac{p+p'}{2}=\frac{PH+QK}{2} \text{ in Fig. 147,}$$

and increase in volume $= HK$

Therefore, total work $$=\frac{PH+QK}{2}\times HK$$

$=$ area of slip PK.

The total work from A to B is evidently the sum of all such slips, or the area AMNB.

This work represents the energy given out by the body merely through expansion against the outside pressure. If the body contracts, moving, say, from B to A, the same area will represent the work done on the body, or the energy given into it through contraction under outside pressure. If the point representing the condition of the body moves from A to B and back again along the same course, evidently, on the whole, no external work is done, the two quantities being equal and opposite.

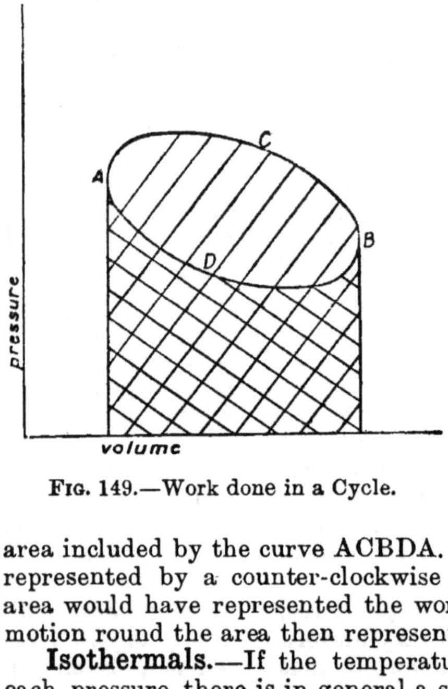

FIG. 149.—Work done in a Cycle.

But if the return course is not the same, we may have a balance left over.

Suppose that the body goes along ACB (Fig. 149), returning along BDA. Marking work done *by* by shading thus ///, and work done *on* by shading thus \\\, where there is a cross shading the two neutralise each other, and the balance of work done by the body is represented by the area included by the curve ACBDA. Had the change of condition been represented by a counter-clockwise motion round the curve, the same area would have represented the work done on the body. A clockwise motion round the area then represents the giving out of so much energy.

Isothermals.—If the temperature of a body is kept constant, for each pressure there is in general a single definite value for the volume, and the relation between volume and pressure will be represented by a curve. For instance, in the case of a gas $p\times v$ is nearly constant for a given temperature, and the curve will be nearly an equilateral hyperbola. Such a curve is termed an isothermal. We may draw a series of isothermals, each corresponding to a different temperature. The reader can easily plot the isothermals for a gas at 0° C., 100° C., and 200° C on the assumption that $pv = RT$ when T is the temperature on the gas scale and R is a constant for a given gas.

The isothermals for liquids and solids are in general nearly parallel to the pressure axis. Thus for water an isothermal as it rises will only approach the pressure axis by $\frac{1}{22000}$ of the volume for an increase of 1 atmosphere, so that it is impossible to represent it to scale on an ordinary diagram.

Adiabatics or Isentropics.—If a body is not allowed to gain or lose heat by conduction, and if the volume and pressure alter, but in such a way that no kinetic energy is acquired by the body or its parts, the change is said to be *adiabatic*. When, for example, a mass of gas is enclosed in a non-conducting cylinder under a loaded piston, a very gradual change in the load will produce an adiabatic change in the gas. But a sudden change, such as a sudden finite decrease of load, will result in rapid motion of the piston and rapid motion of the gas. Some of the energy of this internal motion of the gas will be converted into heat by viscosity, and the body will thus receive heat, though from itself. In the celebrated experiment of Gay-Lussac and Joule on the expansion of a gas when no external work is done (p. 120), no heat is given to or taken from the outside, but the gas acquires considerable kinetic energy which is ultimately converted into heat, each element receiving heat from the surrounding elements through the viscosity. Hence the change is not adiabatic. As we shall show later, the temperature would have fallen about 70° had the change been truly adiabatic, the gas losing energy through external work. This cooling with adiabatic expansion shows that the adiabatics of a gas on the indicator diagram are steeper than the isothermals, for the adiabatic through a given point must move down to a lower isothermal. We shall show later that the adiabatics are steeper than the isothermals in all cases. For a reason to be given later the adiabatics are also termed isentropics.

Isopiestics and Isometrics.—Lines parallel to the horizontal axis indicate a change of volume at constant pressure and are termed *Isopiestics*. The lines of equal volume parallel to the vertical axis are termed *Isometrics*.

Heat Engines.—Any arrangement for the transformation of heat into mechanical energy is termed a heat engine.

By considering the most familiar instance, the steam engine, we may see what are the essentials of a heat engine.

There is a *working substance*, the water and steam. This is made to expand on conversion into steam by the communication of heat from the sides of the boiler—the *source* of heat. But only a part of the heat communicated by the source is turned into work. In order to make the transformation continuous, the steam is ejected from the cylinder and is either allowed to expand into the cooler air, carrying off with it much of the heat originally given, or it is passed into a cold condenser, where it yields up heat to the condenser and is turned into cold water. The general nature of the process then consists in the communication to the working substance of heat from a hotter body—the source—the transformation of some of this heat to work by the expansion of the working substance and the communication of heat not transformed to a colder body, either the outer air or a condenser. In thermodynamics the colder body is termed the *Refrigerator*.

It is of the utmost practical importance to find what is the maximum fraction of the heat leaving the source which can be transformed to work when the temperatures of the source and refrigerator are known. For a comparison of the actual fraction transformed in a given engine with the maximum enables the engineer to determine how far the engine is working economically.

The steam engine is a complicated engine unsuitable for the investigation of the general theory of heat transformation. We shall use, instead, a much simpler ideal machine, first imagined by Carnot, the founder of this part of thermodynamics, and described in a celebrated essay, *Réflexions sur la Puissance Motrice du Feu*, published in 1824. This engine is outside the range of practical experience, but inasmuch as we can see how it would work if the practical details of construction and manipulation were surmounted, the abstract conception is perfectly legitimate.

Carnot's Reversible Heat Engine.—Let us suppose that we have a cylinder (Fig. 150) with absolutely non-conducting walls, and containing a

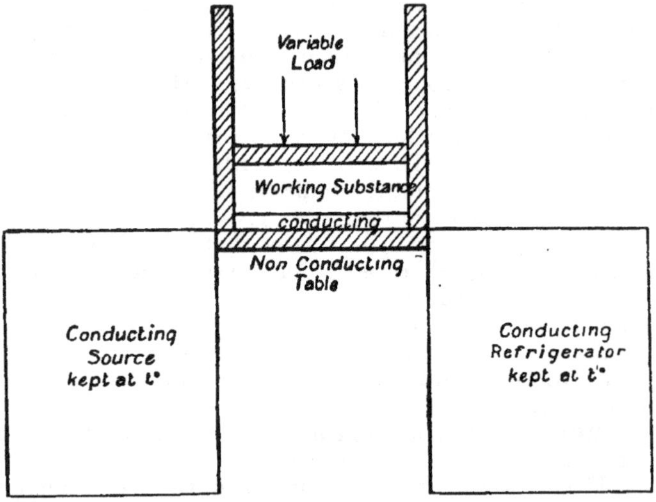

FIG. 150.—Carnot's Engine. Shaded parts non-conducting.

non-conducting piston perfectly fitting yet without friction. Let the cylinder contain air or other working substance under pressure applied by the piston, which we may suppose loaded to any desired extent. Let the bottom of the cylinder be perfectly conducting. As source and refrigerator let us have two bodies arranged as stands, as in Fig. 150, with a non-conducting table between them, their temperatures being maintained constant somehow at $t°$ and $t'°$ respectively. We shall take the working substance through a series of changes such that it ends in its initial condition. This series, known as *Carnot's Cycle*, is represented on the indicator diagram by the four-sided figure, ABCD (Fig. 151), bounded by BC and AD, the t and t' isothermals, and two adiabatics BA and CD. We suppose the substance in the cylinder initially in the condition represented by the point A, Fig. 151, at the temperature t'. Placing it on the central table (Fig. 150), we gradually increase the load, decreasing the volume adiabatically until we arrive at the point B, when the temperature has risen to t. Now, sliding the cylinder along the table on to the source we gradually decrease the load, allowing slow expansion. In general a substance tends to cool on expansion, but the substance is here in contact with a conductor maintained at t, from which it will take in heat, and if the change is slow enough, the temperature of the working substance will never fall sensibly below t. When some arbitrary point C is reached we slide the cylinder back on to the non-conducting table, and then gradually unload the piston, the substance expanding adiabatically till the temperature has fallen again to t', the state being now represented by D. Finally we slide the cylinder on to the refrigerator and increase the load. In general, compression tends to raise the temperature of a substance, but we must here effect it so gradually that all excess of heat is conducted away to the refrigerator and the temperature never rises sensibly above t'. When the pressure arrives at its original value the

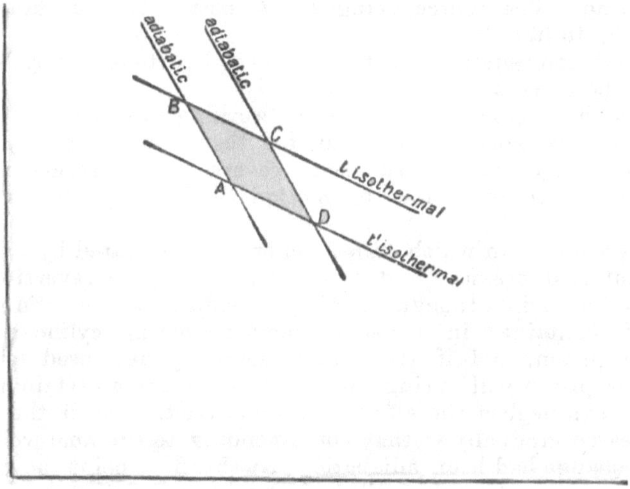

FIG. 151.—Carnot's Cycle.

volume is also at its original value since the temperature is the same, or the state of the body is again represented by A.

The balance of external work done by the working substance is represented by the area ABCD, and this is the mechanical equivalent of the excess of the heat taken from the source over that given to the refrigerator, for the substance in the cylinder has exactly the initial amount of energy and the only changes are in the source and refrigerator.

The process is reversible, *i.e.* we might work counter-clockwise round the cycle ABCD, first sliding the cylinder on to the refrigerator and taking in heat from it during a sensibly isothermal expansion, this heat being equal in amount to that given up in the direct working, then sliding it on to the table and compressing adiabatically till the temperature is t, then compressing isothermally till B is reached, giving up to the source heat equal to that taken in in direct working, and finally expanding adiabatically till we have the initial temperature, volume, and pressure. In each part of the cycle the work done is equal and opposite to the work done in the same part during the direct working. Then by doing work on the substance represented by ABCD, a certain quantity of heat has been taken from the refrigerator and a quantity exceeding this by the heat equivalent of ABCD has been given to the source.

Conditions for Reversible Working.—In order that the same figure may represent the direct and reverse working, that is, in order that indefinitely small changes in the external conditions shall reverse the order of change, certain conditions are necessary. These conditions are—

1. That the working substance shall never differ sensibly in temperature from the bodies to which it is giving or from which it is receiving heat; for suppose that while the working substance goes along BC in the diagram, the source is at a higher temperature T, then it is impossible to make the working substance return along CB merely by a gradual increase of load. The source being $T^\circ - t^\circ$ hotter, the substance cannot yield heat up to it.

This condition requires that the isothermal parts of the cycle BC and DA should be traversed exceedingly slowly.

2. That the pressure exerted by the working substance on the piston shall be sensibly equal to the load, for then, and then only, will infinitesimal changes of load suffice to reverse the direction of motion. This condition requires that the motion shall always be exceedingly slow.

Some processes, in which kinetic energy is generated by a difference between internal pressure and external load, may be reversible at the initial and final points though not at intermediate points. For example, if a gas is contained in a vertical non-conducting cylinder under a frictionless piston, and if the load is suddenly decreased by a finite amount, the piston will spring up and move out to a certain point, the same, if we can neglect the effect of viscous friction, as if the load had been decreased gradually to that corresponding to the final volume, and the whole change had been adiabatic. At this final point the piston will itself reverse the motion, and will in fact continue to move harmonically. The work done between the extreme points is equal to load × distance moved out. But at intermediate points, the motion evidently is not reversible by an infinitesimal change of load.

3. That the machinery moves without friction. If there is friction the load will have to be altered by a finite amount to reverse the motion. The heat generated by the friction is also "irreversible," for it will be given up by the working substance to the surroundings, whether the motion is direct or reverse.

Examples of Reversible Processes.—It will be realised from the foregoing that exactly reversible processes are ideal, in that exact reversibility requires exact equilibrium with surroundings, that is, requires a stationary condition, while a process is necessarily a changing condition. But we can approximate as closely as we like to the conditions of reversibility, by making the conditions as nearly as we like those required, and lengthening out the time of change.

As an illustration, imagine a quantity of water at 100° C. under a piston loaded to 1 atmosphere. Imagine the cylinder to be kept exactly at 100° C. by, say, a current of steam round its outside. There will be equilibrium. But if we decrease the load ever so little, a minute bubble of steam will suffice to start evaporation, which will continue until all the water has become steam at 100°. At any point we may restore equilibrium by holding the piston while we restore the infinitesimal load removed; and at any point we may reverse the motion and condense the steam by adding an extra infinitesimal load so that the pressure is just over 1 atmosphere. As another illustration, consider a quantity of ice and water under atmospheric pressure at 0° C. Keep the surrounding temperature 0° and the proportion of the two remains unchanged. But lower the temperature ever so little below 0° and the ice increases; raise it ever so little above 0° and the water increases. The change from water to ice or ice to water may be reversed by indefinitely small temperature changes.

Examples of Irreversible Processes.—We may contrast with these the case in which water at 100° is under a piston with a load of, say, half an atmosphere. Evaporation with expansion will ensue, and the piston will ultimately rise about twice as far as in the previous case, and will then be in equilibrium. But an indefinitely small increase of load will now only lead to an indefinitely small compression, and not to a reversal of the whole process. Clearly, too, the process is very far from reversible at intermediate points. Or suppose that a lump of ice is thrown into water at 10°. The ice will melt, but no known process will suffice to make the water at 10° change back to ice.

We conclude, then, that in Carnot's reversible cycle the machinery must be free from friction, the working substance must never differ appreciably in temperature and pressure from its surroundings, and that the process must be indefinitely slow.

Efficiency of an Engine.—The fraction of the heat received from the source which an engine converts into work is termed the *efficiency* of the engine.

It is convenient to denote any quantity of heat in calories by H, and its work equivalent by Q = JH when J is the mechanical equivalent of 1 calory.

If an engine receives H units of heat from the source and produces W units of work its efficiency is

$$E = \frac{W}{Q} = \frac{W}{JH}$$

We shall now prove that

All reversible engines working between two given temperatures and taking in equal quantities of heat from the source are equally efficient.

If possible, suppose that one reversible engine A is more efficient between a given source and refrigerator than another reversible engine B. Sending B through a cycle, let it take S of heat from the source and give up R to the refrigerator producing W of work. Or, in reverse working, when W of work is done on it per cycle, it will take R from the refrigerator and give S to the source. Now, by hypothesis, for every quantity of heat S which A receives from the source, it produces more than W of work, say W′. Allow A to work forwards, and out of every quantity W′ of work which it yields, take W to work B backwards through a complete cycle. Then on the whole the source will neither lose nor gain heat, for B restores what A takes. But after each cycle which B goes through, a balance of work $W' - W$ remains over. This can only come from the refrigerator, which must give up to B more heat than it receives from A. Allowing the process to continue indefinitely we should obtain any quantity of work by abstracting heat from the refrigerator. Now the refrigerator may be arranged to be the coldest body in the system, so that we should be obtaining work by merely extracting heat from the already coldest body. For the working substances are at the end of each cycle in their initial condition in every respect. This is contrary to the experience embodied in the Second Law of Thermodynamics.

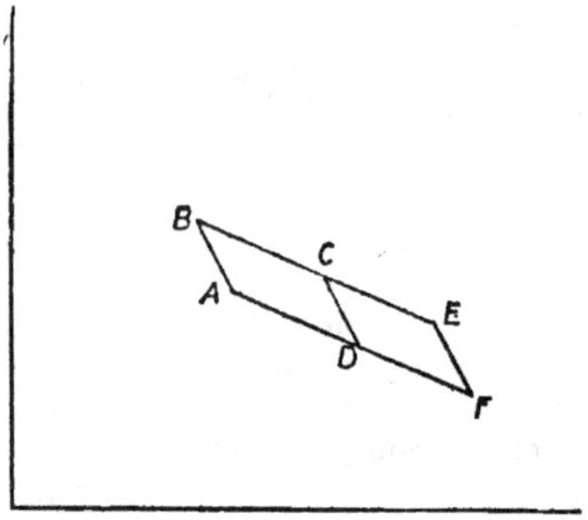

FIG. 152.

Hence we conclude that A cannot be more efficient than B, or that the two reversible engines are equally efficient.

We may further prove that *the efficiency of a reversible engine working between given temperatures is the same whatever quantity of heat is put in at the higher temperature.* That is, if on an indicator diagram (Fig. 152) ABCD represents a Carnot cycle for a reversible engine, the area ABCD is proportional to the heat taken in along BC, so long as BC and AD are given isothermals. For suppose that in a second case double the heat is taken in, the working substance at the higher temperature moving to E along the BC isothermal, and suppose that BEFA now represents the cycle. We may imagine a second engine exactly like the first working round the cycle CDFE. Then from the preceding proposition it has the same efficiency for the same quantity of heat taken in. But the heat given along CE equals the heat given along BC by supposition. Hence the area CDFE equals the area BADC. Or, if the heat given along BE is double the heat given along BC, the work BF is double the work BD. That is, the efficiency is the same whatever quantity of heat is taken in.

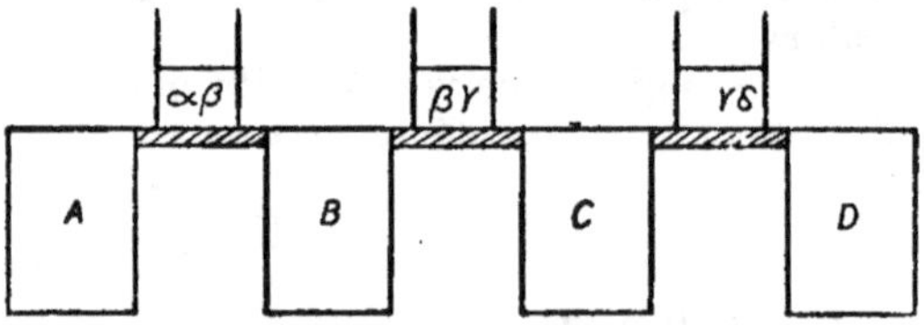

FIG. 153.—Equal Temperature Intervals on the Work Scale.

Absolute or Work Scale of Temperature.—We may express the equal efficiency of all reversible engines between given temperatures, and its independence of the quantity of heat put in, by saying that the efficiency between given temperatures depends only on those temperatures, and not on the nature or conditions of the particular substance used. This independence of the working substance suggested to Lord Kelvin that the efficiency might be used to indicate the temperature on an absolute scale—*i.e.* one in which the given intervals would have the same proportion whatever the substance used. Let ABCD (Fig. 153) be a number of heat reservoirs, arranged in descending order of temperature, and let reversible engines $\alpha\beta$, $\beta\gamma$, $\gamma\delta$ . . . be set working between the successive pairs. Let $\alpha\beta$ go through a cycle, taking Q_A from A and yielding Q_B to B. Let $\beta\gamma$ go through a cycle, taking Q_B from B and yielding Q_C to C. Let $\gamma\delta$ go through a cycle, taking Q_C from C and yielding Q_D to D and so on, each engine taking from its source what its predecessor had yielded to it. The quantities of work yielded will be $Q_A - Q_B$, $Q_B - Q_C$, $Q_C - Q_D$, &c., the heat being expressed in mechanical measures.

Now, adjust the temperatures so that the quantities of work given up by the successive engines are equal, or so that

$$Q_A - Q_B = Q_B - Q_C = Q_C - Q_D = , \&c.$$

Then we define the intervals of temperature as all equal. Or we may say that if a quantity of heat is sent down a succession of temperature steps, the quantity lessening as it goes by reversible transformation into work, the steps are equal when the amount transformed in each step is the same. Though this defines equality of temperature interval, it still leaves it open to us to fix on any desired interval as 1°, and to start from any desired zero point.

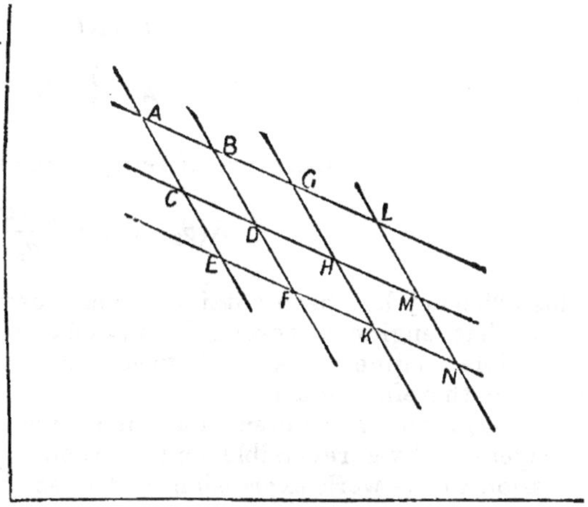

FIG. 154.

We may represent this process on the indicator diagram. Drawing an isothermal AB (Fig. 154) and the adiabatics ACE, BDF, the isothermals drawn so as to make the areas AD, CF, &c., equal, are at equal intervals of temperature.

We may also note that if BG, GL represent distances along which the same amount of heat is taken in as along AB, the adiabatics through G and L must form a series of quadrilaterals equal in area to AD. This, of course, is merely equivalent to saying that the efficiency between given limits is independent of the quantity of heat taken in.

Efficiency expressed on the Absolute Scale.—Any number of the engines $\alpha\beta$, $\beta\gamma$, &c., may be joined together so as to work in concert, and they will take heat from the source of the first, yielding heat to the refrigerator of the last, transforming the difference to work. On the whole, the intermediate reservoirs will be unaffected. Evidently the process may be exactly reversed on putting in the work at the bottom, so that the arrangement forms a compound reversible engine.

If we keep to a given temperature of source indicated by θ_S, on a scale at present arbitrary as to length of degree and zero, and if θ_R be the temperature of the final refrigerator, the efficiency is evidently proportional to the number of engines or steps down, and we may write—

$$\text{Efficiency} = A(\theta_S - \theta_R)$$

where A does not depend on θ_R, but is a function of θ_S alone.

Let us now suppose that it is possible to continue the steps down so far that all the heat put in is converted to work, and that none remains over to put into the lowest refrigerator. The temperature of this refrigerator is the lowest conceivable consistent with the First Law of Thermodynamics, for a still lower temperature would enable us to get more work out of a given quantity of heat than its mechanical equivalent. We choose this temperature, then, as the zero of the new scale, and shall term it the *absolute zero.*

If the refrigerator is at the absolute zero $\theta_R = 0$, and the efficiency is 1, or all the heat put in at θ_S is converted to work. Then we have

$$1 = A\theta_S,$$

or

$$A = \frac{1}{\theta_S}$$

With any other temperature of refrigerator the efficiency is

$$A(\theta_S - \theta_R) = \frac{\theta_S - \theta_R}{\theta_S}$$

This efficiency has been obtained from a peculiar compound engine; but since that engine is reversible, its efficiency equals that of any other reversible engine working between the same temperatures and the result is therefore general.

If Q_S is the heat taken in at the source and Q_R that given out to the refrigerator by a reversible engine, then, equating the temperature expression to the work expression for the efficiency, we have—

$$\text{Efficiency} = \frac{Q_S - Q_R}{Q_S} = \frac{\theta_S - \theta_R}{\theta_S}$$

whence

$$\frac{Q_R}{Q_S} = \frac{\theta_R}{\theta_S}$$

or

$$\frac{Q_S}{\theta_S} - \frac{Q_R}{\theta_R} = 0$$

We may put this result into the following form. If any quantity of heat is allowed to go down a temperature slope, some of it being intercepted and transformed, then if the transformations are reversible, so that by exact reversal the original quantity of heat could be returned to the source, the quantity $\frac{Q}{\theta}$ remains constant down the slope. We may here mention that this quantity, $\frac{Q}{\theta}$, is termed the Entropy put into the system. We shall return to the subject of Entropy later.

Comparison of the Absolute with the Air Thermometer Scale. —To make the formula just found of practical importance, it is necessary to show what relation the absolute bears to known scales. Fortunately it can easily be shown that it nearly coincides with the air thermometer scale. Since all gases give nearly the same temperature indications if far above their condensing points, and the gas scale is therefore nearly independent of the particular gas used, we might perhaps be led to expect this.

Let us work a mass of air in a reversible engine between two temperatures expressed on the air scale. Our knowledge of the properties of air enables us to determine the efficiency of the engine in terms of

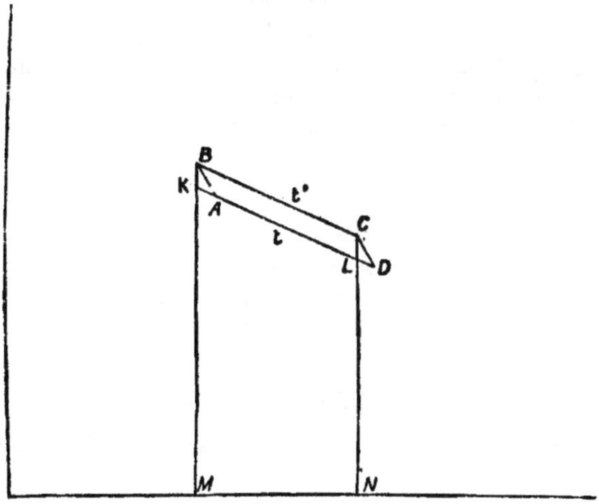

FIG. 155.—Air Scale and Absolute Scale.

the air temperatures, and we may compare the efficiency so found with its expression in terms of the absolute temperatures.

Let AD, BC (Fig. 155) be two neighbouring isothermals of the air, the temperatures measured from $-273°$ C. being t and t' respectively. Let AB, DC be adiabatics, and let AD and $t'-t$ be so small that ABCD is sensibly a parallelogram. Working round the cycle reversibly, heat is taken in along BC and given out along DA, the difference being converted into work represented by the area ABCD.

$$\text{The Efficiency} = \frac{\text{Area ABCD}}{\text{Mechanical Equivalent of Heat taken in along BC.}}$$

But by Joule's experiment on the expansion of gases, a gas expanding and doing no external work remains very nearly at the same temperature without any supply of heat. If it does work and still remains at the same temperature, the heat supplied must be equivalent to the work done. But this is the condition of the air along BC, and the work done is represented by BCNM, which is therefore the mechanical equivalent of the heat taken in along BC.

Then the efficiency—

$$=\frac{\text{ABCD.}}{\text{BCNM.}}$$

$$=\frac{\text{KBCL}}{\text{BCNM}}$$

$$=\frac{\frac{\text{BK}+\text{CL}}{2}\times\text{MN}}{\frac{\text{BM}+\text{CN}}{2}\times\text{MN}}$$

$$=\frac{\text{BK}+\text{CL}}{\text{BM}+\text{CN}}$$

Now BM and KM are the pressures at constant volume corresponding to t' and t, and are therefore proportional to t' and t, this being really the definition of t' and t with the air thermometer.

$$\therefore \frac{\text{KM}}{\text{BM}}=\frac{t}{t'}$$

and

$$\frac{\text{BK}}{\text{BM}}=\frac{t'-t}{t'}$$

Similarly

$$\frac{\text{CL}}{\text{CN}}=\frac{t'-t}{t'}$$

$$\therefore \text{efficiency}=\frac{\text{BK}+\text{CL}}{\text{BM}+\text{CN}}=\frac{t'-t}{t'}$$

If θ θ' are the temperatures on the absolute scale corresponding to t and t'—

$$\frac{\theta'-\theta}{\theta'}=\frac{t'-t}{t'}$$

or

$$\frac{\theta'}{\theta}=\frac{t'}{t}$$

Or the temperatures on the one scale are approximately proportional to those on the other, and we may conveniently choose the degrees so that 0° C. shall be − 273 on either scale, or, more exactly, that the interval between 0° C. and 100° C. shall be 100 degrees on either scale.

This is only an approximate result, for it assumes that in Joule's experiment (p. 120) there was no change of temperature, an assumption not quite exact, as we know by later experiments. It also assumes that the air scale for the volume OM is proportional to that for ON, an assumption exceedingly near the truth, but still probably slightly inexact.

Non-Reversible Cyclical Engine.—The foregoing propositions relate to reversible engines only, but if we have an engine not reversible, say one in which the working substance is at a lower temperature than the source when it takes in heat and is at a higher temperature than the refrigerator when it gives out heat, and yet goes through a cycle so that it recurs to its initial condition, it is easy to show that

No cyclical non-reversible engine working between given temperatures is more efficient than a reversible engine. For let the non-reversible engine work forward, driving a reversible engine backward. If the first is more efficient it can get more work from a given quantity of heat. Put it so that it will be able to drive the reversible one at such rate that the reversible restores to the source the heat taken out and there is a balance of work over, which can only come from the refrigerator. If this is arranged to be the coldest body of the system the result is contrary to the Second Law.

Water-Wheel Analogue of the Reversible Engine.—The reader may be aided in understanding and remembering the foregoing theory of the reversible heat engine by considering an analogous case of energy tranformation.

Let us suppose that we have two reservoirs of water at different levels, and that we wish to transform the potential energy of the water in the higher by means of a water-wheel, allowing the water to fall in the wheel from one reservoir to the other. Evidently we cannot get all the potential energy transformed by the wheel. We put in so much at the top and take out a smaller quantity at the bottom. If the wheel works very slowly, receiving the water at the level of the surface of the higher reservoir, and parting with it at the level of the surface of the lower one, and if there is no leakage and no friction, then the whole of the potential energy lost will be usefully transformed by the wheel, and no wheel between the same levels could be more efficient. We see at once that the conditions of maximum efficiency and reversibility coincide. It is obvious that it is impossible to get mechanical effect out of the lower reservoir if that is already in the lowest available position—the analogue of the Second Law of Thermodynamics. The total potential energy of a mass of water in the upper reservoir is the work which would be done in lifting it from the absolute zero of level—the centre of the earth.

If then we define

$$\text{Efficiency} = \frac{\text{Work got out}}{\text{Energy put in}}$$

the denominator is very nearly the same for the same mass of water at different places, and the efficiency is very nearly proportional to the difference in height, in feet, between the two reservoirs. But the height in feet does not give an absolute scale of efficiency, for, as gravity varies, the work for a fall of one foot depends to some extent on the situation of the wheel. This is analogous to the gas scale of temperature. We may get an absolute scale by discarding the length measure of difference of level and substituting for it a work measure. Dividing up the distance between the centre of the earth and the level of the higher reservoir into steps, so adjusted that a reversible wheel in each step will give the same amount of work from the passage of the same quantity of water, we may define these steps as having equal differences of level. Now the efficiency of any wheel becomes

$$\frac{\text{No. of steps occupied by wheel.}}{\text{Total No. of steps from centre of earth to higher level.}}$$

or if $n =$ No. of steps from centre of earth to higher level
and $n' =$ the No. from centre of earth to lower level.

$$\text{Efficiency} = \frac{n - n'}{n}$$

and the new scale of level is analogous to the absolute scale of temperature.

Since the total quantity of energy is proportional to the total number of steps, and the quantity transformed is proportional to the number descended, the quantity still remaining in the water is proportional to the number of steps remaining.

If then a quantity of water is started downhill, the energy being transformed as it goes by reversible wheels, the quantity

$$\frac{\text{Energy still possessed}}{\text{level of water possessing it}}$$

is constant.

This is really equivalent to saying that the quantity of water remains

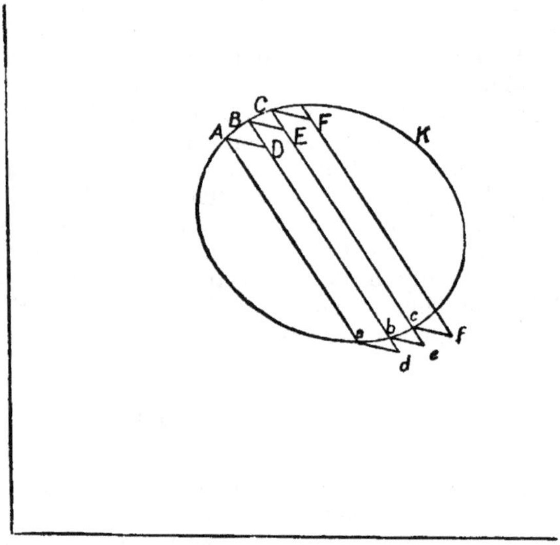

FIG. 156.—Any Reversible Cycle.

constant. Hence, quantity of water is analogous to the quantity $\frac{Q}{\theta}$, which we have termed entropy.

Reversible Cycles in General; with Varying Temperatures of Source and Refrigerator.—We have found that if Q_S of heat is taken in by a reversible engine at a definite temperature θ_S, and Q_R is given out at a definite temperature θ_R, then $\frac{Q_S}{\theta_S} - \frac{Q_R}{\theta_R} = 0$.

But in many cases the heat is not all taken in at one definite temperature and given out at another. We have, therefore, to generalise this result.

Let the working substance go through a cycle represented on the indicator diagram by ABCK*cba*A (Fig. 156).

Take a succession of points ABC . . . very near together, and through them draw adiabatics A*a*, B*b*, C*c* across the area. Draw the isothermals

AD, BE . . . *ad*, *be* . . . &c. Instead of sending the substance round the curve, let it go through the series of steps ADBEC . . . *cebda* . . .

In the limit the heat received along each piece AD is the same as that along each piece AB, for, sending the substance round the cycle ABDA, the area ABD is the mechanical equivalent of the difference between the heat received along AB and that given out along DA, since BD is an adiabatic; and this area is an indefinitely small fraction of the area AD*da*, which is itself only a fraction of the heat received along AD.

We may, therefore, as far as concerns heat taken in or given out, replace the curve ABC . . . by the steps ADBEC . . .

Let Q_{AB} Q_{BC} Q_{CD} . . . be the heats received along AB, BC, CD . . . respectively, or, as we have just shown, along AD, BE, CF . . .

Let Q_{cb} Q_{ba} . . . be the heats yielded along *cb*, *ba* . . . which are sensibly equal to those along *eb*, *da* . . . If now we imagine the substance to be sent round the cycle AD*da*, we see that this is the kind of cycle already considered.

$$\therefore \frac{Q_{AB}}{\theta_A} - \frac{Q_{ba}}{\theta_a} = 0$$

where θ_A and θ_a are the temperatures at A and *a*.

Sending it round the cycle BE*eb*

$$\frac{Q_{BC}}{\theta_B} - \frac{Q_{cb}}{\theta_b} = 0$$

θ_B and θ_b being the temperatures at B and *b* and so on.

If we regard the negative sign as denoting that the substance receives negative heat we may add up all the equations and denote the result by

$$\Sigma \frac{Q}{\theta} = 0$$

where each element of heat received is to be divided by the temperature of the substance when receiving it. Using the notation of the Integral Calculus the result may be written,

$$\int \frac{dQ}{\theta} = 0$$

round a reversible cycle.

This deduction is often quoted as the Second Law of Thermodynamics but it appears preferable to consider the fundamental experience, given at the beginning of the chapter, as constituting the law. In reversible cycles then we have

by the first law $\int dQ = \text{Work done},$

by the second law $\int \frac{dQ}{\theta} = 0,$

the former telling us the relation between the heat disappearing and the work done, and the latter giving a relation between the heat and the temperature.

Entropy.

Definition of Entropy.—If a substance at a temperature θ receives heat Q, it is said to receive entropy $\frac{Q}{\theta}$.

Similarly if it parts with heat Q, when at a temperature θ, it is said to lose entropy $\frac{Q}{\theta}$. If the gain or loss of heat affects the temperature, we may divide the gain or loss into elements each at practically constant temperature, and then the amount of entropy gained or lost is

$$\Sigma\frac{Q}{\theta} \text{ or } \int\frac{dQ}{\theta}$$

The Entropy gained by a body in passing from one state to another depends solely on the initial and final states. Let

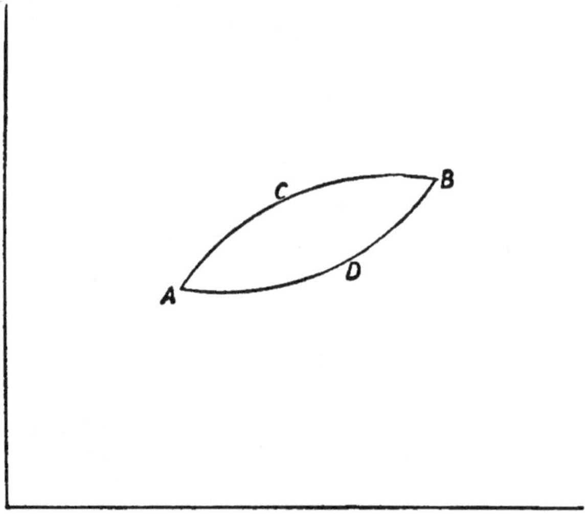

FIG 157.—Entropy. Change in passing from A to B the same by all paths.

A in the indicator diagram (Fig. 157) represent some standard state of the body, B some other state. Then we shall show that, however the body passes from A to B, the entropy received is the same, or the entropy at B depends solely on the position of B and not on the mode in which the state changes from A to B.

Let ACB, ADB, be two paths from A to B. Let the substance be carried round the cycle ACBDA by a reversible process,* then

$$\Sigma\frac{Q}{\theta} \text{ from A through C to B} + \Sigma\frac{Q}{\theta} \text{ from B through D to A} = 0.$$

* Inasmuch as the conditions of reversibility are external conditions, any series of changes of temperature, pressure, and volume in a body can be imagined to be carried out reversibly by suitably adjusting the external surroundings.

Now if BDA is described in the reverse direction ADB, heat taken in becomes heat given out and we must change the sign of Q in the second term of the left side of the equation. Changing the sign and transposing

$$\Sigma\frac{Q}{\theta} \text{ from A through C to B} = \Sigma\frac{Q}{\theta} \text{ from A through D to B}$$

or the entropy received by either path is the same.

The Adiabatics are Isentropics.—If a body is compressed or expanded adiabatically, *i.e.* so that, whatever energy may be transferred as work, none is conducted as heat, the entropy remains constant and the adiabatics on the indicator diagram are lines of equal entropy, or *isentropics.* The different adiabatics may be denoted by the value of the entropy along them, reckoned from some standard condition, just as the different isothermals are denoted by the value of the temperature along them. The entropy of a body is usually denoted by ϕ.

The Entropy Temperature Diagram is a diagram in which the ordinate represents the temperature, and the abscissa the entropy reckoned from some standard condition.

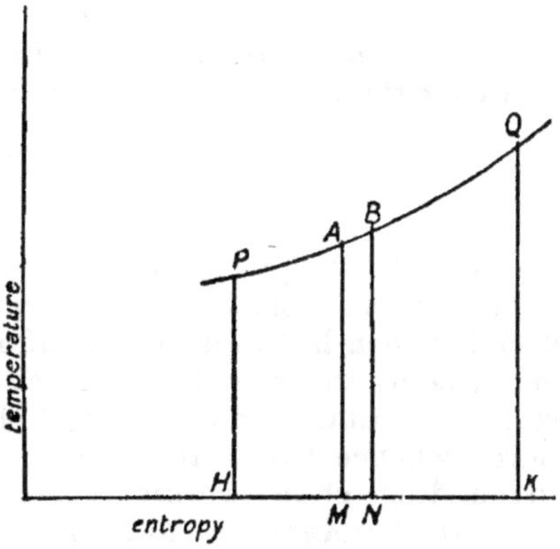

FIG. 158.—Entropy-Temperature Diagram.

If AB (Fig. 158) represents a small change of condition of a body, on an entropy temperature diagram, the temperature changing from AM to BN and the gain of entropy being MN, and if Q be the quantity of heat received in the change from A to B

$$MN = \frac{Q}{\frac{AM + BN}{2}}$$

or

$$Q = MN.\frac{AM + BN}{2} = \text{area ABNM.}$$

Similarly for every element in a finite change. Then the heat received in a change from P to Q is equal to the area PQKH.

Now if we take a body through any cycle, returning to the original state the entropy and temperature both have their original values, and a cycle is therefore represented by a closed curve on the diagram. The area included by the curve is the difference between the heat taken in and the heat given out, and so represents the amount of heat transformed. We shall make use of the diagram later, but without further discussion now, the reader will see that a Carnot's cycle is represented by a rectangle with sides parallel to the axes, and he may easily prove that for a gas the lines of equal pressure and of equal volume are represented by curves of the type

$$\theta = ae^{\frac{\phi}{c}}$$

where, for the one, c is the specific heat at constant pressure, and for the other the specific heat at constant volume.

The total Entropy of a system is unchanged by the performance of reversible cycles between its parts. When a body is worked between two given temperatures we have seen that $\frac{Q_S}{\theta_S}$, the entropy taken from the source, equals $\frac{Q_R}{\theta_R}$, the entropy given to the refrigerator; the total quantity is therefore unchanged.

But if the cycle is a perfectly general reversible one we still have

$$\Sigma\frac{Q}{\theta} = 0$$

or the total entropy received by the working substance is zero. But in order that the cycle should be reversible, all passages of heat must take place between bodies differing infinitely little in temperature, so that Q and θ, being the same for giver and receiver, the entropy gained by the working substance in any part of the system is equal to that lost by the source whence the heat comes. Hence the total entropy lost by the surroundings of the working substance is also zero.

Quantities Analogous to Entropy.—The conception of entropy is somewhat difficult in that, as it does not directly affect the senses, there is nothing physical to represent it. It may assist the reader if we point out some already familiar quantities which are analogous to it. Since

$$\text{Entropy} = \text{heat energy} \div \text{temperature},$$
$$\therefore \text{Heat energy} = \text{entropy} \times \text{temperature}.$$

Also in the Carnot reversible cycle,

$$\text{Energy transformed} = Q_S \times \frac{\theta_S - \theta_R}{\theta_S}$$

$$= \text{Entropy} \times \text{fall of temperature}.$$

Comparing this with gravitation energy—

$$\text{Gravitation energy} = \text{Mass} \times \text{level above zero},$$

the level being measured by work done on unit mass raised from zero level,

and $\qquad \text{Energy transformed} = \text{mass} \times \text{fall in level}.$

Then entropy is analogous to mass and temperature to work measure of level.

Or taking the case of electrical energy,

$$\text{Electrical energy} = \text{charge} \times \text{potential}.$$

If the charge is allowed to pass by an insulated conductor to a lower potential,

Energy transformed = charge × fall in level.

Then charge corresponds to entropy and potential to temperature.

Entropy Tends to Increase.—These analogies are not to be pursued too far, for while mass and electrical charge (taking into account the whole field) are constants, entropy is only constant in the very special case of reversible transformation, when the maximum amount of heat is converted into other forms of energy. In other cases it tends to increase. Whenever there is any conduction between bodies at a different temperature, there is a gain of entropy, for the hotter one loses Q at θ_1, and the cooler gains Q at θ_2; the one loses entropy $\frac{Q}{\theta_1}$, while the other gains entropy $\frac{Q}{\theta_2}$, which is greater. If, then, the source is sensibly hotter than the working substance, or if the refrigerator is sensibly cooler, there is an increase of entropy. Or if friction occurs, some of the mechanical work goes to produce heat and the body receiving it gains entropy, which is not lost by any other body, but which starts into existence through the friction.

Now in actual engines there is always conduction going on through finite differences of temperature and there is always friction, so that there is always gain of entropy.

And more generally in any change of condition of a body, there is no loss and usually a gain in the sum total of entropy. If the body expands or contracts along an adiabatic, then the entropy is unchanged. But if it changes in any other way, heat must be taken in or given out by conduction. The entropy would only remain unchanged in amount (though passed on from one body to another) if the conduction went on between bodies at sensibly the same temperature. In practice there is always a finite temperature slope, and always therefore a gain in the entropy of the heat conducted.

On the whole, then, entropy tends to increase. This principle was first laid down by Clausius.

If we compare transformations from heat with other transformations, we at once notice a marked contrast. In heat transformation there is a constant tendency on the part of the transformed energy to revert to the original heat form, and there is a tendency for it to pass from one body to another by conduction without transformation. But with gravitation energy, for instance, we may allow a mass to fall from one level to another, transforming all the available potential energy to other kinds to the utmost possible extent, though not necessarily to useful forms, and there may be no tendency of the transformed energy to revert to the potential form. Or with electrical energy we may allow a charge to pass from one conductor to another and obtain all the available electrical energy as work or heat. There is nothing in either case corresponding to conduction of heat without transformation. Whenever a rearrangement of mass or of electricity occurs without communication from the outside, there is a loss of potential energy or of electrical energy.

If we could imagine a frictionless machine in which a mass descending to a lower level lifted up another mass from the absolute zero level to meet it, the mass being such that the total potential energy remained unchanged, or if we could imagine an electrical system in which the work gained by passage to a lower potential turned some machine, increasing the charges at the lower potential so that the total electrical energy remained unchanged, we should have cases analogous to the conduction of heat.

The difference really turns on the fact that entropy tends to increase, while its analogues, mass and quantity of charge, are constant.

Dissipation or Degradation of Energy.—Through conduction and radiation there is a continual tendency to reduce differences of temperature in the universe, and therefore to reduce the amount of heat available for transformation. Though not diminishing, but on the contrary increasing in amount, the heat is continually approximating more to a dead level, being "degraded" or "dissipated." The quantity of other forms of energy available for transformation is also diminishing,

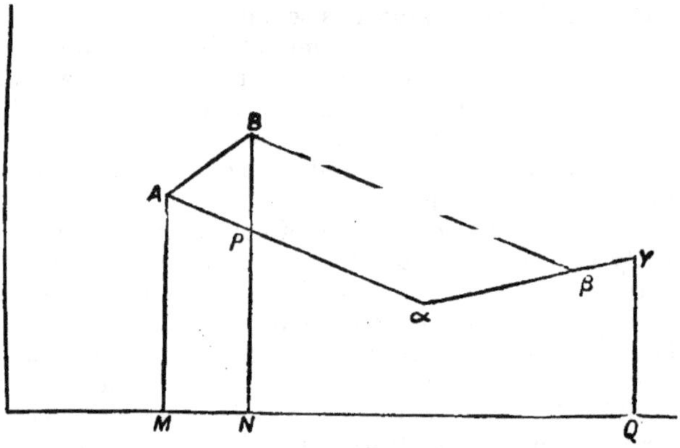

FIG. 159.—Intrinsic Energy.

for in every transformation which occurs some heat is generated, and this can never be wholly retransformed.

There is, as it were, a tax levied on every transformation, the proceeds being put into an untransformable fund. The amount of possible transformation by any method which we can at present conceive is therefore gradually diminishing, a fact which is expressed by saying that the energy is undergoing dissipation or degradation. More and more, energy is assuming the form of heat, which may be regarded as the stable form, and a dead level of temperature is the state which we are gradually approaching.

Intrinsic Energy.—The energy which a substance possesses in itself is termed its intrinsic or internal energy. We have no means of measuring the total quantity of a body's intrinsic energy, but we can find the amount added in changing from one condition to another.

It will obviously be the amount of heat added less the external work done. We may give some attempt at representation on the indicator diagram. Let AB (Fig. 159) be the curve representing the change from the condition A to the condition B Draw the adiabatics $A\alpha$, $B\beta$, down to the absolute zero isothermal $\alpha\beta\gamma$. Of course we have no experimental justification for drawing this isothermal, but for diagrammatic representation we may suppose it to exist.

Taking the substance round the cycle $AB\beta\alpha A$, the whole of the heat taken in along AB will be converted into work since no heat will be given out along $\beta\alpha$, its temperature being zero. Then the area $AB\beta\alpha$ represents the heat taken in along AB. The external work done along AB is ABNM. Then the gain of intrinsic energy along AB is

$$AB\beta\alpha A - ABNM$$
$$= PB\beta\alpha P - APNM$$

or if γ be a point at some distance along $\alpha\beta$ and γQ the ordinate, adding $P\alpha\gamma QNP$ to each, the gain of intrinsic energy is

$$B\beta\gamma QNB - AP\alpha\beta\gamma QMA.$$

The change in the intrinsic energy is represented then by the change in the area $AMNQ\gamma\beta\alpha A$.

Available Energy.—The amount of energy in a body or system which can be transformed to useful work depends entirely on the conditions under which the work is to be done.

If we have, for example, a substance in an envelope under pressure, and we can allow it to expand to a given lower pressure, the work done will depend on the nature of the envelope. If it allows heat to pass we shall get more work by an isothermal than by an isentropic or adiabatic expansion. For the adiabatics on the indicator diagram, as we shall see later, are always steeper than the isothermals, so that, if in Fig. 160, AB is the isothermal, AC the adiabatic through A, the work ABNM is greater than the work ACPM. Of course the body in the first case may have taken in energy through the envelope, and may have converted some of this to work.

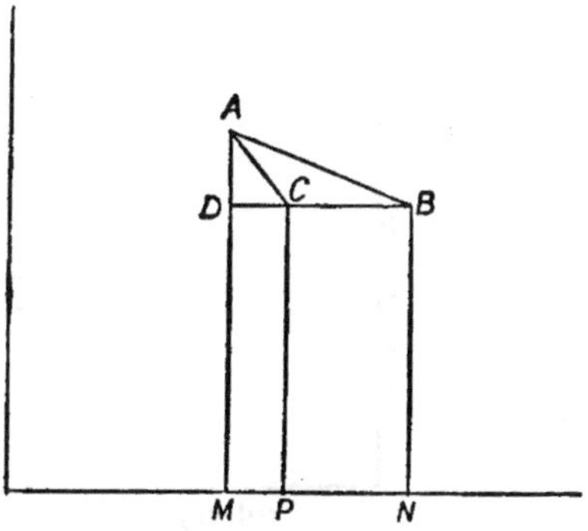

FIG. 160.

If the envelope is non-conducting, then the most work will be done if the external pressure differs only by an infinitely small amount from the internal pressure, so that the expansion will be adiabatic. For if there is a finite difference, kinetic energy will be generated in the substance, which will ultimately tend to become heat, and some of the energy will thus be lost for transformation to work.

Thus, in the gas engine, in which a mixture of coal-gas and air is ignited with transformation of chemical energy to heat, the temperature and pressure of the gas are raised. Obviously the cylinder should be non-conducting, otherwise heat will pass out of the gas which might be transformed. And the expansion should be adiabatic till the pressure is as near as possible to the air pressure, the lowest available.

The Available Energy of an Isolated System.—If we have a system of bodies under such circumstances that they can neither part with heat to, nor receive it from the outside, the entropy of the system cannot diminish. In order to obtain the maximum work from the system, it must evidently be brought to a condition of uniform temperature and of uniform pressure, for if these are not uniform still more work may be obtained by coming to uniformity. There will be some limiting condition given as to pressure or volume. In coming to this, the best course will be by adiabatic transformations, so that the final equals the initial entropy, for if there is any gain, some of the energy which might have been transformed has gone to increase the entropy of the system. Now, if in the final state the entropy and some other quantity, say the pressure, is given, the state of the system is determinate, and therefore its intrinsic energy is determinate. Hence we may calculate the work

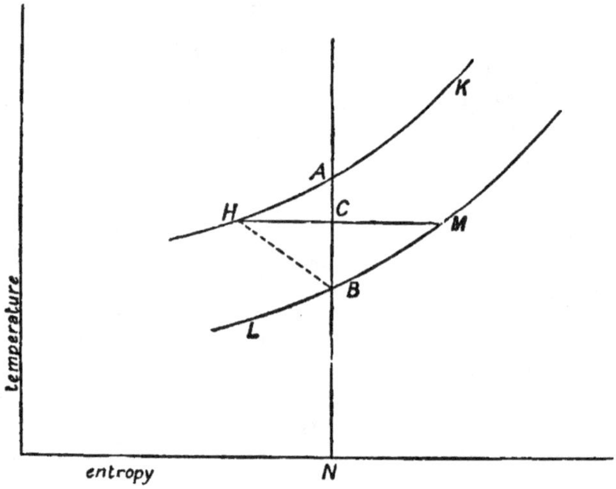

FIG. 161.

done in any convenient method of passing from the initial to the final state, in which we obey the condition of constant entropy. For instance, we may first allow all the bodies to come to the standard pressure adiabatically. We may then use a reversible engine to bring them to the same temperature, first say working between the two coldest as source and refrigerator; then when they are at one temperature, working between the next above as source, and these two as refrigerator, and so on.

We may represent the work obtained in the latter part of the process on the entropy temperature diagram. Let us take the case of two bodies only, at the same pressure but different temperatures. Let HAK (Fig. 161) represent the final pressure line for the hotter substance, AN its temperature; LBM the line for the same pressure of the colder substance, BN its temperature. If we turn the curve LBM round AB as axis till BM is in the plane of the paper on the other side, lying along BH and cutting HAK in H, evidently as the hotter body passes along AH, giving up heat to a reversible engine with the colder as source, the latter passes along BH, the two meeting at the common temperature H. For this secures that the entropy remains constant. The heat converted is given by the area AHB. In other words, if HCM is the horizontal line from one curve to the other, bisected in C by AB, CN is the common temperature and

HAC + BCM is the heat converted.

The Possible Efficiency of a Steam-Engine.—We shall conclude this account of reversible engines, and of the idea of entropy arising therefrom, by a calculation of the possible efficiency of a steam-engine supposing that there is no waste of heat and no friction.

Let us for definiteness take an engine in which the boiler is at 140° C., say 413° absolute, and the condenser at 30° C., say 303° absolute. We shall suppose that the steam is admitted into the cylinder at the temperature and pressure of the boiler till the moment of cut off, that then there is adiabatic expansion till the temperature is reduced to that of the condenser, and that then the contents of the cylinder, which will be partly saturated steam at that temperature and partly water, are let into the condenser and there condensed to water. We shall further suppose, in order that we may take the working substance round a cycle, that the condenser water feeds the boiler.

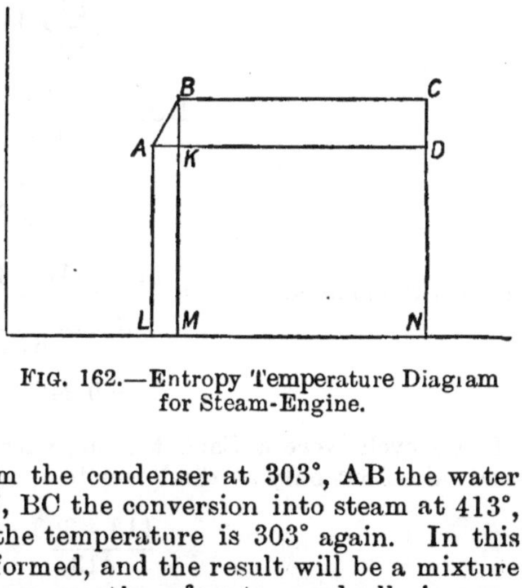

FIG. 162.—Entropy Temperature Diagram for Steam-Engine.

The path may be simply represented on an entropy temperature diagram, Fig. 162. Using absolute or work-scale temperatures, A represents the water returned to the boiler from the condenser at 303°, AB the water as it rises in temperature to 413°, BC the conversion into steam at 413°, CD the adiabatic expansion till the temperature is 303° again. In this expansion water will at once be formed, and the result will be a mixture of saturated steam and water, the proportion of water gradually increasing as the temperature falls.* When 303° is reached, the further condensation into water is represented by DKA.

The heat given during the cycle is represented by the area LABCNL, and that transformed by ABCDA.

$$\text{Then the efficiency} = \frac{\text{ABCDA}}{\text{LABCNL}} = \frac{\text{ABK} + \text{BCDK}}{\text{ABML} + \text{BCNM}}.$$

In calculating the value of this, we shall make very slight error in

* The reader may calculate the proportion of water at 303°—very nearly one-fifth—by assuming that the entropy is the same at the beginning and at the end. We may refer to Perry's *Steam-Engine*, § 200, for the mode of calculating the entropy of water and of steam, or to p. 320 of the same work for tables giving the entropy at different temperatures.

taking AB as a straight line, and the area LABML as 110 per gramme of water. We have the values

$$\begin{aligned} BCNM &= \text{latent heat at } 413^\circ, \text{ or } 140^\circ \text{ C.} \\ &= 606{\cdot}5 - {\cdot}695 \times 140 = 508 \quad \text{(Regnault)} \end{aligned}$$

$$AL = 303\,;\; BM = 413.$$

To find ABK

$$ABK = \frac{1}{2}AK{\cdot}BK = \frac{1}{2}AK(BM - AL),$$

also

$$ABML = \frac{1}{2}AK(BM + AL) = 110,$$

whence

$$\begin{aligned} ABK &= \frac{BM - AL}{BM + AL}\,ABML \\ &= \frac{110}{716} \times 110 \end{aligned}$$

$$\begin{aligned} BCDK &= \frac{BK}{BM} \times BCNM \\ &= \frac{110}{413} \times 508 \end{aligned}$$

Then the efficiency

$$= \frac{\dfrac{110 \times 110}{716} + \dfrac{110}{413} \times 508}{618}$$

$$= 0{\cdot}246 \text{ about.}$$

If the cycle were a Carnot cycle, with heat put in only at the higher and taken out only at the lower temperature, the efficiency would be

$$\frac{413 - 303}{413} = 0{\cdot}266.$$

This is only a particular cycle, and the efficiency might be brought still nearer the Carnot cycle value by allowing, say, isothermal expansion of the steam at 413°, until the pressure was such that, on further adiabatic expansion till the temperature fell to 303°, condensation was then just beginning. But such a cycle would be less like that in the actual steam-engine than the one considered.

The engine we have imagined is a perfect one for the particular cycle, for we have supposed no loss by conduction or friction. Let us now see how much coal would be required per horse-power per hour in such an engine.

We may estimate that one gramme of coal gives a supply of 8000 calories in the source, or $8000 \times 4{\cdot}2 \times 10^7$ ergs.

Of this only 0·246 is transformed by our engine, or 1 gramme of coal yields $8{\cdot}27 \times 10^{10}$ ergs as work.

Now, 1 H.-P. is 76 kgm. metres/sec.,

say 746×10^7 ergs/sec.,

and 1 H.-P. for 1 hour is $746 \times 3600 \times 10^7$,

$= 2680 \times 10^{10}$ ergs say.

This requires $2680 \div 8{\cdot}27 = 324$ grammes of coal,

say 0·71 lb. of coal.

Actual engines have by no means reached this efficiency over such a temperature range. Probably $2\frac{1}{2}$ to 3 lbs. of coal per horse-power hour would be a good result at present.

It may be noted that if the whole of the heat-yield of the coal were transformed, then we should have

1 H.-P. for 1 hour requiring $(2680 \times 10^{10}) \div (8000 \times 4{\cdot}2 \times 10^7)$,

$= 80$ grammes nearly,

or $= 0{\cdot}18$ lb. „

CHAPTER XVIII.

THERMODYNAMICS OF ISOTHERMAL AND ADIABATIC CHANGES.

Heat taken in when a Body expands Isothermally—Heat to a neighbouring Adiabatic the same by all paths—Change in Temperature when a Body undergoes a small Adiabatic Change—Adiabatics steeper than Isothermals—Specific Heats at Constant Pressure and Constant Volume—Their Ratio γ equal to the Ratio of the Isentropic and Isothermal Elasticities—Experimental Determinations of γ for Gases—Adiabatic Gas Equation—Fall of Temperature Upwards with Convective Equilibrium—Internal Energy taken up by a Gas in Expanding—Comparison of Air Scale with Absolute Scale—Generalisation of Indicator Diagram for any Stress and corresponding Strain.

FROM the two laws of Thermodynamics, which, for reversible cycles, may be stated in the forms

$$(1)\ \int dQ = W \quad \text{and} \quad (2)\ \int \frac{dQ}{\theta} = 0$$

we may deduce the temperature changes or the heat developments in many alterations of physical condition, in terms of the known properties of the bodies affected.

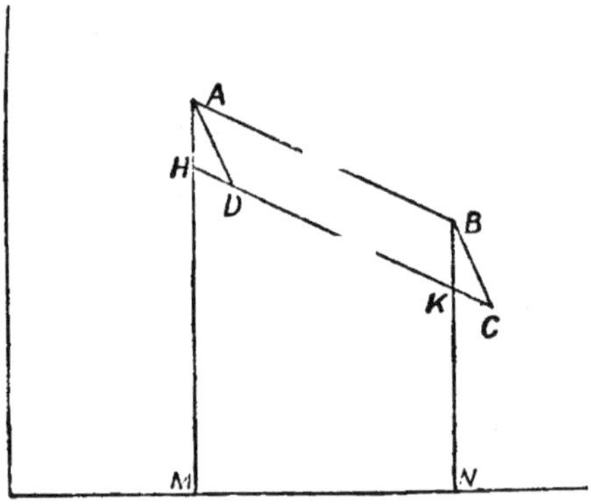

FIG. 163.

Some of the results may be obtained by elementary methods, using the indicator and entropy-temperature diagrams. We proceed to obtain several of great importance.

The Heat taken in when a Body Expands Isothermally.—Let the body expand at temperature θ by a small amount dv, its path being represented on the indicator diagram by the element AB of an isothermal in Fig. 163. Then dv is represented by MN. Let us make AB part of a Carnot cycle, the lower isothermal DC being at temperature $\theta - d\theta$. Produce CD to cut AM in H. If all changes are small, ABCD may be regarded as a parallelogram. We shall equate its area expressed by aid of the Second Law of Thermodynamics to that expressed in terms of the constants of the body.

Let Q be the work measure of the heat along AB.

Then
$$\frac{Q\,d\theta}{\theta} = ABCD = AH \times MN.$$

If we denote the change of pressure at constant volume per degree rise in temperature by ω, then $AH = \omega d\theta$, and putting $MN = dv$,

$$\therefore \frac{Q d\theta}{\theta} = \omega d\theta \cdot dv$$

and $$Q = \omega\theta dv. \qquad (1)$$

The quantity ω is only observed in gases, for which it is approximately $\frac{p}{\theta}$, but we can express it in all cases in terms of e_θ, the isothermal elasticity, and a, the coefficient of expansion at constant pressure. Let AB be an isothermal at θ (Fig. 164), and let AC, CB be drawn parallel to the axes, meeting at C, of which the temperature is $\theta - d\theta$. The definition of e_θ is

$$e_\theta = \frac{v \times \text{small increase of pressure}}{\text{consequent decrease of volume}}$$

where the changes are at constant temperature

$$= v\frac{AC}{CB}.$$

But $AC = \omega d\theta$; $CB = av d\theta$

$$\therefore\ e_\theta = \frac{\omega}{a}. \qquad (2)$$

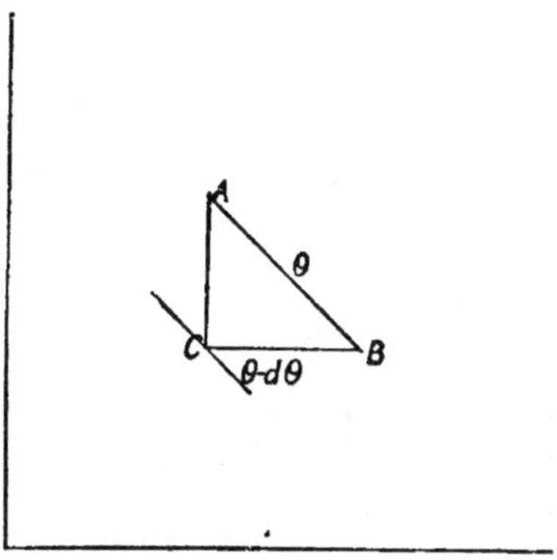

FIG. 164.

Substituting in (1) from (2) we get

$$Q = ae_\theta\theta dv.$$

If we put $e_\theta dv = vdp$, we get the heat in terms of the pressure change, viz.,

$$Q = av\theta dp.$$

The heat taken in or given out by a body in passing from a given condition to a neighbouring adiabatic is the same along all paths. Let H in Fig. 165 indicate the initial condition of the body, and let BA be a neighbouring adiabatic. Let HA, HB be any two different paths to the adiabatic. If the body is worked reversibly round HAB, then $\int \frac{dQ}{\theta} = 0$. Or if Q_1 be the heat taken in along HB, Q_2 that along HA, θ_1 θ_2 the average temperatures at which it is taken in along the two respectively, then since no heat is taken in or given out along AB

$$\frac{Q_1}{\theta_1} = \frac{Q_2}{\theta_2} \text{ or } \frac{Q_1}{Q_2} = \frac{\theta_1}{\theta_2}$$

In the limit when H approaches AB, $\frac{\theta_1}{\theta_2} = 1$, and therefore $Q_1 = Q_2$.

The reader will easily see how this proposition comes out at once by the use of the entropy-temperature diagram.

The change in temperature when a body undergoes a small adiabatic change of volume by a given change of pressure.—Let

AB represent, in Fig. 165, the small adiabatic change due to an increase of pressure HB, a unit mass of the body being dealt with. Let the temperature at A be θ, and that at B be $\theta+d\theta$. Let BC be the $\theta+d\theta$ isothermal, and HAC the line of equal pressure through A. Then by the last proposition the heat along AC equals the heat along BC.

But heat along $BC = av\theta dp$,

while heat along $AC = K_p d\theta$,

where K_p is the specific heat at constant pressure, as expressed in work measure. Equating

$$K_p d\theta = av\theta dp$$

or

$$d\theta = \frac{av\theta dp}{K_p}$$

If e_ϕ is the adiabatic elasticity, *i.e.* the elasticity along AB defined by

$$e_\phi = v\cdot\frac{HB}{HA}$$

$$d\theta = \frac{ae_\phi\theta dv}{K_p}$$

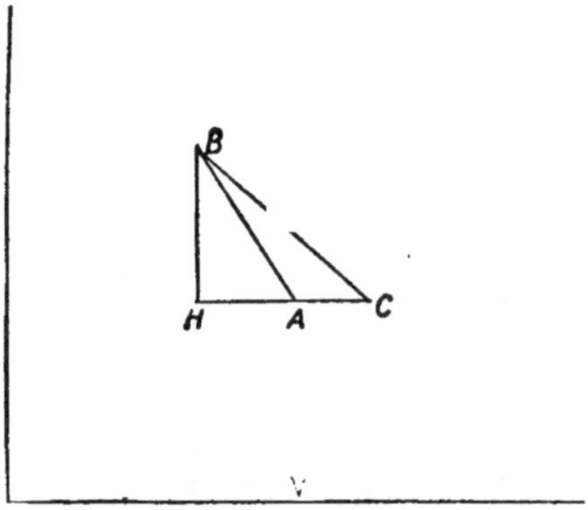

FIG. 165.

Though the area of ABC is of the second order and vanishes in comparison with the heat received along either AC or BC, it may be of interest to show how it may be used to obtain the above result when equated to the corresponding area on the entropy-temperature diagram, the two areas representing the work done, and the heat transformed, respectively.

The area

$$ABC = \frac{1}{2}HB\cdot AC$$

$$= \frac{1}{2}dp\cdot avd\theta$$

On the entropy-temperature diagram, Fig. 166, the corresponding area is

$$abc = \frac{1}{2}ab\cdot bc$$

$$= \frac{1}{2}\,\frac{ab}{bm}\cdot \text{area } bmnc$$

But in the limit

$$bmnc = amnc$$
$$= \text{heat along } ac$$
$$= K_p d\theta$$

and

$$\frac{ab}{bm} = \frac{d\theta}{\theta}$$

$$\therefore\ abc = \frac{1}{2}\,\frac{d\theta}{\theta}\cdot K_p d\theta$$

Equating the two areas, we get as before

$$d\theta = \frac{av\theta dp}{K_p}$$

The adiabatics are steeper than the isothermals on the indicator diagram. If a is positive, *i.e.* if a body expands with rise of temperature at constant pressure, then $d\theta$ is positive if p is increased. The adiabatic, as AB, Fig. 165, rises then from a lower to a higher temperature, and must slope up more rapidly than the isothermal.

If a is negative so is $d\theta$. In this case the isothermals for lower temperatures are above those for higher temperatures, on the diagram, and as the adiabatic AB rises to a lower temperature, again it must slope more rapidly.

The formula $d\theta = av\theta dp/K_p$ has been verified in various cases, first by Joule for water and sperm oil (*Scientific Papers*, vol. i. p. 474), the liquid being suddenly compressed in a copper vessel by the addition of weights on to a piston in a small connected cylinder. The change of temperature was determined by a thermo-electric circuit, of which one junction was in the centre of the copper vessel, the other in a constant temperature water bath outside. To justify the use of the thermoelectric circuit it was necessary to prove that the pressure alone did not affect the E.M.F. This Joule did by having the difference of temperature between the junctions in one case small and in another great, and showing that pressure produced the same effect in each case. Had the E.M.F been altered by the pressure the alteration would have been greater for a greater temperature difference.

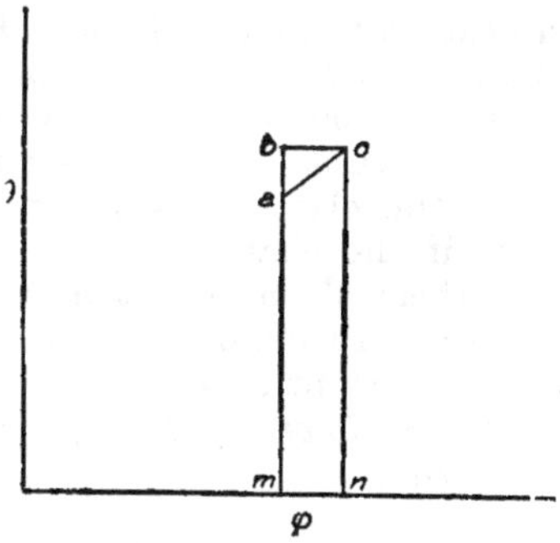

FIG. 166.—Entropy-Temperature diagram.

We may take as a specimen an experiment on water at 30° C. The pressure put on was 53,634 lbs. per sq. ft., equivalent to $2{\cdot}57 \times 10^7$ dynes per sq. cm.

At 30° $a = {\cdot}00032$
$\theta = 303$
$v = 1$
$K_p = 4{\cdot}2 \times 10^7$, the mechanical equivalent of 1 calory.

Then by the formula we should have

$$d\theta = \frac{{\cdot}00032 \times 303 \times 2{\cdot}57 \times 10^7}{4{\cdot}2 \times 10^7}$$

$$= {\cdot}059^\circ$$

Joule found by experiment

$$d\theta = {\cdot}054^\circ$$

According to the formula $d\theta$ should change sign with a and, other things being the same, should be proportional to it. This conclusion was verified by Joule for water. He showed that on compression there was a very slight cooling at 1·2° C., which changed to a very slight warming at 5° C., and the higher the temperature above 4° C. the greater was the warming.

The Specific Heats at Constant Pressure and at Constant Volume.—The specific heat of a substance at a given temperature is not an invariable quantity but depends to some extent on the conditions under which the heat is communicated, this dependence being most marked in the case of a gas. It is usual to consider especially two kinds of specific heat, that when the pressure is maintained constant, and that when the volume is maintained constant. For gases (as we have already seen in chap. vi.) these two specific heats differ very nearly by the heat equivalent of the external work done in the expansion, and the difference is a large fraction of either. For solids and liquids the difference, though appreciable, is not nearly so important. We shall denote the specific heats at constant pressure and constant volume, respectively, by C_p and C_v when in heat units, and by K_p and K_v when in work units. The elasticity of a substance, *i.e.* the ratio of a small change of pressure to the consequent change of volume per unit volume, is also a variable quantity depending on the circumstances of the change. There are two especially important cases, the elasticity when the change is adiabatic—a condition which holds if the change is very sudden; and the elasticity when the change is isothermal—a condition which holds if the change is sufficiently slow. The two are denoted respectively by e_ϕ and e_θ. We shall now prove a very important relation connecting the specific heats and the elasticities.

The ratio of the two specific heats γ is equal to the ratio of the two elasticities,

$$\text{or} \qquad \frac{K_p}{K_v} = \frac{e_\phi}{e_\theta}$$

In Fig. 165 let BA and BC represent adiabatic and isothermal elements through B, cutting a line of equal pressure in AC, and let BH be perpendicular to AC

$$e_\phi = v \times \frac{\text{change of pressure along BA}}{\text{change of volume along BA}}$$

$$= v \times \frac{\text{BH}}{\text{HA}}$$

Similarly

$$e_\theta = v \cdot \frac{\text{BH}}{\text{HC}}$$

Then

$$\frac{e_\phi}{e_\theta} = \frac{\text{HC}}{\text{HA}}$$

Now $$\frac{\text{HC}}{\text{HA}} = \frac{\text{heat received along HC}}{\text{heat received along HA}} \quad \text{(since both are at constant pressure)}$$

$$= \frac{\text{heat along HC}}{\text{heat along HB}} \quad \text{(since AB is an adiabatic)}$$

$$=\frac{K_p d\theta}{K_v d\theta} \quad \text{(where } d\theta \text{ is the temperature difference between H and BC)}$$

$$=\frac{K_p}{K_v}$$

Then

$$\frac{e_\phi}{e_\theta}=\frac{K_p}{K_v}=\gamma$$

This relation only involves the Second Law of Thermodynamics in the assumption of the smallness of HBA, and it was obtained by Laplace before the law was known.

The ratio is usually denoted by γ.

The Difference between the Two Specific Heats.—Taking unit mass of a substance round HBC, and using the value for the heat along BC found on p. 285, we have

$$K_v d\theta + a\theta e_\theta dv - K_p d\theta = \text{area of HBC}$$

$$=\frac{avd\theta dp}{2}$$

$$=0 \text{ in the limit,}$$

whence

$$K_p - K_v = \frac{a\theta e_\theta dv}{d\theta}$$

Putting $dv = avd\theta$ we have

$$K_p - K_v = a^2 e_\theta v\theta$$

Dividing by K_p we get at once

$$\gamma = \frac{1}{1-\frac{a^2 e_\theta v\theta}{K_p}}$$

From this we see that γ is constant when $a^2 ve_\theta/K_p$ is constant.

The Numerical Value of γ, the Ratio of the Two Specific Heats.—The last formula at once gives us γ, when the quantities on the right are known.

Thus let us take :—For air at 0° and 760 mm.,

$a = \cdot 00367$
$v = 773\cdot 4$
$\theta = 1/a$
$e_\theta = p = 1$ atmosphere $= 1013600$ dynes per cm^2
$K_p = \cdot 2389 \times 4\cdot 19 \times 10^7$

whence $\gamma = 1\cdot 41$ nearly.

If we take Swann's value (p. 86) as $0\cdot 2414$ we get $\gamma = 1\cdot 396$.

For water at 30°.

$a = \cdot 00032$
$v = 1$
$\theta = 303$
$e_\theta = 1\cdot 014 \times 10^6 \times 22000$
(since the decrease in volume is 1 in 22000 for 1 atmo.)
$K_p = 4\cdot 2 \times 10^7$

whence $\gamma = 1\cdot 017$

For iron at 30° we may take the quantities as approximately :—

$$
\begin{aligned}
& a = \cdot 000033 \\
& v = \frac{1}{7\cdot 8} = \cdot 128 \\
& \theta = 303 \\
& e_\theta = 1 \times 10^{12} \\
& K_p = \cdot 112 \times 4\cdot 2 \times 10^7 \\
\text{whence} \quad & \gamma = 1\cdot 009
\end{aligned}
$$

Experimental Determinations of γ for Gases.—This ratio has been the subject of much research since Laplace first showed that the formula for the velocity of sound, found by Newton, must be multiplied by the square root of γ, and it has been determined from that velocity, from comparisons of the two elasticities, and from measurements of the two specific heats. We shall give short accounts of some of the methods used.

1. The velocity of sound.

The velocity of propagation of waves of longitudinal disturbance in air was shown by Newton to be $U = \sqrt{\text{elasticity} / \text{density}}$. He supposed the elasticity to be e_θ.

Laplace pointed out that the elasticity used in the propagation of sound waves in air is the adiabatic elasticity e_ϕ, for the alternations of compression and rarefaction are so rapid and the conductivity of air is so low that practically no conduction of heat takes place out of the compressions or into the rarefactions. This supposition is confirmed by the consideration (Lord Rayleigh's *Sound*, vol. ii. p. 26) that if heat were partially diffused out of and into the waves by conduction, the entropy would, on the whole, increase, and the operations would then not be reversible. When, therefore, a rarefaction followed a compression, the work done in expansion outwards would be less than that done in compression inwards, and with the dissipation of energy the sound would die away. Lord Rayleigh has shown that if the elasticity were only slightly below the adiabatic elasticity, the rate of dying away would be very rapid. This consideration only applies when there is partial conduction. With condensations and rarefactions taking place so slowly that the temperature remained constant, all heat transfers would take place at the same temperature and the entropy would remain constant. The operations would therefore be reversible, and the waves would be propagated without dissipation of energy. As the waves of sound are propagated with very little dissipation of energy, the elasticity is either isothermal or adiabatic, and not between the two.

The former supposition, that it is isothermal, gives

$$
\begin{aligned}
U &= \sqrt{\text{pressure} \div \text{density}} \\
&= \sqrt{1013600 \times 773\cdot 4} \\
&= 28{,}000 \text{ cm. per sec.}
\end{aligned}
$$

But experiment gives between 33,100 and 33,300, say 33,200, as the value at 0° and 760 mm. Then the latter supposition, that the elasticity is adiabatic, is alone admissible.

But $e_\phi = \gamma e_\theta = \gamma p$ for air

$$\therefore 33200 = \sqrt{\gamma \times \text{pressure} \div \text{density}}$$
$$= \sqrt{\gamma} \cdot 28000$$

whence $\gamma = 1{\cdot}406$

2. By comparison of the two elasticities on change of volume of the gas.

This method was devised by Clément and Desormes. They used a large glass reservoir A (Fig. 167), furnished with a tap B and a water manometer aa' to indicate the pressure of the contained air.

To begin with, A was partially exhausted, and when its contents were at the temperature of the surroundings, the pressure indicated by the level a of the manometer was read. The tap B was now turned on for a moment to establish equilibrium with the external pressure, and then turned off again. The temperature having risen by this compression, which was taken as adiabatic, it was allowed to fall again to its original value, and the final level a' of the manometer was read.

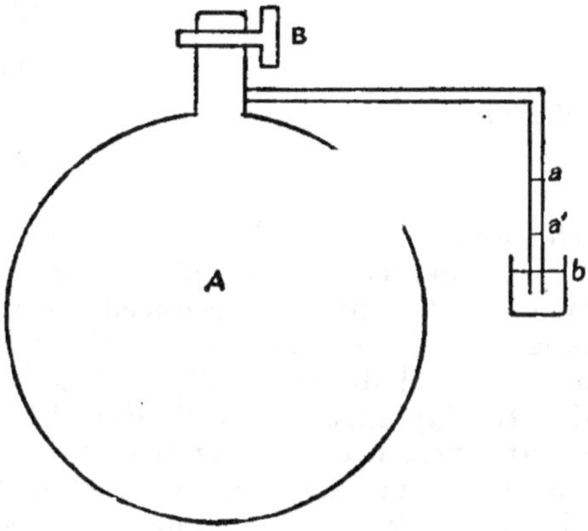

FIG. 167.

The experiment may be regarded as one to determine the ratio of the elasticities. The air originally filling A was regarded as crushed down to a fraction of the volume of A, the same at both readings of the manometer. But if we make the *same* change of volume, first suddenly and adiabatically, and then slowly and isothermally,

$$\gamma = \frac{e_\phi}{e_\theta} = \frac{\text{sudden increase of pressure}}{\text{slow increase of pressure}}.$$

Now, in this experiment the volume of the gas is altered by a definite amount *suddenly*, and the pressure increase noted. The volume being kept the same, the temperature falls to that of the surroundings, and then the increase of pressure above its original value is the same as if the same change of volume had been made slowly, so that

$$\gamma = \frac{ab}{aa'}.$$

Taking one of their results as a specimen—

$$\left.\begin{array}{ll}\text{Atmospheric pressure} = 766{\cdot}5 \\ \text{Initial} \quad ,, \quad = 752{\cdot}7 \\ \text{Final} \quad ,, \quad = 762{\cdot}9\end{array}\right\} \quad ab = 13{\cdot}8, \quad aa' = 10{\cdot}2$$

$$\therefore \gamma = \frac{138}{102} = 1{\cdot}353.$$

This assumes that the pressure changes are exceedingly small, which is obviously not exact.

Taking p_o, P, and p as the initial, atmospheric, and final pressures, and V_o, V, and V as the corresponding volumes, the first change being adiabatic, we have (p. 295)

$$p_o V_o^\gamma = PV^\gamma \qquad (1)$$

The second change being isothermal,

$$p_o V_o = pV \qquad (2)$$

Then

$$\frac{p_o}{P} = \left(\frac{V}{V_o}\right)^\gamma \text{ from (1)}$$

$$= \left(\frac{p_o}{p}\right)^\gamma \text{ from (2)}$$

whence

$$\gamma = \frac{\log \frac{p_o}{P}}{\log \frac{p_o}{p}}$$

This gives $\gamma = 1{\cdot}348$ in the above example.

The method, as originally devised, has a serious fault in the assumption that the newly introduced air will have the same temperature the moment after introduction as the air already in the receiver, and that all will cool down equally. This fault was avoided in a modification due to Gay Lussac and Welter, in which, to begin with, A contained air at a pressure p_o higher than the atmospheric pressure P. On turning the tap the pressure fell to P, while the air which had previously occupied a part V_o of the volume expanded to fill the whole volume V.

Finally the pressure rose to p.

Then

$$p_o V_o^\gamma = PV^\gamma$$

and

$$p_o V_o = pV$$

whence as above

$$\gamma = \frac{\log \frac{p_o}{P}}{\log \frac{p_o}{p}}$$

In one of the experiments (Baynes's *Thermodynamics*, p. 136), the pressures on an arbitrary scale were—

$$P = 1{\cdot}0096,\ p_o = 1{\cdot}0314,\ p = 1{\cdot}0155.$$

whence

$$\gamma = 1{\cdot}3745.$$

There are still sources of error in the method. The change of volume is not accurately adiabatic, for momentum is produced in the issuing gas, and if the tap is turned off at the first moment of equalisation of pressure with that of the atmosphere, the air being still in motion, its kinetic energy will be gradually converted to heat, and so the entropy will be increased. If, however, the tap be left on for a longer time the outrush will cease and be followed by an inrush, and there will be a series of oscillations, the energy of the oscillations being gradually converted to heat, and the entropy being thereby increased. M. Cazin showed the existence of these oscillations, and was able to turn off the tap at the end of the first outrush. The expansion up to that point might be considered as equivalent to an adiabatic one, for nearly all the work done by the gas has gone out of it, the amount retained through viscosity being so far negligible. M. Cazin's determinations were much nearer those given by the velocity of sound than the earlier results.

Röntgen used as a pressure indicator a corrugated plate like that in an aneroid barometer, fixed over an opening in the reservoir. Any movement of the plate was communicated to a mirror in which was viewed the reflection of a scale.

He obtained the following values for γ:—

Air	1·4053
Carbon dioxide . . .	1·3052
Hydrogen	1·3852

The value for hydrogen is no doubt too small. The gas so rapidly regains the temperature of the surroundings that the adiabatic change of pressure cannot be observed accurately.*

3. By direct comparison of the two specific heats.

If a quantity of heat H is given to a unit mass of gas at constant pressure it raises the temperature by $d\theta$ where

$$H = C_p d\theta.$$

If V is the initial volume and dv the increase, a the co-efficient of expansion,

$$dv = aV d\theta,$$

$$\therefore H = \frac{C_p dv}{aV}$$

If the same quantity H is given to the unit mass with constant volume it raises the temperature by $d\theta'$ where

$$H = C_v d\theta'.$$

If P is the initial pressure, dp its increase, β the co-efficient of pressure increase,

$$dp = \beta P d\theta',$$

$$\therefore H = \frac{C_v dp}{\beta P}$$

Equating the two values of H, we get

$$\frac{C_p dv}{aV} = \frac{C_v dp}{\beta P}$$

and

$$\frac{C_p}{C_v} = \frac{aV dp}{\beta P dv}$$

* Among later researches we may refer to a paper by Capstick, "On the Ratio of the Specific Heats of the Paraffins and their Monohalogen Derivatives," *Phil. Trans.* A. 1894, Pt. I., p. 1.

Jamin and Richard carried out a method based upon this result, in which a definite quantity of heat H was communicated to the gas contained in a large vessel by means of a wire heated by an electric current.

In one case the gas was allowed to expand at constant pressure, and dv was measured, and in the other the volume was kept constant and dp was observed. No absolute measures could be made owing to the loss of heat from the wire by radiation, which was unknown. But it was the same in both parts of the experiment, and so the quantity H remaining in the gas was the same.

They obtained as values of γ—

Air	1·41
Hydrogen	1·41
Carbon dioxide	1·29

Joly has succeeded in making direct measurements of the specific heat at constant volume by the steam calorimeter, as already described in chap. vi. His result for air at 0° and 760 mm. is

$$C_v = 0{\cdot}17154.$$

If we take Wiedemann's value for that at constant pressure, $C_p = 0{\cdot}2389$,

we get
$$\gamma = \frac{C_p}{C_v} = 1{\cdot}393.$$

With Swann's value taken as 0·2414 at 0° C. we get $\gamma = 1{\cdot}407$.

The Determination of the Absolute Temperature at 0° *C. from the Value of* γ.—If in the equation for the difference of specific heats

$$K_p - K_v = a^2 v e_0 \theta$$

we substitute for K_v its value $\dfrac{K_p}{\gamma}$

we get
$$K_p = \frac{\gamma}{\gamma - 1} a^2 v e_0 \theta$$

Rankine, taking the value of γ given by the velocity of sound, used this equation to determine K_p and C_p, but if we use the value of K_p found by direct experiment, it may also be regarded as an equation giving θ, the absolute temperature at 0° C., the length of the degrees of the absolute and air scales so nearly coinciding at 0°, that a may be taken as the same for each.

Putting the equation in the form

$$\theta = \frac{\gamma - 1}{\gamma} \times \frac{K_p}{a^2 v e}$$

if we take

$$K_p = {\cdot}2389 \times 4{\cdot}2 \times 10^7$$
$$a = {\cdot}00367$$
$$v = 773{\cdot}4$$
$$e_0 = 1{\cdot}0136 \times 10^6 \text{ (Paris atmosphere)},$$

and use the value of γ given by the velocity of sound as 1·406 we get

$$\theta = 274{\cdot}4$$

But if we use the value of γ obtained by Joly, viz. 1·393, then we get

$$\theta = 268{\cdot}1$$

Inasmuch as a difference of ·002 in the value of γ makes a difference of about 1° in θ, and as the values of γ obtained by different methods differ by far more than this, the value of θ thus obtained can only be regarded as an approximation justifying the use of the gas scale for the absolute scale when no great exactness is required. We shall see later how a much closer approximation may be found.

The Relations between Volume, Pressure, and Temperature of a Gas expanding Adiabatically.—We may use the approximate equation—

$$\begin{aligned} pv &= p_0v_0(1+\alpha t) \\ &= p_0v_0\alpha\left(\frac{1}{\alpha}+t\right) \\ &= \mathrm{RT} \end{aligned}$$

where $\mathrm{R}=p_0v_0\alpha$ and T is the gas temperature measured from $\frac{1}{\alpha}$ below 0° C.

Now the elasticity of a substance is, by definition,

$$-v\frac{dp}{dv}$$

If T is constant

$$vdp+pdv=0$$

or $$e_\theta=\frac{vdp}{dv}=p$$

But $$e_\phi=\gamma e_\theta=\gamma p$$

Hence if any small adiabatic changes of p and v occur

$$-\frac{vdp}{dv}=\gamma p$$

or $$\frac{dp}{p}=-\gamma\frac{dv}{v}$$

whence $$\log p=\log Cv^{-}$$

or $pv\ =\mathrm{C}$ where C is a constant. If we eliminate v by means of the equation $pv=\mathrm{RT}$ we have

$$p^{\gamma-1}=\frac{\mathrm{R}^\gamma\mathrm{T}^\gamma}{\mathrm{C}}=\mathrm{AT}^\gamma$$

Or if we eliminate p we have

$$v^{\gamma-1}=\frac{\mathrm{C}}{\mathrm{R}}\cdot\frac{1}{\mathrm{T}}$$

As an illustration let us find the fall of temperature if a mass of air at 15° C. expands adiabatically to double its volume, the value of γ being 1·4. If the new temperature is T′ we shall have the two equations

$$v^{\gamma-1}=\frac{\mathrm{C}}{\mathrm{R}}\cdot\frac{1}{288}$$

$$(2v)^{\gamma-1}=\frac{\mathrm{C}}{\mathrm{R}}\cdot\frac{1}{\mathrm{T}'}$$

and dividing

$$T' = \frac{288}{2^{\gamma-1}} = \frac{288}{2^{0 \cdot 4}}$$

$$= \frac{288}{1 \cdot 32} = 218$$

a fall of 70°.

Decrease of Temperature Upwards with the Atmosphere in Convective Equilibrium.—As already pointed out (p. 216), the limiting condition of equilibrium for the atmosphere is one of "convective" equilibrium, *i.e.* one in which the temperature decreases upwards in such a way that a mass of air in ascending will expand and cool in doing work so as always to be at the temperature of its surroundings. Such expansion is adiabatic. Hence the temperature and pressure should be connected in this condition in air in which there is no condensation by the equation

$$p^{\gamma-1} = AT^{\gamma}$$

Let us apply this to find the decrease per 100 feet rise from the surface, which we will suppose to be at 0° C.

Taking logarithms we have

$$(\gamma - 1) \log p = \log A + \gamma \log T$$

whence

$$(\gamma - 1)\frac{dp}{p} = \gamma \frac{dT}{T}$$

or

$$dT = \frac{\gamma - 1}{\gamma} \cdot \frac{dp}{p} T.$$

If dp is the decrease due to a rise of 100 feet,

$$\frac{dp}{p} = \frac{100}{26000}$$

since 26,000 feet is the height of the homogeneous atmosphere.

$$\gamma = 1 \cdot 41$$
$$T = 273$$
$$\therefore \ dT = \frac{\cdot 41}{1 \cdot 41} \times \frac{100}{26000 \times 273}$$
$$= \cdot 3^{\circ}$$

Or, when the atmosphere is just in equilibrium, the temperature decreases by about ·3° per 100 feet rise, when the mean is about 0° C. With a more rapid decrease there will be circulation.

The Internal Energy taken up by a Gas in Expanding.—We have already mentioned (chap. viii.) Mayer's assumption that a gas in expanding does no work against internal forces, and have described the experiment, first made by Gay Lussac and later made by Joule, to test the truth of the assumption. In Joule's experiment air was compressed in one vessel to a pressure of 22 atmospheres, and was then allowed to expand into another equal vessel previously empty, so that no energy was given to or taken from the air by external work. If in mere expansion the gas had done work against its own cohesion, it would have drawn on its own heat for the supply, and the temperature would have fallen.

Joule could not detect any alteration of temperature on the whole. But the heat capacity of the gas was so small compared with that of the calorimeter and its contents that a very small change in the temperature of the gas would not have affected the temperature of the calorimeter sensibly. The result of the experiment, therefore, only showed that Mayer's assumption was not far from the truth.

Lord Kelvin afterwards proposed a method of experiment giving a much more delicate test for the presence or absence of internal work in gaseous expansion. This method was put into practice by himself and Dr. Joule.* Their researches showed that in Joule's original experiment there must have been a slight cooling effect, *i.e.* that some work must have been done against internal forces. The method of experiment also gave a means of comparing the absolute scale of temperature with the indications of the air thermometer.

In order to understand the principle of the method, let us suppose that a long pipe AB (Fig. 168) is obstructed in the middle by a narrow passage C, and let air be gently driven towards C by a piston at A, on which the pressure is P. On the other side of C let the air expand against a less presssure p, applied by a piston at B. Let the volume of unit mass be V on the A side and v on the B side. Then the work done

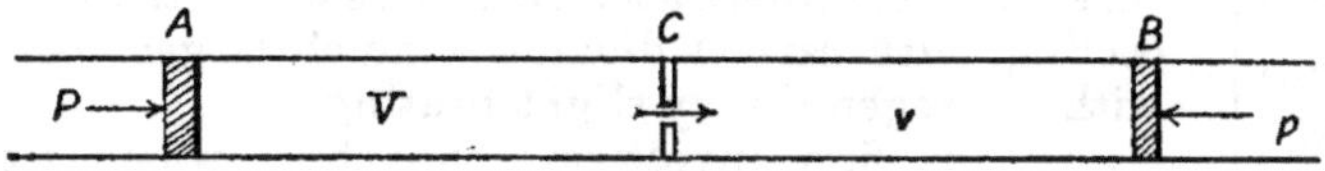

FIG. 168.

on the gas on the A side is PV per unit mass, and the work done by the gas on the B side is pv for the same mass. Let the walls of the enclosure be either non-conducting or everywhere at the temperature of the gas in contact with them.

It may appear at first sight that this is an adiabatic transaction, since no heat is communicated from the outside as heat; but if, as we suppose, the passage at C is very small, there will be a "rapid" there, the air passing through it acquiring a considerable amount of kinetic energy. But this kinetic energy is quickly lost on the B side through the viscosity of the surrounding air, *i.e.* some of the work done on the gas gives rise to kinetic energy and this kinetic energy is transformed to heat, so that there is, on the whole, a gain of entropy.

Let us first suppose that the temperature is found to be the same on the two sides of C at a sufficient distance away from the "rapid." Let us also suppose that Boyle's law is true.. Then

$$PV = pv,$$

or the work done on the gas on the A side is equal to that done by the gas on the B side. On the whole, then, no energy is given to the gas from the outside, and since the temperature is unaltered, we conclude that, while the volume has increased from V to v, the gas has not transformed any of its own heat into energy of molecular separation. In other

* Joule's *Scientific Papers*, vol. ii. p. 217, *or* Thomson's (Lord Kelvin's) *Papers*, vol. i. p. 333.

words, the internal or intrinsic energy is the same for the same temperature whatever the volume.

In such a case, therefore, Mayer's assumption would be justified.

But if, as was actually the case in Thomson's and Joule's experiments, there is an alteration of temperature on the passage through C, this alteration may indicate either that Boyle's law is untrue or that Mayer's assumption is untrue, or that both are untrue. We know already that Boyle's law is only an approximation, and it remains therefore to determine whether the alteration of temperature indicates that Mayer's assumption is also only an approximation.

FIG. 169.

In order to measure the change of temperature accurately, Thomson and Joule forced the gas through a porous plug consisting of a boxwood cylinder *bb* (Fig. 169), 1·5 in. internal diameter, containing two perforated brass plates A, B, 2·72 in. apart, the space between these plates being filled with cotton or silk, more or less compressed. The gas was driven up to A through a long pipe immersed in a constant-temperature bath, and flowed out from B round a thermometer, being allowed there to come to the atmospheric pressure. When a steady state of temperature was everywhere attained, there was with air a sensible cooling, with carbon dioxide a much larger cooling, and with hydrogen a very slight heating.

Let us first investigate the effect which we might expect if Boyle's law does not exactly express the relation between volume and pressure, neglecting meanwhile the effect of change in intrinsic energy.

According to Amagat's researches, in the case of air the product *pv* decreases by about 12 in 1000 for an increase of pressure from 1 to 31 atmospheres. We are perhaps not far wrong in assuming that the decrease is nearly uniformly at the rate of 1/2500 per atmosphere within these limits. With this assumption, if the air is kept at the same temperature on the two sides of the plug, and if it expands to the atmospheric pressure *p* on the further side, then

$$PV = pv\left(1 - \frac{P-p}{p} \cdot \frac{1}{2500}\right)$$

and

$$PV - pv = -\frac{P-p}{p} \cdot \frac{pv}{2500}$$

or

$$pv \text{ exceeds } PV \text{ by } \frac{P-p}{p} \cdot \frac{pv}{2500}$$

And on the whole external work $\frac{P-p}{p} \cdot \frac{pv}{2500}$ is done *by* the air, and in order to keep the temperature constant this amount of energy must be supplied from without.

If it be not supplied, the air will cool by an amount nearly given by

$$\frac{P-p}{p}\cdot\frac{pv}{2500}\cdot\frac{1}{K_p}$$

where K_p is the mechanical value of the specific heat at constant pressure. In fact the cooling will be rather less, because with lowering of temperature the expansion against the atmospheric pressure is rather less, pv is diminished, and therefore $pv-PV$ is less.

In some of the experiments P was about 130 inches of mercury and p was 30 inches.

pv is about 780×10^6 in C.G.S. units, while K_p is $\cdot 2375\times 4\cdot 2\times 10^7$.

Substituting these values in the expression for the cooling, we find that it should be about 0·1° C. But the actually observed cooling was about 0·9° C. Then some of the energy was also taken up in the separation of the molecules, or there was an increase of intrinsic energy on expansion.

In a similar manner it may be shown that the cooling in the case of carbon dioxide was chiefly due to change in intrinsic energy.

In the case of hydrogen the product *pv increases* with the pressure. Then, if the intrinsic energy is independent of the volume, energy must be subtracted from the gas to keep the temperature the same in the expanded condition. If it be not subtracted, the temperature will rise. The actual effect observed was a slight heating, and of the order due to the increase of pv. Hence it was concluded that the change in intrinsic energy due to change in volume is very small.

Use of the Porous Plug Experiment for the Comparison of the Air Scale with the Absolute Scale.—We may express the change in intrinsic energy of the gas in two ways; (1) in terms of the energy communicated from the outside and (2) in terms of the observed changes of physical condition, temperature, pressure, &c., and equating these two expressions we obtain a value for the absolute temperature.

In considering the energy received from the outside, we remark that on the whole there is very little change of temperature in passing through the plug, though no doubt there are very great local variations in the "rapids"—the fine streams issuing from the pores. The gas is therefore very nearly in temperature equilibrium with the surrounding walls and the temperature slopes in these walls are very gradual. This is still more nearly true since the gas was allowed to flow for so long a time before observations were made that affairs had arrived at a steady state and the wall at each point was at practically the same temperature as the gas passing it. The walls were bad conductors so that the heat conducted into or out of the gas might be neglected. The kinetic energy in the gas was also small when it had settled down to steady motion beyond the "rapids." Then the gain of intrinsic energy is practically due to the difference between the works done on and by the gas on the two sides of the plug or if E is the initial, and e the final, intrinsic energy,

$$e-E=PV-pv.$$

In finding the value of $e-E$ in terms of the observed changes in physical condition, we may suppose the change made in any convenient way, for the intrinsic energy corresponding to a given state is quite independent of the manner in which that state is reached. Let A (Fig. 170) represent the state of unit mass of the gas on the entrance side, and B that of the same mass on the exit side of the plug.

We shall suppose the change from A to B to take place along the isothermal AC, and the equal pressure line CB, and to avoid consideration of the variation of the different quantities, such as coefficient of expansion and specific heat, we shall suppose A and B so near together that these quantities are practically constant.

The heat put in along AC is

$$a\theta V(P-p) \text{ or } a\theta V dP.$$

The heat put in along CB is $K_p\, d\theta$

where $d\theta$ is the rise in temperature. The external work done is ACBNL.

Then $$e-E=a\theta V dP+K_p d\theta-\text{ACBNL}.$$

But we also know that $$\begin{aligned} e-E &= PV-pv \\ &= \text{AHOL}-\text{BKON} \end{aligned}$$

Then $$\begin{aligned} a\theta V dP+K_p d\theta &= \text{AHOL}-\text{BKON}+\text{ACBNL} \\ &= \text{ACKH} \\ &= V dP \end{aligned}$$

and $$\theta=\frac{1}{a}-\frac{K_p}{aV}\cdot\frac{d\theta}{dP}. \qquad (1)$$

If $d\theta=0$ then $\theta=\frac{1}{a}$ so that if a substance had the same temperature on the two sides of the plug, equal expansions of that substance would indicate equal intervals of temperature on the absolute scale. Hydrogen approaches most nearly to this condition, hence its use in the standard gas thermometer.

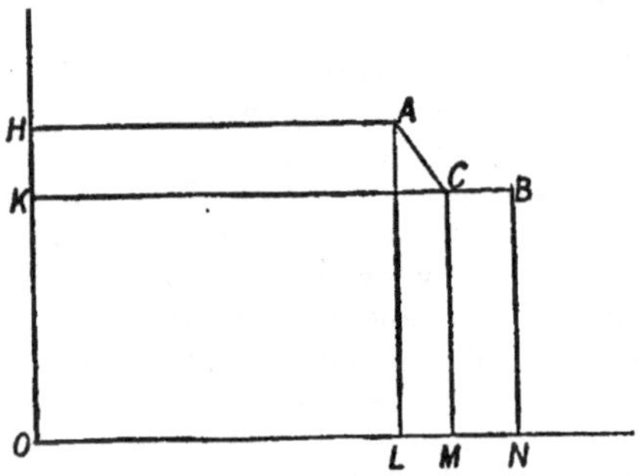

FIG. 170.

In their experiments Thomson and Joule found that $d\theta$ was proportional to dP for a given temperature, throughout the range of pressures employed, and inversely as the square of the absolute temperature, so that we may put

$$\frac{d\theta}{dP}=-\frac{A}{\Pi}\cdot\frac{\theta_o^2}{\theta^2}$$

where θ_o is the absolute temperature of 0° C., Π the atmospheric pressure, and A a constant for each gas, being the cooling per atmosphere difference of pressure at 0° C.

The observed values for A were

Air	+ ·275
Carbon dioxide	+ 1·388
Hydrogen	− 0·03 about

Though $\frac{d\theta}{dP}$ was obtained with finite differences of pressure, it is evident that, since it is independent of the pressure, we may use it in equation (1), where the difference of pressure is supposed to be very small. We obtain

$$\theta=\frac{1}{a}+\frac{K_pA\theta_o^2}{aV\Pi\theta^2} \qquad (2)$$

We may at once obtain from (2) the absolute temperature at 0° C., assuming that the degrees have the same value as those of the air scale about that temperature. Thus, for air, since at 0° C. $\theta_o=\theta$,

$$\frac{K_pA\theta_o^2}{aV\Pi\theta^2}=\frac{\cdot 2389\times 4\cdot 18\times 10^7\times \cdot 275}{\cdot 00367\times 773\times 1013600}$$

$$=\cdot 95$$

or

$$\theta_o=\frac{1}{a}+\cdot 95$$

Or the absolute temperature does not differ from the air temperature by as much as 1° C.

We refer the reader to Thomson's article on "Heat," *Ency. Brit.*, 9th ed., § 60–2, for the mode of comparing the absolute with the gas scale not assuming that the degrees have the same length about 0° C., but taking the interval from 0° to 100° C. as the same on both.

It will be observed that equation (2) implies that a, the coefficient of expansion, is not strictly independent of the density of the gas, for the second term on the right is not constant, but when A is positive it increases as V decreases or as the density increases.

Then $\frac{1}{a}$ must decrease and a increase with the density for gases such as air and carbon dioxide. For hydrogen, on the contrary, a decreases as the density increases.

Starting from equation (2) Thomson (*loc. cit.* § 63) has calculated the expansion of air and other gases at different temperatures and obtains results agreeing very closely with those found experimentally by Regnault.*

The change in internal energy of a gas on expansion has been brought into greater prominence by the "regenerative" method of liquefying gases, in which a gas is forced from a pipe, where it is under great pressure, out through a fine nozzle into the air, the process corresponding very nearly to that of the "porous plug" experiment. At low temperatures

* Maxwell (*Heat*, 5th ed., p. 214) gives an equation obtained by integrating equation (1), which is equivalent to

$$\theta=\frac{1}{a}-\frac{K_p d\theta}{pva\log\frac{P}{p}}$$

In the integration it is virtually assumed that a is constant, an assumption which is shown to be unjustifiable by the consideration that, if a were constant, $d\theta$ would be proportional not to $P-p$ but to $\log\frac{P}{p}$. The proper treatment of the equation after its reduction to the form (2) will be found in the Article "Heat," *Ency. Brit. loc. cit.*

and expansion from high densities there appears to be great cooling in all cases, even in that of hydrogen. And this we might expect, for with the high densities the molecules are, in much larger proportion, within each other's sphere of action, and work is needed to separate them.*

Generalisation of the Indicator Diagram for any Stresses and the Corresponding Strains.—If we take abscissæ to represent strains and ordinates stresses, so measured that areas represent work done on or by the quantity of matter considered, we may use the indicator diagram for any corresponding stresses and strains and draw isothermals and adiabatics just as with pressure and volume.

As an illustration let us suppose we have a wire under an end pull and let the relation between stretch ds per length 1 and pull P per square centimetre section be given by

$$P = Y.ds$$

where Y is Young's modulus. This implies that we are dealing with unit cube of the material.

If Y is constant for a given temperature, the isothermal for, say, 0° C. may be represented by a straight line going through the origin as OH (Fig. 171), the tangent of the slope representing Y. Now as the temperature rises the length under zero pull in general increases. Let α be the ordinary coefficient of length expansion. Drawing $OP = \alpha$, PK will represent the 1° isothermal, and since Y in general decreases with rise of temperature it is at a slightly less slope than OH. Similarly QL will represent the 2° isothermal.

It may be noted that these isothermals, if straight lines, will meet beyond O, at a point in the lower quadrant to the left, say at T. Draw TM to represent the pressure. If Young's modulus at any temperature is given by

$$Y = Y_o(1 - \lambda t)$$

then

$$Y_o = \frac{TM}{OM}$$

$$Y_1 = \frac{TM}{OM + OP} = \frac{TM}{OM + \alpha}$$

and

$$\frac{Y_o}{Y_1} = 1 + \lambda = \frac{OM + \alpha}{OM} = 1 + \frac{\alpha}{OM}$$

whence

$$OM = \frac{\alpha}{\lambda}.$$

With iron α is of the order $\frac{11}{10^6}$, while λ is of the order $\frac{2}{10^4}$, so that $\frac{\alpha}{\lambda} = \frac{11}{200}$, a compression which is never approached. But if it could be reached without disintegration, and if the isothermals were straight lines as supposed, the interpretation would be that at this compression the increase in yielding due to rise of temperature would just neutralise the expansion due to the same rise of temperature.

* For a full discussion see Callendar, "On the Thermodynamical Correction of the Gas Thermometer," *Phil. Mag.* v., 1903, p. 48.

Let us now see how the wire can be taken round a Carnot cycle. Let AB, DC represent neighbouring isothermals at θ and $\theta - d\theta$. Let AD, BC represent adiabatics. We draw these steeper for we may expect the change in tension for given stretch to be greater if no heat is allowed to pass, *i.e.* if we introduce a constraint and the results obtained confirm the supposition. We may take ABCD to represent a Carnot cycle. But we must note that since work is done by the body in contracting we must go round counter-clockwise to get a balance of work done by the body in returning to the starting-point. Heat is taken in

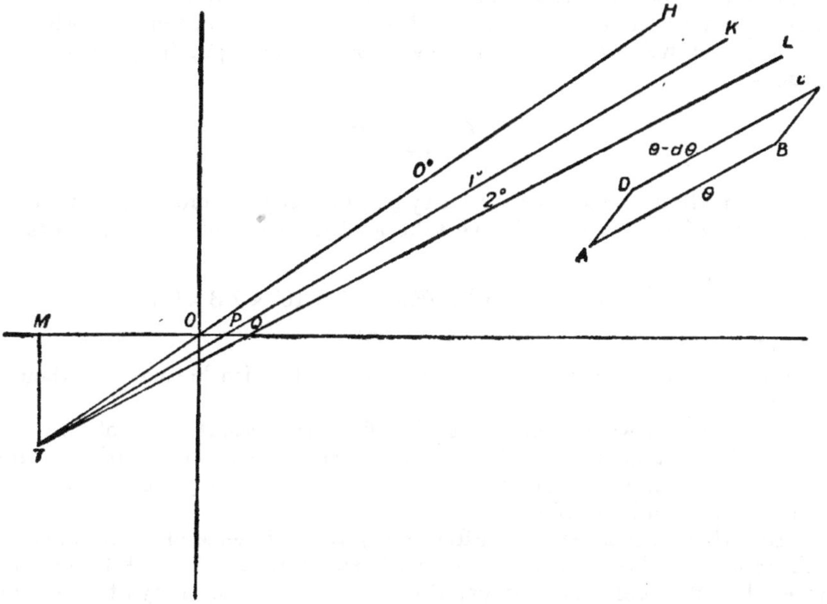

FIG. 171.

along AB and given out along CD, and an adiabatic stretch along AD cools the wire.

Just as in the case of the pressure-volume diagram, it can be shown that the heat along AB is

$$Q = \beta Y_{\theta} \theta ds$$

where ds is the change in length along AB, and β is the change in length under constant load per 1° rise in temperature. If α is the ordinary coefficient of expansion, λ the temperature coefficient of Young's modulus, and s the total stretch up to A, we may put $\beta = \alpha + \lambda s$.

Similarly the change in temperature under adiabatic change of pull dP is

$$d\theta = -\frac{\beta\theta dP}{\rho K_P} = -\frac{\beta Y_{\phi} \theta ds}{\rho K_P}$$

when ρ is the density, K_P the specific heat under constant pull—probably nearly the same as the specific heat under no pull—and Y_{ϕ} is the adiabatic

Young's modulus. We must give the negative sign, since the lower isothermals are higher in the diagram.

In the case of india-rubber β is negative, for if a weight be hung on to a rubber cord, the weight rises if the cord be heated. Hence $d\theta$ is positive for a sudden increase of stretch. This may be verified by suddenly pulling out an india-rubber band and applying it to the upper lip—a sensitive thermoscope—when we can easily detect the warming. Keeping the band extended, it soon falls to the temperature of the air. Now allowing it to contract suddenly, it is very appreciably cooled. The ratio of the specific heats under constant stress and constant strain is easily seen to be equal to the ratio of the adiabatic and isothermal moduli, and we can obtain as the formula corresponding to that on p. 289—

$$\gamma = \frac{1}{1 - \frac{\beta^2 Y_\phi \theta}{K_P \rho}}$$

For small stresses $\beta = a + \lambda s$ may be generally assumed equal to a, the coefficient of expansion. If we take as the values of the constants for steel

$$\beta = \frac{11}{10^6} : Y_\phi = 2 \times 10^{12} : \rho K_P = 4.2 \times 10^7 \times 7.8 \times 0.112$$

we get

$$\gamma = 1.002$$

or the adiabatic elasticity is greater than the isothermal by about 1 in 500.

The term λs would only seriously affect this result when of the order 10^{-6}. Since λ is of the order 10^{-4}, this requires a stretch of the order 10^{-2}, which cannot be given without permanent set, when our assumptions are all inadmissible.

But though not seriously affecting γ the existence of λs implies that the adiabatic elasticity increases with the strain. It will follow that when longitudinal waves travel along a wire, the elasticity at the points of greatest tension will be greatest, and if we represent the waves by the curve of tension, the crests in the curve will move most rapidly and will therefore gain on the hollows.

Similar formulæ can at once be worked out for the case of shear strains and stresses involving the rigidity modulus. If abscissæ represent angles of shear and ordinates tangential stresses per unit area, areas on the diagram will represent work done with unit cube of the substance. We shall not now have anything corresponding to the ordinary coefficient of expansion a, but if we put for the modulus of rigidity

$$n = n_0(1 - \lambda t)$$

a shear ϵ at constant stress will increase by $\lambda\epsilon$ for a rise of 1°, so that we must use $\lambda\epsilon$ instead of β in the last case.

The adiabatic change of temperature for change $d\epsilon$ in shear is

$$d\theta = -\frac{\lambda \epsilon n_\phi \theta d\epsilon}{K_P \rho}$$

that is, there is a cooling for increase of shear. But in general it is of the second order. since both λ and ϵ are usually small.

The consideration of the ratio γ is especially important in this case since the torsional resistance of wires is so frequently used to measure small torques, the torque being deduced from the angle of twist and the time of vibration of the system acted on. Now the angle of twist may be produced slowly and the condition of the wire may be isothermal; the vibrations, on the other hand, may be rapid and the conditions may be somewhere between isothermal and adiabatic conditions. If the adiabatic and isothermal elasticities differed seriously, error might arise in such experiments as that of Cavendish on the constant of attraction.

The error would be the greatest if the conditions in vibration were truly adiabatic and those in steady deflection by the torque truly isothermal.

We may show that

$$\gamma = \frac{1}{1 - \frac{\lambda^2 \epsilon^2 n_\theta \theta}{\rho K_P}}$$

For steel λ is not far from $\frac{2}{10^4}$. Suppose a wire 1 mm. radius and 1 metre long is turned through 1 radian at its lower end, then the extreme value of ϵ is $\frac{1}{10^3}$; n is about 10^{12}.

Substituting the numerical values we get

$$\gamma = \frac{1}{1 - 10^{-6}}$$

and the ratio will be still more nearly 1 for less shear. This implies that the two rigidities are for all practical purposes identical for steel wire.

If we have any other type of stress and strain we have the results

$$d\theta = \pm \frac{\beta \theta dp}{\rho K_P}$$

(+ if the effects of rise of temperature and increase of stress are opposite)

and
$$\gamma = \frac{1}{1 - \frac{\beta^2 E_\theta \theta}{K_P \rho}}$$

where β is the change of strain under constant stress for 1° rise in temperature, E_θ is the isothermal modulus of elasticity, K_P is the specific heat under constant strain, and ρ is the mass in the volume so strained. Thus in a spiral spring we may deal with unit length measured along the axis of the spiral, when ρ will be the mass of the spring per unit length. The reader will find that for such a spring γ is practically 1, as might be expected when we remember that the strain is chiefly torsional.

CHAPTER XIX.

THERMODYNAMICS OF CHANGE OF STATE AND OF SOLUTIONS.

First Latent Heat Equation—Volume of Saturated Steam—Triple Point and Difference of Vapour-Pressures of Ice and Water below 0° C.—Second Latent Heat Equation—Alteration of Vapour-Pressure with Curvature of Liquid Surface—Connection with Change in Melting-Point by Pressure—Solutions—Vapour-Pressure less than that of the Solvent—Osmotic Pressure—Raising of Boiling Point—Lowering of Melting-Point—Semi-Permeable Membranes—Van't Hoff's Application of Thermodynamics—Molecular Theory of Osmotic Pressure.

The First Latent Heat Equation.—If a substance is at such a temperature, that with suitable pressure and volume it can exist in two

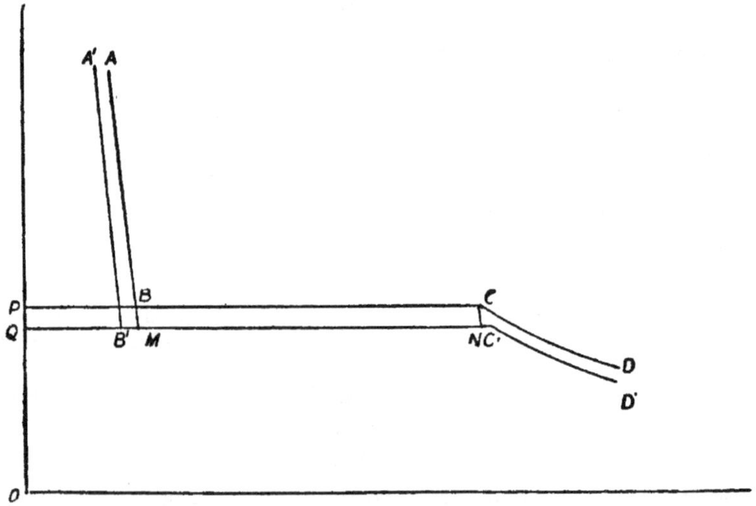

FIG. 172.

states at once, the two being in equilibrium with each other, then the isothermals on the indicator diagram will have a horizontal position. For example steam and water at 100° C. may coexist and the isothermal for 1 gramme at 100° will be represented by a curve, of which the general course only (it is quite out of scale) is represented by ABCD, Fig. 172. AB is water, BC is steam and water at 100° and 1 atmosphere, and CD is steam. If the volume is fixed, say at QM, then the proportions of steam and water will be fixed, and they will be in equilibrium with each other at 1 atmosphere, as long as the temperature is 100°. If the volume increases, then more steam is formed till at C all is steam with a volume about 1640 c.c. On the other hand, if the volume diminishes more water is formed till at B all is water with a volume about 1·04 c.c. The process is reversible at every point.

The general course of a substance at the melting-point may also be represented by a curve like ABCD. For ice and water at 0° DC will represent ice. At C the ice begins to melt, having then a volume about 1·09 c.c. At B it is all water having a volume about 1 c.c., and BA represents water.

If the temperature is not that of the normal boiling-point or the normal melting-point the curve will have the same general character, but the horizontal portion will be at a different level. For steam and water at 99° C., for example, it is at 733·21 mm. instead of 760 mm. For ice and water at $-1°$ C. it is, as we shall see below, raised up to about 134 atmospheres. We do not, in ordinary experience, meet with ice and water in equilibrium except at or very near the normal melting-point. The possibility of such equilibrium at other temperatures was indeed first discovered as a result of thermodynamical reasoning by Professor James Thomson (*Trans. R.S. Edin.*, xvi., 1849, p. 575), and was first experimentally demonstrated by his brother shortly afterwards (see chap. xii.).

The following investigation applies to either change of state. Let ABCD, A′B′C′D′, Fig. 172, be isothermals at θ and $\theta - d\theta$ respectively for 1 gramme of the substance. Let B represent volume v_1 and let C represent volume v_2. Let L be the latent heat taken in along BC in changing state. Through B and C draw adiabatics BM, CN, where M and N are on the $\theta - d\theta$ isothermal. Then BCNMB may be taken as a Carnot cycle, and we have

$$\frac{\mathrm{L}d\theta}{\theta} = \text{external work} = \text{area MBCN}.$$

If the isothermals are sufficiently near to each other, any want of parallelism of BM and CN may be neglected, and the area of MBCN may be taken as

$$\mathrm{BC} \times \mathrm{PQ} = (v_2 - v_1)dp$$

Hence

$$\frac{\mathrm{L}d\theta}{\theta} = (v_2 - v_1)dp$$

or

$$\frac{\mathrm{L}}{\theta} = (v_2 - v_1)\frac{dp}{d\theta} \qquad (1)$$

If then v_2 is greater than v_1, as in the case of steam and water, $\frac{dp}{d\theta}$ is positive, or the boiling-point rises with the pressure.

If, on the other hand, v_2 is less than v_1, as in the case of water and ice, $\frac{dp}{d\theta}$ is negative, or the pressure at which equilibrium is possible increases with fall of temperature. That is, pressure lowers the melting-point.

In change from solid to liquid, in which there is an increase of volume, a rise in pressure corresponds to a rise in temperature, or pressure raises the melting-point.

The Volume of Saturated Steam.—The latent heat equation at once gives us the volume of a saturated vapour at any temperature at which we know the latent heat, the liquid volume and the rate of change of the vapour-pressure. Thus, for water and steam at 100° we have

$$L = 537 \times 4{\cdot}19 \times 10^7$$
$$\theta = 373$$
$$v_1 = 1{\cdot}043$$
$$p_{100\cdot5} = 77{\cdot}371 \text{ cm. of mercury}$$
$$p_{99\cdot5} = 74{\cdot}650 \quad ,, \quad ,,$$

whence
$$\frac{dp}{d\theta} = 2{\cdot}721 \text{ cm. of mercury}$$
$$= 2{\cdot}721 \times 13{\cdot}596 \times 981 \text{ dynes per sq. cm.}$$

Substituting in equation (1), we have

$$v_2 = 1{\cdot}043 + \frac{537 \times 4{\cdot}19 \times 10^7}{373 \times 2{\cdot}721 \times 13{\cdot}596 \times 981}$$
$$= 1663 \text{ c.c.}$$

Fairbairn and Tate (*Phil. Trans.*, cl., 1860) found by direct experiment that $v_2 = 1616$. While there is some uncertainty about the thermodynamic result depending on the uncertainty in the values of the constants, it is easily seen that we cannot accept so low a result as that obtained by Fairbairn and Tate, which was probably vitiated by the adherence of steam to the sides of the containing vessel. The thermodynamic method is preferable here to direct methods.*

It is interesting to compare the value $v_2 = 1663$ with that which steam would have if it followed the same law of expansion as air. We know from Regnault's researches † that the density of steam at low temperatures and pressures is 0·623 of the density of air—almost exactly the value obtained on the supposition that two volumes of hydrogen unite with one volume of oxygen to form one volume of steam. If it maintained the same relative density at 100° C. and 760 mm., its volume would be

$$\frac{1}{0{\cdot}001293 \times 0{\cdot}623} \times \frac{373}{273} = 1696$$

which is certainly greater than the actual volume. Hence steam does not expand according to the laws of Boyle and Charles at higher temperatures.

The Change of Melting-Point under Pressure.—Taking the latent heat equation we have, for water-ice—

$$L = 80 \times 4{\cdot}19 \times 10^7$$
$$\theta = 273$$
$$v_1 = 1{\cdot}09$$
$$v_2 = 1$$

whence
$$\frac{dp}{d\theta} = -\frac{80 \times 4{\cdot}19 \times 10^7}{273 \times 0{\cdot}09} = -136500000 \text{ dynes/cm.}$$

* Ramsay and Young, "The Properties of Water and of Steam," *Phil. Trans.* A., 1892, p. 107.

† *Ann. de Chim. et de Phys.*, 3rd series, t. xv. p. 141.

Taking 1 atmosphere as 1014000 dynes/cm., this gives 134 atmospheres as the pressure required to produce a lowering of 1° C., or the lowering per atmosphere is ·0074° C.

This result, due to James Thomson, was confirmed by Lord Kelvin (*Phil. Mag.* [3] xxxvii., 1850, p. 123). He used a strong glass piezometer, in which was a mixture of ice and water. A manometer and a sensitive thermometer enclosed in an outer tube were inserted in the piezometer, and when the pressure was increased a mean lowering of ·00735° per atmosphere was obtained.

Other experiments have already been described in chap. xii.

The Triple Point and the Difference of Vapour-Pressures of Ice and Water below 0° *C.*—When ice and water are at the melting-point under the pressure of their vapour alone, that point is above 0° C. by about 0·0074°, since the vapour-pressure is less than 5 mm. or nearly an atmosphere less than 1 atmosphere. At this point it is easily shown that the vapour-pressures of ice and water are exactly equal, by taking the substance through a "one temperature" reversible cycle, or a cycle in which the temperature is throughout the same. Since the cycle is reversible, $\int \frac{dQ}{\theta} = 0$, and if θ is constant $\int dQ = 0$, or the total heat given is zero. Then the total external work done is zero. Now imagine some such apparatus as that in Fig. 173, where ice and water at this melting-point, + 0·0074° C., are in contact through the pipe represented in the lower part of the figure, and let all be maintained at one temperature. Suppose, if possible, that ice-vapour has a greater pressure than water-vapour. Then, in allowing ice-vapour to expand into water-vapour, it will do work. Allow this expansion to take place isothermally and reversibly through an engine in the pipe represented in the upper part of the figure. Then the ice will continually give off vapour which can be passed through the engine, where its pressure is reduced to that of water-vapour. It will then condense on the water surface, and the excess of water can freeze on to the lower surface of the ice. Here is a reversible cycle which can continue endlessly, giving off work through the engine. But this is contrary to the principle that the total work must be zero. Then our supposition is wrong, and the vapour-pressure of the ice cannot be different from that of the water.

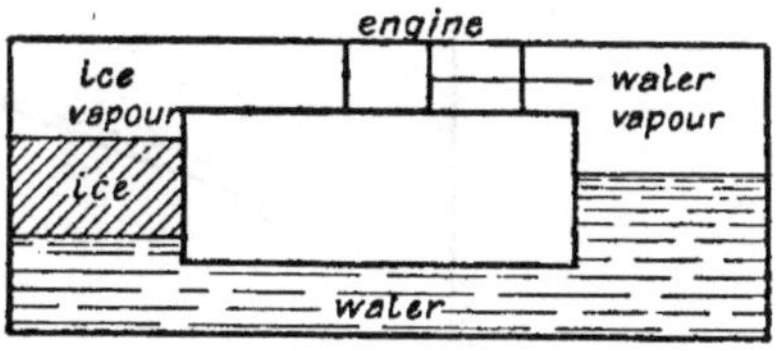

FIG. 173.

At this point, ·0074° C., ice, water, and their vapour are all in equilibrium with each other at the same pressure. The three states or the three "phases" can co-exist without changing from one to another. The point is therefore called the *triple point.*

If we use a temperature-pressure diagram, as in Fig. 174, T will represent the triple point, TA the vapour-pressure of water above the temperature of the triple point, TB the vapour-pressure of ice below that temperature, TC the line giving the relation between melting-point and pressure—a line going up very nearly vertically, sloping in fact only ·0074° per atmosphere.

The three included regions may then be marked vapour, water, ice respectively. If we have all three states or phases present, T is the only point of equilibrium, and the system is said to be *non-variant.*

If, however, we have two phases only, say water and vapour, we may have equilibrium anywhere along TA. We can vary the temperature, but given any one value of the temperature the pressure is fixed. The system is therefore said to be *monovariant.*

If we have only one phase present, as water, we can vary both temperature and pressure so long as they give a point within the region CTA, and the system is said to be *divariant.*

When Regnault made his celebrated researches on the vapour-pressure of ice and water, he supposed that the line AT was continuous

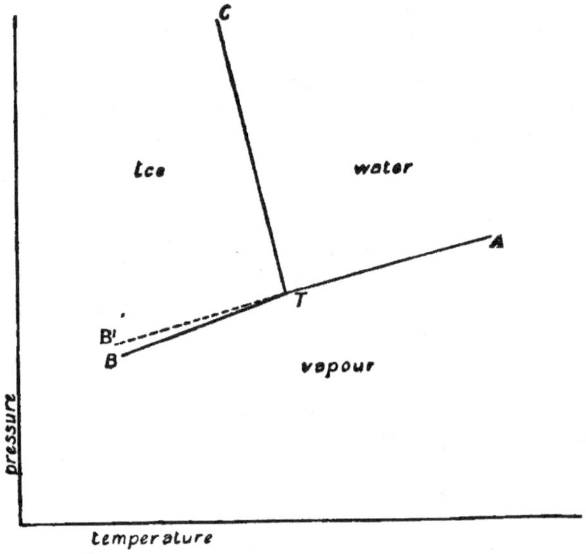

FIG. 174.

through T. But Kirchhoff (*Pogg. Ann.*, ciii. p. 206) showed that the slope of AT continued was different from that of TB, a conclusion which follows at once from the latent heat equation. Putting that equation in the form

$$\frac{dp}{d\theta}=\frac{L}{\theta\,(v_2-v_1)}$$

p may represent the saturation pressure of the vapour of either ice or of water, while v_2-v_1 is the change in volume from the denser to the rarer state. Let ω represent the pressure of water-vapour, ω' that of ice-vapour. Let L be the latent heat from water to steam at 0°, L′ the latent heat of fusion of ice. Then the latent heat from ice to steam is L + L′. We may evidently neglect v_1, the volume in the denser state, since it is exceedingly small compared with v_2, and if we put $\frac{1}{v_2}=\sigma$, the vapour density

$$\frac{d\omega}{d\theta}=\frac{L}{\theta v_2}=\frac{L\sigma}{\theta} \qquad (1)$$

$$\frac{d\omega'}{d\theta}=(L+L')\frac{\sigma}{\theta} \qquad (2)$$

whence
$$\frac{d\omega'}{d\theta}-\frac{d\omega}{d\theta}=\frac{L'\sigma}{\theta} \qquad (3)$$

or dividing by (2)

$$\left(\frac{d\omega'}{d\theta}-\frac{d\omega}{d\theta}\right)\div\frac{d\omega'}{d\theta}=\frac{L'}{L+L'} \qquad (4)$$

These equations obviously apply to any substance at its triple point. In the case of water at 0°

L = 606·5 from Regnault's formula (p. 180)

$L'=80$; $\frac{d\omega'}{d\theta}=4{\cdot}60-4{\cdot}26=0{\cdot}34$ mm. from Regnault's researches on vapour-pressure,

whence
$$\frac{d\omega'}{d\theta}-\frac{d\omega}{d\theta}=\frac{80\times{\cdot}34}{686{\cdot}5}={\cdot}04\text{ mm.}$$

That is, the vapour-pressure of ice below 0° falls down more rapidly than that of water by about 0·04 mm. per degree.

Then the line AT, if continued beyond the triple point, lies above TB; see the dotted line TB′ in Fig. 174.

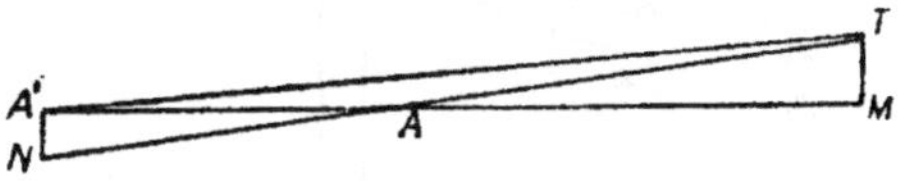

FIG. 175.

The difference in vapour-pressures of ice and water just below 0° is too small to measure accurately, though Regnault's work on re-examination indicates its existence. But if, instead of taking the difference of vapour-pressures for the same temperature, we take the difference of temperatures for the same vapour-pressure, we get a quantity large enough to be measured. For let T (Fig. 175) be the triple point on a pressure-temperature diagram. Let TA, TA′ represent the ice-vapour and water-vapour lines. Let A′AM be a line of equal pressure. Then AA′ is the difference of temperature of ice and water in equilibrium with their vapours at that pressure.

But
$$\frac{AA'}{AM}=\frac{A'N}{TM} \text{ and } AA'=\frac{A'N\,.\,AM}{TM}$$

$$=\frac{\omega-\omega'}{\frac{d\omega'}{d\theta}}=\frac{{\cdot}04\theta}{{\cdot}34}$$

(where θ is the temperature of the water below the triple point)

$$=0{\cdot}116\times\theta$$

or 0·116° per degree below the triple point.

This has been verified by Ramsay and Young ("On the Influence of Change from the Liquid to the Solid State on Vapour-Pressure," *Phil. Trans.*, Part II., 1884, p. 461). They evaporated water and ice in separate vessels at the same pressure below that of the triple point, and found that the water when in equilibrium was colder than the ice.

While it is possible to supercool water in contact with its vapour, *i.e.* to have it to the left of the line TC, Fig. 174, so that TB′ has an actual existence, no method of superheating ice has yet been devised. The prolongation of BT beyond T, therefore, represents nothing physical. But ice can be obtained in a condition represented by points slightly to the right of TC. For instance, if a block of ice at the triple point is subjected to pressure, it is raised in temperature slightly (from the formula of p. 286 about 0·002° C. per atmosphere if we take its coefficient of expansion as 0·00015 and its specific heat as 0·5), while it only melts at its surfaces.

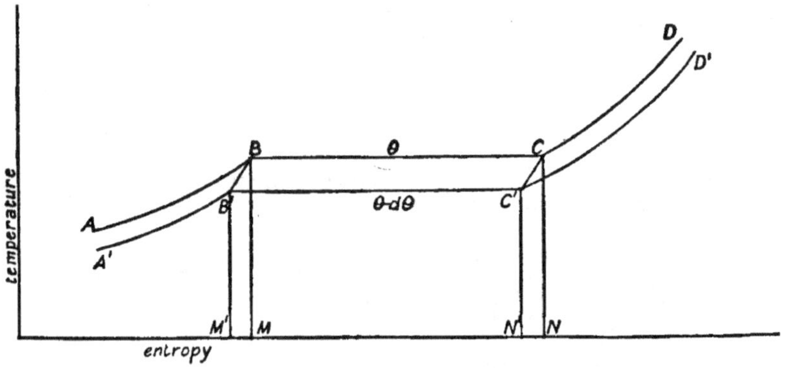

FIG. 176.—Equal Pressure Lines on the Entropy-Temperature Diagram.

Thus it will be represented on Fig. 174 by a line from T nearly vertical but sloping slightly outwards as it rises. And no doubt it would be possible to obtain ice under tension in a condition represented by the prolongation of this line downwards into the vapour region. In the case of water substance there are only three phases and one triple point. But there are substances, such as sulphur, with four phases, *i.e.* two solid forms as well as the liquid and gaseous, and for these substances there are three triple points. For an account of these triple points and for the general phase rule of Willard Gibbs, giving the condition of equilibrium of phases, we refer the reader to Whetham's *Theory of Solution*, chap. ii., or to Findlay's *The Phase Rule*.

The Second Latent Heat Equation.—Another equation connecting latent heat and its change with the specific heats of the two phases may be readily obtained from the entropy temperature diagram. Let ABCD, A′B′C′D′ (Fig. 176) be equal pressure lines, the horizontal portions representing the mixture undergoing change of state with the receipt of latent heat at θ and $\theta - d\theta$ respectively.

Taking the substance round the cycle BCC′B′B, the heat given along BC is L. That given along CC′ is $-C_2 d\theta$ where C_2 is the specific heat of the body in the second state, always just on the point of changing into the first state. That given along C′B′ is

$$-\left(L-\frac{dL}{d\theta}d\theta\right),$$

and that given along B′B is $+C_1 d\theta$, where C_1 is the specific heat of the body in the first state, always just on the point of changing into the second. Thus the total heat is the sum of these,

or
$$\left(\frac{dL}{d\theta}+C_1-C_2\right)d\theta.$$

But this is equal to the area BCC′B′, which is practically equal to the rectangle with base BC and height $d\theta$. Now BC × BM = total heat along BC = L; then $BC=\frac{L}{\theta}$.

Then the area $BCC'B'=\frac{Ld\theta}{\theta}$.

Equating the two values of the heat given we have

$$\frac{dL}{d\theta}+C_1-C_2=\frac{L}{\theta}.$$

In the case of water and steam at 100° C.,

$$\frac{dL}{d\theta}=-0{\cdot}695 \text{ from Regnault's equation (p. 180)},$$

$$C_1=1 \quad L=537 \quad \theta=373$$

whence
$$C_2=-1{\cdot}135,$$

or the specific heat of steam when kept saturated is negative.

This means that if we have a quantity of saturated steam at 100°, *i.e.* at a pressure of 760 mm., and we increase the pressure adiabatically to 787 mm., which is the saturation pressure at 101°, the work done will raise the temperature above 101°, and heat must be abstracted to keep it down to that temperature. Or if we decrease the pressure adiabatically to 733 mm., the saturation pressure corresponding to 99°, the temperature will fall below 99°, and heat must be given to restore it to that point.

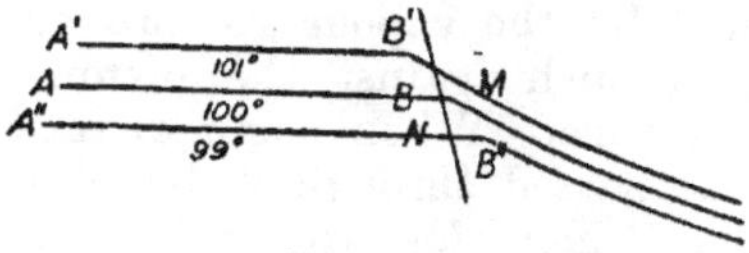

FIG. 177.

In this latter case of adiabatic expansion, if dust nuclei are present, condensation will occur, for as the temperature is reduced the pressure will always be above the saturation pressure. For example, when the temperature has reached 99° the pressure will not yet have fallen to 733 mm., and the excess of vapour will condense on the dust nuclei. This agrees with the common observation that the adiabatic expansion of saturated water vapour produces cloud. Let A′B′, AB, A″B″ (Fig. 177) represent the three isothermals at 101°, 100°, and 99° on an indicator diagram, and let MBN be the adiabatic through B; then the above result implies that BM cuts the upper isothermal in the vapour part to the right of B′, while BN cuts the lower isothermal in the mixture part to the left of B″.

Usually for saturated vapours the specific heat is negative, but for ether vapour (Clausius's *Mechanical Theory of Heat*, p. 136) it is positive at ordinary temperatures. Hence an adiabatic expansion of ether vapour does not produce condensation.

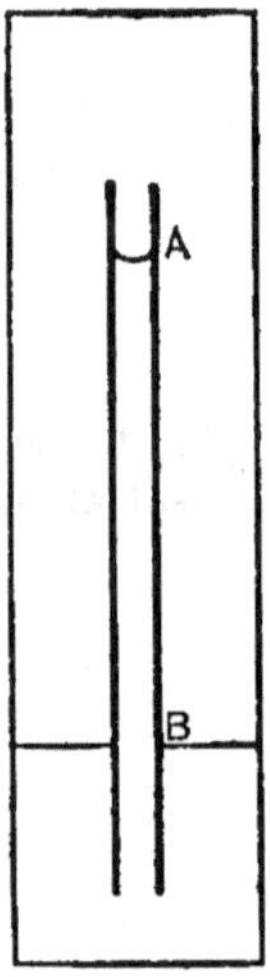

FIG. 178.

Alteration of Vapour-Pressure with Curvature of Liquid Surface.—Lord Kelvin (*Proc. R.S.E.*, February 7, 1870) pointed out that the vapour pressure of a liquid with a curved surface must be different from that of the same liquid with a plane surface. Suppose that a capillary tube (Fig. 178) dips in a liquid contained in a closed vessel, and that no gas other than the vapour of the liquid is present above the free surface. Further, suppose that the liquid rises in the tube, and is in equilibrium at the level A. Then there must be equilibrium between liquid and vapour at A, as well as at the plane surface B; that is, the concave curved surface is in equilibrium with the vapour at a pressure less than that at the plane surface B by the weight of unit column of vapour of height AB. For if not, suppose the vapour rising from A can have a pressure greater than the pressure of the vapour at the same level outside the tube. The space above A will not be saturated, and circulation will take place through continuous evaporation from A and condensation at B, and work can be obtained. We may make the idea more definite by imagining a small engine above the capillary surface, as in Fig. 179, the whole being maintained at one constant temperature. If the pressure of the vapour from A exceeds the pressure of the surrounding vapour, let evaporation take place from A at the maximum pressure, and let the vapour go through the valve V and push against the piston P with this pressure. When some quantity of vapour has passed through V, let V be closed and let P move forward with isothermal expansion of the vapour between it and V till there is the same pressure on both sides of the piston. The excess of vapour pushed in front of P will produce condensation at B. Then let a valve be opened in P, and let P move back to V. It is now ready for another stroke. This process can be carried on endlessly, and the piston does work while the substance is carried through a cycle which can be made perfectly reversible by merely reversing the action of the valves. But in such a cycle no balance of work can remain over, so that our supposition of difference of pressure between the vapour from A and its surroundings must be false.

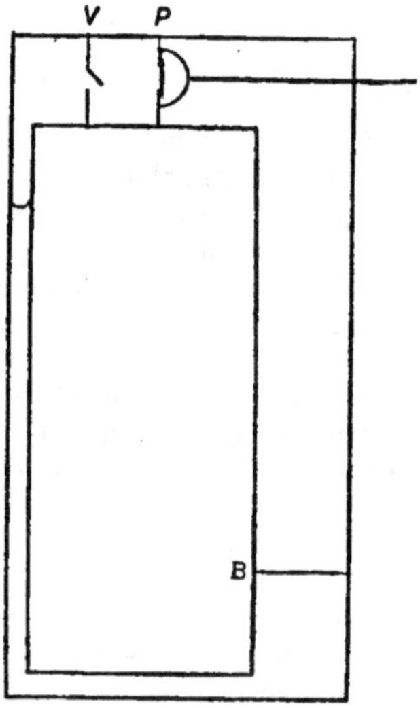

FIG. 179.

If ω be the pressure of vapour at B and ω' be its value at A, and if h, the intervening height, is so small that we may regard the density of the vapour σ as uniform, we have

$$\omega - \omega' = g\sigma h.$$

We may express h in terms of the surface tension T and the radius of curvature r of the surface. The pressure in the liquid just under the surface A is

$$\omega' - \frac{2T}{r},$$

therefore the pressure in the liquid at the level of the plane surface is

$$\omega' - \frac{2T}{r} + g\rho h,$$

where ρ is the density of the liquid. But this is equal to the pressure of the vapour outside at the same level, viz.,

$$\omega = \omega' + g\sigma h$$

equating we get

$$\frac{2T}{r} = g(\rho - \sigma)h$$

whence $$\omega - \omega' = \frac{2T}{r}\frac{\sigma}{\rho - \sigma}.$$

If the surface is convex, so that the liquid is depressed in the tube, we have

$$\omega' - \omega = \frac{2T}{r}\frac{\sigma}{\rho - \sigma}.$$

The expression is simpler if we introduce the difference of hydrostatic pressures P under the curved and under the plane surfaces. This is

$$P = g\rho h = g\sigma h\frac{\rho}{\sigma},$$

whence $$\omega - \omega' = \frac{P\sigma}{\rho}.$$

If the height h is great we cannot take σ as uniform. We must then integrate the change in pressure with the varying density as we descend from A to B.

Let dh be a small step down, and let $d\omega$ be the change in pressure in the step, σ being the density.

Then $$d\omega = g\sigma dh$$

and $$\omega = K\sigma \qquad \text{(Boyle's law)}$$

whence $$\frac{d\omega}{\omega} = \frac{g\,dh}{K}$$

and integrating from ω' to ω, and $h=0$ to $h=h$

$$\log \frac{\omega}{\omega'}=\frac{gh}{K}$$

$$=\frac{gh\sigma}{\omega}$$

$$=g\rho h \cdot \frac{\sigma}{\omega\rho}$$

$$=\frac{P\sigma}{\omega\rho}$$

where P is the difference of hydrostatic pressures under the two surfaces.

We may apply this result to the case of a small spherical drop, radius r surface tension T. The hydrostatic pressure just within the drop will be greater than that under a plane surface at the same level by $P=\frac{2T}{r}$. Hence the vapour-pressure from the drop will exceed that from the plane surface by $\frac{2T}{r} \cdot \frac{\sigma}{\omega\rho}$ if P is not very great or r is not very small. If we cannot make these limitations, then we must use the logarithmic formula, and put

$$\log\frac{\omega}{\omega'}=\frac{2T}{r} \cdot \frac{\sigma}{\omega\rho}$$

Hence a space over a plane liquid surface and saturated for that surface, is not yet saturated for a small drop, and such a drop if formed in any way will tend to evaporate and disappear. We can see, then, how in a dust-free space vapour can exist without condensation, though supersaturated as far as a plane surface is concerned.

If, however, dust is present, a particle of it may have such small curvature that if any condensation occurs on it the vapour pressure of the liquid is practically that of a plane surface. The liquid may spread round the particles, and from the beginning form drops of such size that the space is saturated for them and they continue to grow.

It is to be noted that the alteration of vapour-pressure is very small until the radius of curvature is exceedingly minute.

Thus, taking water vapour at 0°, we have its density $\frac{5}{8}$ density of air.

Now, for air at 0° and 760 mm.

$$\frac{\omega}{\sigma}=\frac{1014000}{\cdot 001296}$$

whence for water vapour

$$\frac{\omega}{\sigma}=\frac{1014000}{\cdot 00081}=1\cdot 25\times 10^9$$

Putting $\rho=1$ and $T=80$

$$\log_e \frac{\omega}{\omega'} = \frac{1.60}{1{\cdot}25 \times 10^9} \cdot \frac{1}{r}$$

$$= \frac{1{\cdot}28 \times 10^{-7}}{r}$$

or

$$\log_{10} \frac{\omega}{\omega'} = \frac{1{\cdot}28 \times 10^{-7}}{2{\cdot}30 \times r}$$

$$= \frac{{\cdot}56 \times 10^{-7}}{r}$$

If

$$r = 10^{-5} \text{ cm.} \quad \frac{\omega}{\omega'} = 1{\cdot}013$$

$$r = 10^{-6} \text{ cm.} \quad \frac{\omega}{\omega'} = 1{\cdot}138$$

$$r = 10^{-7} \text{ cm.} \quad \frac{\omega}{\omega'} = 3{\cdot}631$$

or it is only when the drop is of the order of a millionth of a centimetre that its vapour-pressure is sensibly greater than that from a plane surface, and if the dust particles are greater than this, a very small excess above normal saturation may make drops form on them.

If no dust particles are present, we may perhaps still see how drops are formed if the pressure is great enough (chap. x.). For there may be groups of the vapour molecules formed here and there by collisions which are virtually small liquid masses. Now the radius of the sphere of action of a molecule is of the order 10^{-8} (chap. ix.), so that if a group of size 10^{-7} is formed it may find the space saturated if the pressure is three or four times its normal value and it may continue to grow.

The Connection between Alteration of Vapour-Pressure by Pressure on the Liquid and the Change in Melting-Point by Pressure.—An alteration in the hydrostatic pressure to which a liquid or a solid is subjected, affects the vapour-pressure, whether the alteration is produced by capillarity, as in the rise or fall of a liquid in a tube, or by increasing or decreasing the atmospheric pressure on the surface, as when ice or water is contained in a vessel into which more or less air may be pumped.

If we have a mixture of ice and water in such a vessel with their vapour only above the mixture, then their equilibrium point is the triple point. But if air be pumped in till the pressure is P, the equilibrium temperature of the ice and water falls by

$$d\theta = \frac{P(v_1 - v_2)\theta}{L'}$$

where v_1 is the volume of 1 gm. of ice, v_2 that of 1 gm. of water, and L' is the latent heat of fusion.

At this temperature we may show that the two vapour-pressures of solid and liquid are so changed by the pressure P that they are again equal, or we have, as it were, another triple point. This may be proved by imagining an arrangement to take the substance through a one temperature reversible cycle. That used on p. 309 for the ordinary triple point is no longer applicable. Suppose that the two vapour-pressures at the new equilibrium pressure are different, that of water say being the greater, then if there be water in a vessel (Fig. 180) containing air at pressure P, at the level of the water surface the vapour in the air would be supersaturated for ice. But if the vessel be sufficiently lofty, then the pressure both of air and of water vapour will decrease from below upwards, and at some height A, ice will be in equilibrium with the water vapour present. Then arrange a side tube of this height containing a cylinder of ice and supported a little above its base by upward pulls T. If now the cylinder is slowly pulled up by forces applied at the level C, evaporation will take place at A, condensation will go on at the water surface, and solidification at the base of the ice. Or if the cylinder is slowly let down, doing work against T, everything is reversed, and we have a one-temperature reversible cycle from which work may be obtained. Our supposition of different vapour pressures must therefore be false.

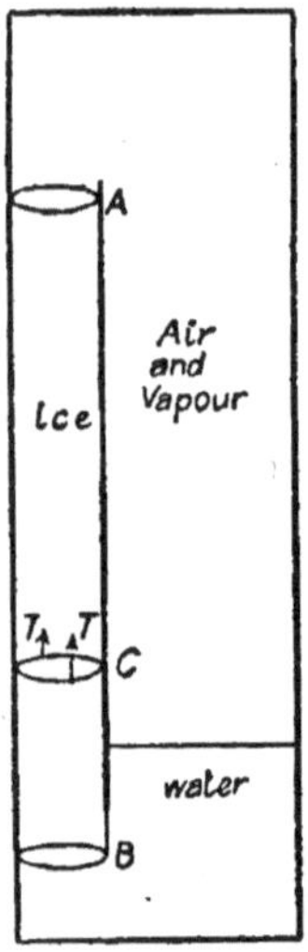

FIG. 180.

It may be shown that the alteration given by the formula of p. 315 just accounts for this new equality of vapour-pressures. For if TA TA′ (Fig. 181) represent the vapour-pressures of ice and water, respectively, below the triple point T, a pressure P on the surface alters the ice vapour pressure by $TI = \frac{P\sigma}{\rho_1} = P\sigma v_1$, where v_1 is the specific volume of ice. That is, the new vapour-pressure line is IT′, $P\sigma v_1$ above TA. Similarly, the new water vapour pressure line is represented by WT′, $P\sigma v_2$ above TA′ where v_2 is the specific volume of water. W is lower than I, since v_2 is less than v_1, and $IW = P\sigma(v_1 - v_2)$. If WT′, IT′ meet at T′, and if T′ be $d\theta$ below the triple point, evidently IW is equal to the difference of height of TA, TA′ for the same temperature. But by p. 311 (3) this is

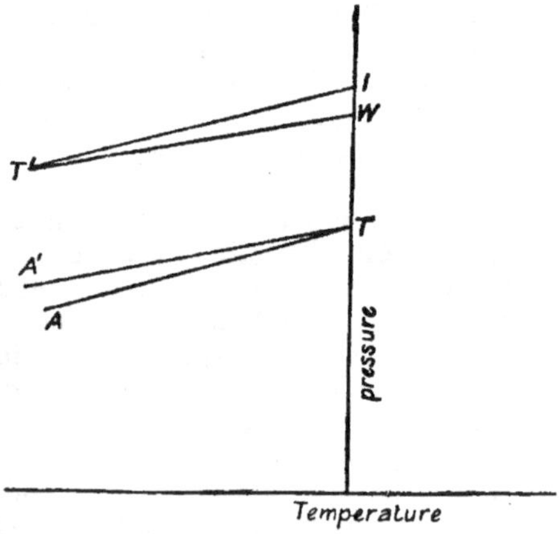

FIG. 181.

$$\omega' - \omega = \frac{L'\sigma d\theta}{\theta}$$

whence $$P\sigma(v_1 - v_2) = \frac{L'\sigma d\theta}{\theta}$$

and $$d\theta = \frac{P(v_1 - v_2)\theta}{L'}$$

or the two vapour-pressures are equal at the new melting-point.

The point T′ is really above T, or as the melting-point falls by pressure the vapour-pressure rises. For if we take the water vapour pressure $d\theta$ below T when the water is under pressure P, it is

$$\omega' = \omega - \frac{d\omega}{d\theta}d\theta + P\sigma \cdot v_2$$

where ω is the vapour-pressure at the triple-point.

But $$\frac{d\omega}{d\theta} = \frac{L\sigma}{\theta}$$

and $$P = \frac{L'd\theta}{(v_1 - v_2)\theta}$$

then $$\omega' = \omega + \frac{\sigma d\theta}{\theta}\left(\frac{L'v_2}{v_1 - v_2} - L\right)$$

$$= \omega + \frac{L\sigma d\theta}{\theta}\left\{\frac{L'v_2 - L(v_1 - v_2)}{L(v_1 - v_2)}\right\}$$

but $$\frac{L'}{L} = \frac{80}{606{\cdot}5} \text{ and } \frac{v_1 - v_2}{v_2} = {\cdot}09$$

then the second term on the right is positive, and ω' is greater than ω.

The results just obtained enable us to give a molecular account of the phenomena of melting under pressure. The vapour-pressure of a substance depends on the number of molecules escaping from the surface per second, or upon the "mobility" of the molecules, and we may take the vapour-pressure as measuring this mobility. At the triple point the three states are apparently in equilibrium; but this is only a "mobile equilibrium" due to the equality of the number of molecules passing in each direction across each separating surface. In other words, the mobility of each state is the same. Below the triple point the vapour-pressures differ, and the mobilities differ also. Thus the liquid gives to the solid more molecules than it receives in return and the solid grows. The liquid evaporates more than the solid, and would distil over on to the solid in the space above their surface. But if pressure is put on, the mobilities are increased, though not to the same extent, the denser state being less increased than the less dense. Hence at the triple point equilibrium is destroyed, the ice having a greater mobility than the water, and the new point of equal mobility is below the triple point.

We have applied the foregoing equations to the case of ice and water; but it is evident that they apply to any case of a substance capable of existing in the three states or phases at the same time.

Solutions.

Vapour-Pressure of a Solution always Less than that of the Solvent.—It has long been known that when a salt is dissolved in a liquid the vapour-pressure of the solution is less than that of the pure solvent, the reduction of pressure increasing with the proportion of salt dissolved.

It follows that if we have two vessels in an enclosure, as in Fig. 182, one containing pure solvent and the other containing a solution, the rest of the space being filled with vapour of the solvent — the salt being supposed to have no appreciable vapour — then the solvent will distil over into the solution, and the level of the latter will tend to rise till its surface is so high that the pressure of the vapour actually present at the new level is equal to the vapour-pressure of the solution.

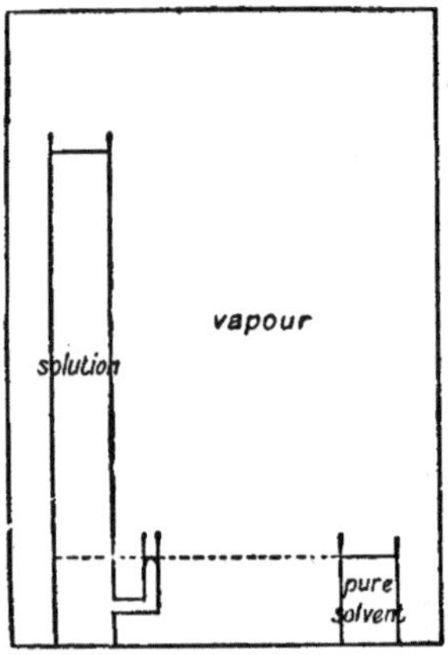

FIG. 182.

Let h be the difference in levels of the surfaces when equilibrium is attained, let ω be the vapour-pressure of the pure solvent, ω' that of the solution, and σ the density of the vapour. We suppose that the solution is very dilute so that h will not be too great to suppose σ uniform in the column intervening between the levels of the two surfaces. Then we have

$$\omega = \omega' + g\sigma h$$

$$= \omega' + g\rho h \cdot \frac{\sigma}{\rho}$$

where ρ is the density of the solution, and if this is very dilute it is equal practically to that of the solvent. Let us put $P = g\rho h$, then P is the excess of hydrostatic pressure in the solution over that in the solvent at the level of the surface of the latter.

Then
$$\omega - \omega' = \frac{P\sigma}{\rho} \qquad (1)$$

But (p. 315) if we put pressure P on to the solution its vapour-pressure is increased by $\frac{P\sigma}{\rho}$, or since

$$\omega = \omega' + \frac{P\sigma}{\rho},$$

it is increased to the vapour-pressure of the solvent. Hence the solution will be in equilibrium with the vapour at such a height above the surface of the solvent that the hydrostatic pressure due to that height of liquid will make the vapour-pressure of the solution equal to that of the pure solvent.

We can see this also by imagining a capillary tube, which the solution does not wet, to be fixed in the side of the solution vessel, as in Fig. 182, and of such narrow bore that the solution is depressed in it to the level of the surface of the solvent. Evidently the vapour-pressure at this surface must be equal to that of the solvent, otherwise a one-temperature reversible cycle giving work could be arranged. Or, again, the vapour-pressure of the solution is made equal to that of the solvent by putting pressure on it, $P = g\rho h$, where h is the height at which the solution is in equilibrium with the vapour above the solvent.

The pressure P which must be put on to the solution to equalise the vapour-pressures is, for a reason to be given later, called the **osmotic pressure.**

If the difference of vapour-pressures is too large to allow us to assume the vapour-density constant, we must use the logarithmic formula, whence

$$\log_e \frac{\omega}{\omega'} = \frac{P\sigma}{\omega\rho}$$

or

$$P = \frac{\omega\rho}{\sigma} \log_e \frac{\omega}{\omega'},$$

This may be obtained, of course, exactly as on p. 316.

Raising of the Boiling-Point.—One effect of the lowering of the vapour-pressure by solution is a rise in the boiling-point. For, if the solvent vapour has a pressure of 1 atmo. at a given temperature, the solution has a pressure less than 1 atmo. at that temperature, and it must therefore be raised to a higher temperature before it will boil.

We can easily calculate the rise of boiling-point in terms of the osmotic pressure.

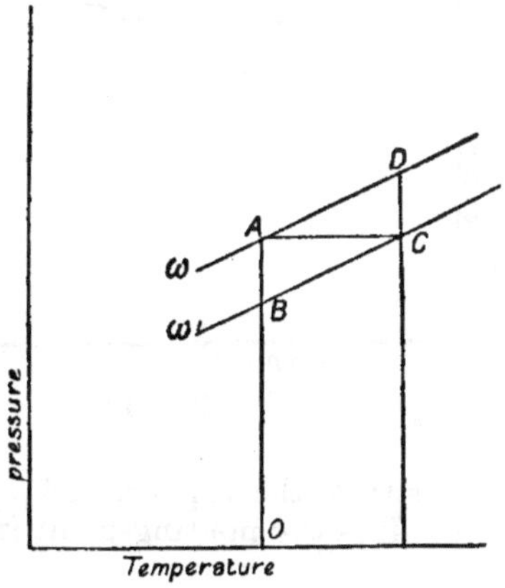

FIG. 183.

For if AO, BO, Fig. 183, represent the pressures of solvent and solution vapours at the normal boiling-point of the solvent, and AD, BC, the vapour-pressure lines as the temperature rises, if we draw AC parallel to the temperature axis to meet BC in C, AC is the rise in boiling-point, say $d\theta$. As we suppose the solution to be dilute, AD and BC will be near together and may be regarded as parallel straight lines. This follows also from Raoult's researches (see below) confirming earlier work, which showed that for a given solution $\frac{BO}{AO}$ is independent of the temperature.

Then AD is practically parallel to BC if $\frac{AB}{AO}$ is very small.

Then
$$d\theta = AC = CD. \frac{AC}{CD} = \frac{AB}{\frac{CD}{AC}}$$

$$=\frac{\omega-\omega'}{\frac{d\omega}{d\theta}}$$

$$=\frac{P\sigma}{\rho}\div\frac{L\sigma}{\theta} \qquad \text{by p. 320 and p. 311}$$

$$=\frac{P\theta}{L\rho} \qquad (2)$$

where L is the latent heat of the solvent, and θ is the absolute temperature of the normal boiling-point.

Lowering of the Melting-Point.—Another effect of salt in solution which has long been known is a lowering of the melting-point. When the solution does begin to freeze, the solid which crystallises out is pure solvent. At the new melting-point, when solution and solid solvent are in equilibrium with each other, their vapour-pressures must be equal. For if not, if, for example, the vapour of the solution were the greater, a reversible one-temperature cycle could be arranged giving work.

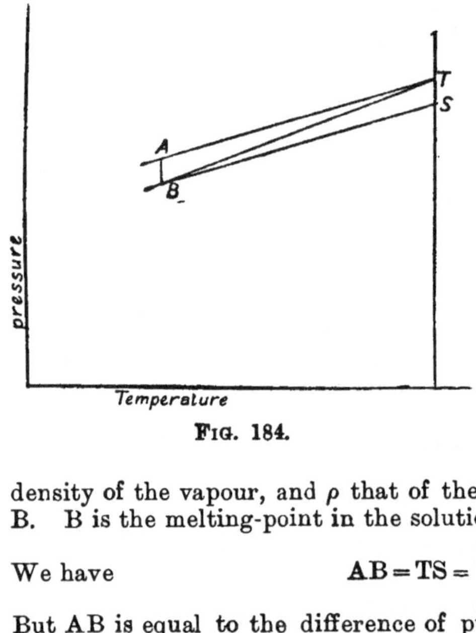

FIG. 184.

Let T, Fig. 184, represent the triple point of the pure solvent, TA the liquid solvent vapour-pressure line, TB the solid solvent vapour-pressure line, SB the vapour-pressure line of the solution, lower than TA by $\frac{P\sigma}{\rho}$ where P is the osmotic pressure of the solution; σ the density of the vapour, and ρ that of the liquid, and let TB, SB meet in B. B is the melting-point in the solution. Let it be $d\theta$ below T.

We have $$AB=TS=\frac{P\sigma}{\rho}.$$

But AB is equal to the difference of pure solid and pure liquid solvent vapour-pressures $d\theta$ below the melting-point, whence, as on p. 311,

$$AB=\omega-\omega'=\frac{L\sigma}{\theta}d\theta$$

where L is the latent heat from solid to liquid. Equating the two values of AB

$$\frac{L\sigma d\theta}{\theta}=\frac{P\sigma}{\rho}$$

or $$d\theta=\frac{P\theta}{L\rho} \qquad (3)$$

The quantity P, the osmotic pressure, or as we are now regarding it, the hydrostatic pressure which must be put on the solution to restore its vapour-pressure to equality with that of the solvent, can be found from either of these effects, lowering of melting-point or raising of boiling-point, or going back to equation (1) it may be found from the difference of vapour-pressures.

Though much work had been done by earlier observers * the effect of solution on vapour-pressure, and on melting-points was first put into a satisfactory form by Raoult, who worked not only with salts in water, but with solutes in other solvents. Herein lay the success he attained in formulating laws. For aqueous solutions are generally electrolytic and for these the osmotic pressure is irregular, probably for a reason which we shall give below. But with other solvents it is easy to make great numbers of non-conducting solutions, and these give more regular values leading to definite laws. Raoult showed † that

1. $\frac{\omega - \omega'}{\omega}$ or the proportionate lowering of the vapour-pressure, is independent of the temperature for a particular solution, ether being the solvent.

2. $\frac{\omega - \omega'}{\omega}$ is proportional to the amount of salt or solute dissolved.

3. If $n = \frac{\text{grammes dissolved in 100 of solvent}}{\text{molecular weight of solute}}$

= "gramme molecules" of salt per 100 grammes of solvent

then in the case of ether as solvent

$$\frac{\omega - \omega'}{\omega} \div n \text{ is constant,}$$

or the lowering per gramme molecule is the same for different substances. Later he showed ‡ that if n is the number of molecules of solute in N of the solvent, then for very dilute solutions

$$\frac{\omega - \omega'}{\omega} = \frac{n}{N} \text{ very nearly.}$$

Still later § he found that

$$\frac{\omega - \omega'}{\omega} = \frac{cn}{N + n}$$

would hold for much stronger solutions if c were 0·9, and that as the dilution increased c tended towards 1.

For electrolytic solutions, c may be very much more than 1, rising towards 2 for such salts as NaCl or KCl in water, and towards 3 for such salts as $Ca\,Cl_2$ or $Ba\,Cl_2$, in water.

The law $\frac{\omega - \omega'}{\omega} = \frac{n}{N + n}$ implies that the lowering of the vapour-pressure

* See Ostwald's *Solution*, chap. vii., for a history of the subject.
† *C.R.*, ciii. p. 1125, 1886.
‡ *C.R.*, civ. p. 976, 1887.
§ *Zeit. für Phys. Chem.*, ii. p. 353, 1888

is proportional to the number of molecules of solute present, whatever their nature. When we have to multiply by c, where c is nearly 2, it is taken to imply that the molecules of salt are for the most part dissociated each into 2. When we have to multiply by a number nearly 3, it is taken to imply that the dissociation is from one into three molecules and so on.

For very dilute non-electrolytic solutions n, the number of salt molecules, is very small compared with N, the number of solvent molecules, and we may put

$$\frac{\omega - \omega'}{\omega} = \frac{n}{N} \tag{4}$$

Substituting in (1), p. 320, we get for the osmotic pressure

$$P = \frac{\omega\rho}{\sigma} \cdot \frac{n}{N} \tag{5}$$

If we put the relation between the pressure ω and the density σ of hydrogen as

$$\frac{\omega}{\sigma} = R\theta$$

we may put that for any rarefied vapour of molecular weight M as

$$\frac{\omega}{\sigma} = \frac{2}{M} R\theta$$

whence
$$P = \frac{2\rho R\theta}{M} \cdot \frac{n}{N} \tag{6}$$

Further, if s is the density of the salt in the solution or the number of grammes per c.c., and if S is its molecular weight, then $\frac{s}{S}$ is the number of gramme molecules per c.c. Similarly $\frac{\rho}{M}$ is the number of gramme molecules of the solvent per c.c. if the solution is sufficiently dilute.

Hence
$$\frac{n}{N} = \frac{s}{S} \div \frac{\rho}{M}$$

and substituting in (6) we get

$$\frac{P}{s} = \frac{2}{S} R\theta \tag{7}$$

Comparing this with $\frac{\omega}{\sigma} = \frac{2}{M} R\theta$ we see that the osmotic pressure P is the pressure which the salt would exert if it were gaseous, and of the density at which it actually exists in the solution. This most remarkable result was first obtained in a very different way by Van t'Hoff, and is known as Van t'Hoff's law of osmotic pressure. We shall give some account of his theory later.

It will be useful to collect the results which we have obtained. We have for very dilute solutions

Osmotic pressure $$P=(\omega-\omega')\frac{\rho}{\sigma} \qquad (1)$$

Raising of boiling-point *or* lowering of freezing-point of liquid density ρ

$$d\theta=\frac{P\theta}{L\rho} \qquad (2) \text{ and } (3)$$

Raoult's experimental value of the difference of vapour-pressure,

$$\frac{\omega-\omega'}{\omega}=\frac{cn}{N+n}=\frac{cn}{N} \text{ nearly}$$

where c is nearly 1 for non-electrolytes. For such non-electrolytes by substituting in (1)

$$P=\frac{\omega\rho}{\sigma}\cdot\frac{n}{N} \qquad (5)$$

Whence Van t'Hoff's result $$P=\frac{2s}{S}\cdot R\theta \qquad (7)$$

For the purpose of illustration let us calculate the actual values for 1 gramme molecule in solution in water. It must be remembered that as a rule aqueous solutions are electrolytic, and it is only for non-electrolytic solutions, such as one of cane sugar, that the results hold.

We may determine the value of R from the density of hydrogen ·0000896 at 0° C. and 1 atmo. or 1014000 dynes/sq. cm.

Putting $$\frac{1014000}{\cdot 0000896}=273R$$

we get $R=4{\cdot}15\times 10^7$ when ω is in dynes/sq. cm.

or $=40{\cdot}9$ when ω is in atmos.

When the solution is 1 gramme molecule per litre of water, $\frac{s}{S}$ which is the number of gramme molecules per c.c., is $\frac{1}{1000}$. Then (7) becomes

$$P=8{\cdot}3\times 10^4\times\theta \text{ dynes/sq. cm.}$$

or $$\cdot 0818\theta \text{ atmos.,}$$

and if the temperature is 0° C., putting $\theta=273$, this becomes

$$P=22{\cdot}3 \text{ atmos.}$$

If there are n gramme molecules per litre

$$P=22{\cdot}3\times n \text{ at } 0^\circ$$

or $$=22{\cdot}3n\,(1+at) \text{ at } t^\circ \text{ C.}$$

It must be noted that this only holds for dilute solutions, so that in general n must be small, indeed a small fraction.

The lowering of the vapour-pressure per 1 gramme molecule per litre is

$$\frac{\omega-\omega'}{\omega}=\frac{n}{N}=\frac{1}{\frac{1000}{18}}=\frac{18}{1000}=\frac{1}{55 \cdot 6}$$

since N, the number of gramme molecules of water in a litre, is 1000/18. Thus at the boiling-point the lowering per gramme molecule in mm. of mercury is

$$\frac{760}{55 \cdot 6}=13 \cdot 6 \text{ mm.}$$

The raising of the boiling-point is obtained from (3) and (7) by putting $\theta=373$; $L=537 \times 4 \cdot 19 \times 10^7$; $\rho=1/1 \cdot 043$

$$d\theta=\frac{8 \cdot 3 \times 10^4 \times 373^2 \times 1 \cdot 043}{537 \times 4 \cdot 19 \times 10^7}$$

$$=0 \cdot 54^\circ.$$

The lowering of the melting-point is obtained from the same formulæ by putting $\theta=273$; $L=80 \times 4 \cdot 19 \times 10^7$; $\rho=1$

$$d\theta=\frac{8 \cdot 3 \times 10^4 \times 273^2}{80 \times 4 \cdot 19 \times 10^7}$$

$$=1 \cdot 85^\circ.$$

Thus for cane sugar, $C_{12}H_{22}O_{11}$, S = 342, and a solution containing 342 grammes per litre of water will boil at 100·54° C., will freeze at − 1·85° C., and will have a vapour-pressure less than that of pure water by 1 in 55. These results agree very closely with direct observation.* Observations have been made with a large number of salts and solvents, and when the solutions are non-electrolytic the agreement is equally close.

If the lowering of the freezing-point or the raising of the boiling-point is greater than the values obtained from Van t'Hoff's formula—if, in fact, *c* in Raoult's formula is greater than 1 we can reconcile the observations with the theory that each gramme molecule has the same effect by supposing that dissociation of the salt molecules has taken place, and that the dissociated ions are acting independently, each producing its own effect. Thus, if we have such a salt as KCl, and if a fraction λ of it is dissociated, then of the total n molecules originally present only $n(1-\lambda)$ remain single, while $n\lambda$ have split each into two, so that we have $n(1-\lambda)+2n\lambda=n(1+\lambda)$ molecules actually in solution.

Hence the results just obtained must be multiplied by $1+\lambda$, where $1+\lambda$ may rise to 2 if dissociation is complete.

Thus with KCl in water a solution of 1 gm. molecule per 10 litres of water gives a depression not of 0·185°, but about 1·86 times as much; whence we may conclude that 86 per cent. of the salt is dissociated. As the dilution increases so does the molecular depression of the freezing-point, whence it is concluded that the dissociation also increases. This supposition of dissociation is confirmed by observations on the electric

* Whetham, *Solution*, chap. vi.

conductivity of the solution, where the percentage dissociated comes out to nearly the same value as that given by the depression of the freezing-point.

The theory of osmotic pressure which we have just given appears to be the most satisfactory mode of treating the subject, in that it approaches it through the phenomenon of vapour-pressure and the effects of change of vapour-pressure, which are most easily subjected to direct measurement. But historically the idea of osmotic pressure was arrived at through the study of diffusion through membranes, and the difference of vapour-pressure was regarded as a secondary phenomenon. We shall give a brief account of the subject from this point of view, as it gives some excellent examples of the application of thermodynamic reasoning.

Semi-Permeable Membranes.—It has long been known that if a mixture or solution of spirit and water be separated by bladder from pure water then the water passes into the solution more rapidly than the spirit passes out, and will maintain an excess of pressure on the spirit side. Dutrochet, who first carefully investigated this passage through membranes, gave to it the name osmosis. Graham in 1854 (*Phil. Trans.*, 1854, p. 177) showed that osmosis took place rapidly into alkaline solutions; and in 1861 he published his celebrated memoir on Dialysis (*Phil. Trans.*, 1861, p. 1183), in which he divided substances into two classes: the crystalloids, which in solution will pass through membranes, and which include bodies susceptible of crystallisation; and colloids, or glue-like substances, such as gum, gelatine, albumen, which will not pass, or only with great slowness.

A new turn was given to the subject by Pfeffer's discovery in 1877 that membranes could be made through which a salt could not pass, while water could pass freely. One such membrane he made by filling a porous pot with a solution of potassium ferrocyanide and immersing it in a solution of copper-sulphate. Where the two solutions met in the pores of the clay an insoluble gelatinous precipitate was formed, supported by the clay and constituting the membrane. This membrane will allow water to pass through it easily, but is impermeable to sugar. It is therefore termed a *semi-permeable membrane.*

Pfeffer working with sugar solution inside a semi-permeable membrane and water without, found that the water passed in more rapidly than it passed out, until a certain definite excess of pressure existed within the vessel over that without. This excess of pressure, *the osmotic pressure*—we shall see later that it is identical with osmotic pressure as defined above—is found to be proportional to the amount of sugar in solution and to rise with the temperature. Thus the osmotic pressure was 53·1 cm. of mercury for 1 per cent. of sugar dissolved at 14·2° C., and it rose from 50·5 cm. at 6·8°C. to 54·8 cm. at 22°. Pfeffer also worked with some other solutions. In 1884 De Vries (Pringsheim's *Jahrbücher*, xiv. p. 427, 1884) found that when certain vegetable cells, those of the epidermis on the under side of the midrib of the leaves of *Tradescantia discolor* being the best, are placed in concentrated salt solutions the protoplasmic contents contract and shrink away from the cell walls, the protoplasm being apparently covered with a semi-permeable membrane, which allows the water to pass more freely from the dilute solution in the protoplasm than in the opposite direction. If the external solution is gradually diluted a point is at last reached at which the protoplasm tends to fill the cell again, or the osmotic pressure of the contents of the cell equals that of the solution. De Vries called the solutions at this stage *isotonic.* The same cells could be used for solutions of different salts, and he found that equimolecular solutions of similar salts are isotonic.

Van t'Hoff's Application of Thermodynamics.—In 1885 Van t'Hoff (*Phil. Mag.*, xxvi., 1888, p. 81) pointed out that the osmotic pressure of sugar solution as obtained by Pfeffer was the same as would be exerted by the sugar in the gaseous form at the same density. Thus at 14° a 1 per cent. solution of sugar, or $\frac{1}{34 \cdot 2}$ of a gramme molecule should produce a pressure

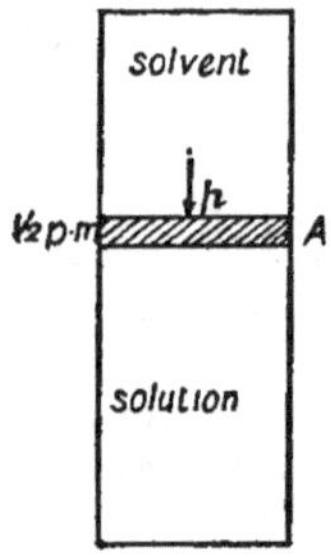

FIG. 185.

$$P = \frac{22 \cdot 3}{34 \cdot 2}\left(1 + \frac{14}{273}\right) \times 76 \text{ cm. of mercury.}$$

$$= 52 \cdot 2 \text{ cm.}$$

almost exactly that found by Pfeffer.

Van t'Hoff also showed that in the variations of the pressure with temperature as found by Pfeffer the pressure is proportional to the absolute temperature. He then applied thermodynamic reasoning to show that in any case the osmotic pressure is proportional to the absolute temperature if a certain assumption is made which will appear in the course of the proof. He used the conception of a semi-permeable membrane which will allow the solvent to pass, but which is impermeable to the solute, even though no such membrane may have been found in practice.

The following is equivalent to Van t'Hoff's work. Suppose that we have a cylinder containing a quantity of solution with osmotic pressure p, separated from the pure solvent by a semi-permeable piston, A, as in Fig. 185. If there is a pressure p on the piston A so that the solution is under excess pressure equal to the osmotic pressure, there will be equilibrium between solvent and solution through the membrane, and the piston can be moved either way reversibly.

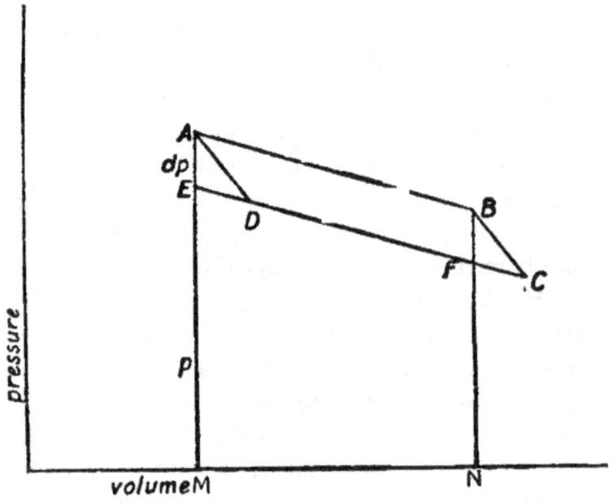

FIG. 186.

Let us take the piston through a reversible cycle represented on the diagram, Fig. 186, by ABCD, where abscissæ represent volumes, ordinates osmotic pressures, *i.e.* pressures by the piston on the solution. Beginning at temperature θ at A, allow $MN = dv$ of solvent to flow into the solution, the temperature being maintained at θ by the addition of heat H. Then allow a further flow of solvent, with no further addition of heat, represented by the adiabatic BC, the temperature falling to $\theta - d\theta$. Then let the piston be pushed back at $\theta - d\theta$ along CD, and finally back along the adiabatic DA to its initial position.

Then the work done is ABCD = AEFB,

$$= \frac{dp}{d\theta} d\theta \cdot dv,$$

and by the second law this is

$$= \frac{\mathrm{H} d\theta}{\theta},$$

or

$$\frac{dp}{d\theta} \cdot dv = \frac{\mathrm{H}}{\theta}.$$

Now, if the dilution is sufficiently great, further dilution neither gives out nor absorbs heat. Whatever in the action of solution does produce heat changes, is all finished by the time the solution is sufficiently diluted. We may picture the action perhaps as a separation of the salt molecules, and as a surrounding of each salt molecule by solvent molecules. When this is once done, it does not matter how many more solvent molecules we add, for we shall not alter the condition of the salt molecules. Hence H is, as with a gas isothermally expanding, equivalent to the external work done, and is 0 if no such work is done.

We put $\mathrm{H} = p dv,$

whence $\frac{dp}{d\theta} = \frac{p}{\theta}$

or $p = c\theta.$

When the dilution has not proceeded so far that no heat is evolved or absorbed on further dilution, we cannot assume that $\mathrm{H} = p dv$, and we cannot draw the conclusion that $p = c\theta$.

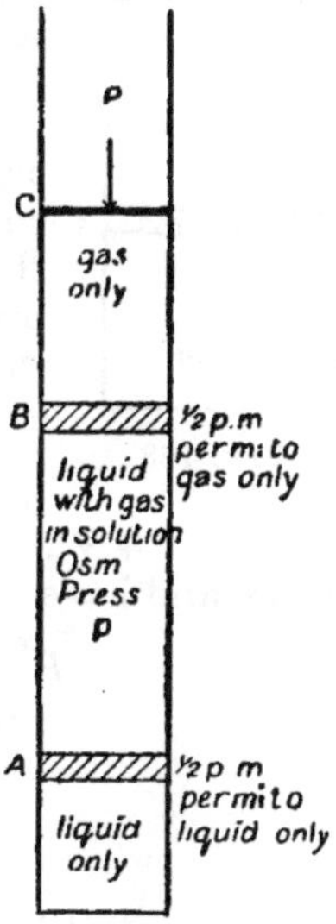

FIG. 187.

Having thus shown that osmotic pressure follows one of the gas laws, Van t'Hoff then showed that when the solute is a gas, it follows from Henry's law that the osmotic pressure is actually equal to the pressure the dissolved gas would exert if it occupied the same volume in the absence of the solvent. Henry's law states that a liquid at a given temperature always dissolves the same multiple of its own volume of a given gas whatever the pressure of the gas. Or, put in another way—if a space contains a liquid and a gas, the liquid dissolving some of the gas, say n volumes at a given pressure, then if the volume and pressure be varied the relation between them will be given by Boyle's law, on the supposition that the one volume of the liquid counts for n volumes. To apply Henry's law to osmotic pressure, imagine a cylinder containing a quantity of liquid with gas dissolved in it (Fig. 187). Let A represent a movable piston permeable to the liquid only. Let B represent a fixed partition permeable to the gas only, and C a solid piston against which any desired pressure can be exerted. Let us begin with A at the bottom and C close against B, and let the pressure of the gas necessary to keep the solution at its actual strength be P; let the osmotic pressure be p. Then A and C are in equilibrium if the pressures on them are p and P. Let V be the volume of the liquid, and let the dissolved gas have volume nV at pressure P. Now move the piston A upward through the solution, pure liquid being left behind it, and at the same time allow the gas to pass through B, and C to move forward so that the solution is always of the same strength between A and B. Then C must move n times as far as A, and ultimately when A has reached B the volume of gas between B and C is nV. The work done is $pV - PnV$. Now, allow the gas between B and C to expand isothermally to some new volume $nV + V'$ and pressure P′. Let the osmotic pressure of the gas in solution under pressure $P' = p'$. Now let the pressure on A be made equal to p', and let C be moved down under pressure P′ while A is moved down under pressure p'. C moves through nV while A moves back through V. Then we have nV volumes of gas dissolved under pressure P′ and V′ volumes still outside. If we now force the piston C down till all the gas is dissolved, at every pressure n volumes at that pressure will be in the liquid, so that the gas compresses into the liquid just as if we started with $nV + V'$ volumes at P, and ended with nV at P. Hence the work done on the gas in this last compression is just that done by the gas in the expansion from nV to $nV + V'$. We have now gone through a one-temperature reversible cycle, and have total work zero, or

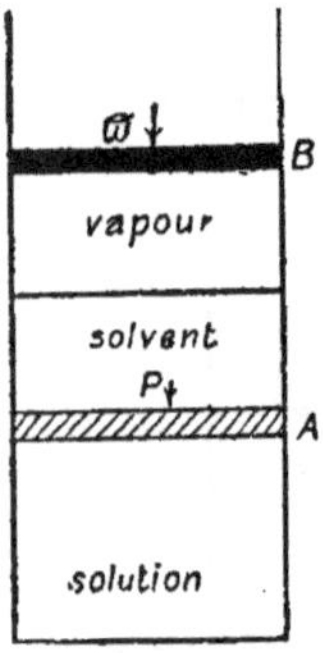

FIG. 188.

$$pV - PnV - p'V - P'nV + \text{equal and opposite}$$
$$\text{works in change from } nV \text{ to } nV + V' = 0.$$
$$\therefore\ pV - PnV = p'V - P'nV,$$
$$\text{or } p - nP = \text{constant}.$$
$$\text{But evidently if } P = 0,\ p = 0, \text{ so that the constant} = 0$$
$$\text{and } p = nP.$$

No method has yet been found of extending this proof to dilute solutions of solids and liquids, and in the absence of anything corresponding to Henry's law we cannot expect such extension. But we might perhaps expect the result to hold, inasmuch as in dilute solutions of solids and liquids the molecular distances of the solutes are of the order of gaseous molecular distances, even if we did not know it from the direct investigation of osmotic pressures by Pfeffer.

It now only remains to show that Raoult's value of the lowering of the vapour pressure in solutions follows from Van t'Hoff's law of the gas value of osmotic pressure.

Let us suppose that we have a solvent of density ρ, vapour pressure ω, and density of vapour σ. Let it form a solution in which the osmotic pressure is P while the vapour pressure is ω' and its density σ'. Let the solution be under a semi-permeable piston A (Fig. 188), and let solvent be above the piston, and let a movable piston B be above the solvent. Allow volume v mass ρv of the solvent to go through a one-temperature cycle, thus—

(1) Let A be pushed down under pressure P till volume v of the solvent has passed out of the solute, work Pv being done.

(2) Let volume v, mass ρv, of solvent evaporate at pressure ω, pushing B up, forming volume $\frac{\rho v}{\sigma}$ of vapour and doing work $\frac{\omega \rho v}{\sigma}$.

(3) Let the vapour expand isothermally from volume $\frac{\rho v}{\sigma}$ to $\frac{\rho v}{\sigma'}$ at pressure ω', doing work $\frac{\omega + \omega'}{2} . \rho v \left(\frac{1}{\sigma'} - \frac{1}{\sigma}\right)$ approximately since ω and ω' are for dilute solutions not very different.

(4) Let the vapour condense back on to the solution by some suitable contrivance, the solution now being under pressure ω', the work done being $\frac{\omega' \rho v}{\sigma'}$.

The cycle is now complete, and the total work is zero, or

$$Pv - \frac{\omega \rho v}{\sigma} - \frac{\omega + \omega'}{2} \rho v \left(\frac{1}{\sigma'} - \frac{1}{\sigma}\right) + \frac{\omega' \rho v}{\sigma'} = 0.$$

But $\frac{\omega}{\sigma} = \frac{\omega'}{\sigma'}$ by Boyle's law, whence

$$P = \frac{\omega + \omega'}{2} \rho \left(\frac{1}{\sigma'} - \frac{1}{\sigma}\right)$$

$$= \frac{\omega + \omega'}{2} \frac{\rho}{\sigma} \frac{\omega - \omega'}{\omega'},$$

or $$\omega - \omega' = \frac{P\sigma}{\rho} \text{ since } \omega + \omega' = 2\omega' \text{ nearly.}$$

If s be the density of the solute, and S its molecular weight, Van t'Hoff's gas value of P gives

$$P = \frac{2s}{S} \cdot R\theta.$$

while if M be the molecular weight of the solvent

$$\omega = \frac{2\sigma}{M} R\theta.$$

Hence, substituting for P and σ

$$\frac{\omega - \omega'}{\omega} = \frac{s}{S} \div \frac{\rho}{M}$$

$$= \frac{n}{N}$$

where $n : N$ = number of gramme molecules of salt : number of gramme molecules of solvent.

It is not necessary to show how the lowering of the freezing-point follows from the value of the osmotic pressure.

Molecular Theory of Osmotic Pressure.—Since a solvent and a solution are in equilibrium on the two sides of a semi-permeable membrane when the solution is under pressure equal to what we have termed the gas pressure of the solute, we are naturally led to look upon the pressure of the solution as consisting of two parts, that of the pure solvent and that of the solute, the latter being equal to its gas pressure. Hence the solvent alone has the same pressure on the two sides of the membrane. If the solute were actually gas, with the gas velocity assumed in the kinetic theory, and if its molecules moved among those of the solvent, quite independent of them, it would produce the observed osmotic pressure by its impacts against the membrane.

But it is difficult to imagine how this independent motion exists, and it is remarkable that if, instead of supposing that the molecules of solute move freely among those of the solvent, we go to the other extreme and suppose that they enter into some sort of combination with them, we can obtain the same value of the osmotic pressure.*

Let us suppose that there are n molecules of solute to N of solvent, and let the vapour pressure of the pure solvent be ω. Now, if each of the n molecules of solute attached itself to a molecule of solvent, and loaded it so that it had no tendency to evaporate, there would be $N-n$ evaporating molecules in the solution to N in the solvent. Then the rate of escape from the liquid surface would be as $N-n : N$. If the rate of return were equally hindered, so that with equal density of vapour over solution and solvent the return of vapour molecules were also in the ratio $N-n : N$, then we should have equilibrium with the same vapour pressure in each case—just as we have equilibrium with the same vapour pressure over an open surface and over a surface partially covered with a perforated plate. But if, as is more natural in this case, we suppose the rate of return the same for the same vapour pressure, that is, if we suppose a vapour molecule to run the same chance of getting entangled whether it descends into solution or pure solvent, then over the solution the return will balance the reduced escape when the pressure of the vapour is less in the ratio $N-n : N$. Hence

$$\omega : \omega' = N : N-n$$

or

$$\omega : \omega - \omega' = N : n$$

and

$$\frac{\omega - \omega'}{\omega} = \frac{n}{N}$$

the approximate form of Raoult's law.

* *Phil. Mag.*, xlii., 1896, p. 289.

CHAPTER XX.

THERMODYNAMICS OF RADIATION.

General Principle—The Pressure of Radiation—The Normal Stream of Radiation, the Total Stream, and the Energy Density—The Pressure on a fully Radiating Surface—The Relation between E and θ in full Radiation, the Fourth Power Law—Full Radiation remains full Radiation in any Adiabatic Change—Relation between Volume and Temperature in an Adiabatic Change—Entropy—Application of Doppler's Principle—Change of Energy in a given Wave Length—Change of Energy of each Wave Length in an Adiabatic Expansion of full Radiation—Maximum Value of Energy for given Range of Wave Length—Form of the Function expressing the Distribution of Energy in the Spectrum.

General Principle.—The thermodynamic theory of radiation starts with the idea that a space containing radiation resembles the working substance in a heat-engine, in that it exerts pressure on its boundary, that it can be compressed or extended, that it can receive more energy as radiation from the boundary and can part with energy to the boundary by absorption, and finally that it can be taken round a cycle. This most fruitful idea is due to Bartoli, who in 1875 * showed that it was possible by means of radiation to transfer heat from a colder to a hotter body in a manner which will be understood from what follows in this chapter. Though the details of his calculation require amendment, he showed that the possibility of such transfer implies that radiation shall press against any surface upon which it falls.†

In 1884 Boltzmann ‡ took up Bartoli's method. He used Maxwell's electromagnetic theory of radiation,§ according to which a pencil of light or radiation, incident normally on a surface, exerts a pressure on that surface equal to the energy per c.c., or equal to the energy density in the pencil. This pressure has now been shown to exist by Lebedew, and by Nichols and Hull.‖ Taking an enclosure full of radiant energy from a fully radiating surface, round a Carnot cycle, Boltzmann showed that the density of the energy in a uniform temperature enclosure containing full radiation is proportional to the fourth power of the temperature, that is, he deduced Stefan's law. The subsequent development of the theory is largely due to Wien.¶

* Bartoli's most important paper was not fully published at first. It is given in full in Exner's *Repertorium der Physik*, xxi., 1885, p. 198.

† Another proof of such pressure is given by Larmor, *Radiation*, *Encyc. Brit.*, 10th ed., xxxii. It depends on the energy in a train of waves of given amplitude being inversely as the square of the wave length.

‡ *Wied. Ann.*, xxi., 1884, p. 364.

§ Maxwell's *Electricity and Magnetism*, 1873, vol. ii., contains the first account of this pressure.

‖ Lebedew, *Congrès Int. de Phys.*, ii. p. 133 ; Nichols and Hull, *Am. Acad. Arts and Sciences*, xxxviii., 1903, p. 559 ; *Pressure of Light*, S.P.C.K.

¶ *Congrès Int. de Phys.*, ii. p. 23.

The Pressure of Radiation.—We shall assume Maxwell's value for the pressure of radiation. He showed that a pencil of electromagnetic waves exerts pressure each way in the line of propagation numerically equal to the energy per c.c. or energy density in the pencil. On his theory there is no pressure at right angles to the line of propagation. If, then, E is the energy density of the radiation falling normally on a receiving surface, it exerts pressure E per sq. cm. on that surface. If the radiation is incident at θ (Fig. 189), let AB be the trace of 1 sq. cm. of the surface, BC the trace of area AB cos $\theta \times 1$ perpendicular to the pencil. The force on BC is E cos θ, and resolving, this gives a normal pressure on AB equal to $E\cos^2\theta$. There is also a tangential force on AB equal to E cos θ sin θ, but this does not concern us here. If the pencil is totally reflected, the reflected ray also exerts pressure $E\cos^2\theta$. It may be noted that the tangential force in this case is equal and opposite to that due to the incident pencil, but that if there is absorption the tangential force is not wholly neutralised.

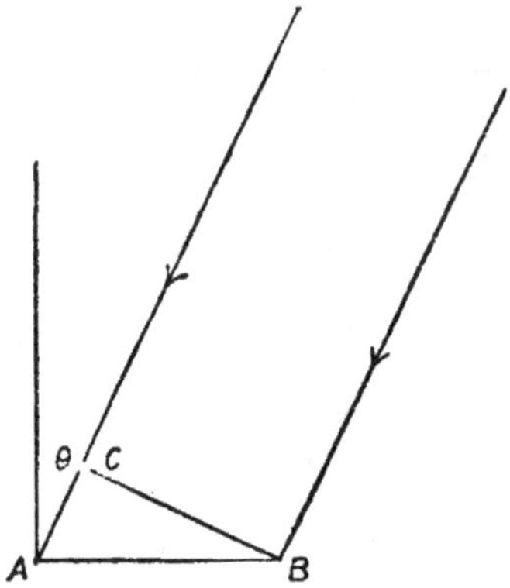

FIG. 189.

The Normal Stream of Radiation, the Total Stream, and the Energy Density.—We can describe the stream of energy from a full radiator by three quantities, between which there are necessary relations. These are

1. *The Normal Stream of Radiation* N.

If a sq. cm. is placed parallel to an emitting sq. cm., at a distance r from it along the normal, then if the energy from the emitting sq. cm. passing through the other sq. cm. is $\frac{N}{r^2}$ per second, N is the normal stream of radiation from the radiator. Or if dS is an element of a radiator, dS' an element along the normal parallel to dS and r from it the energy from dS passing through dS' is $\frac{NdSdS'}{r^2}$ per second.

2. *The Total Stream of Radiation* R.

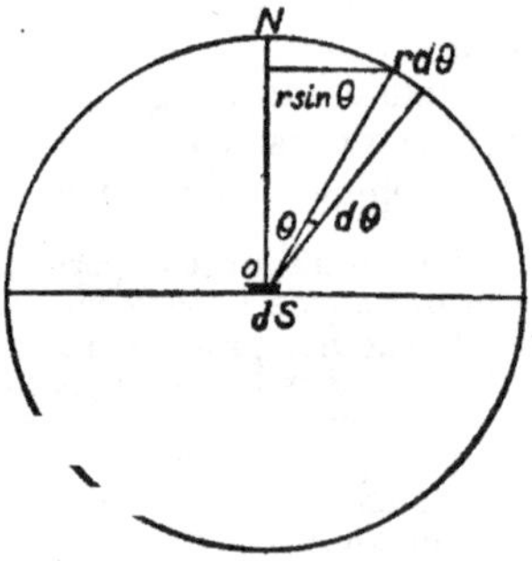

FIG. 190.

The total stream of radiation is the energy emitted from 1 sq. cm. per second round a hemisphere. Let dS be an element of the radiating surface (Fig. 190). Draw a sphere radius r round dS and let ON be the normal radius. The stream from dS passing through an element dS' of the sphere at N is $\frac{NdSdS'}{r^2}$ per second. If dS' is at the end of a radius making θ with ON, the stream through it is, by the law of inclined pencils, $\frac{NdSdS'\cos\theta}{r^2}$. Divide the hemisphere above dS into rings round

ON as axis. A ring between θ and $\theta + d\theta$ will have area $2\pi r\sin\theta . rd\theta$, and the radiation passing through it will be

$$\frac{\mathrm{N}d\mathrm{S}\cos\theta \cdot 2\pi r\sin\theta \cdot rd\theta}{r^2} = 2\pi \mathrm{N}d\mathrm{S}\sin\theta\cos\theta d\theta$$

The total radiation $\mathrm{R}d\mathrm{S}$ is obtained by integrating this over the hemisphere, *i.e.* from $\theta = 0$ to $\theta = \frac{\pi}{2}$, and we get

$$\mathrm{R}d\mathrm{S} = 2\pi \mathrm{N}d\mathrm{S}\int_0^{\frac{\pi}{2}}\sin\theta\cos\theta d\theta$$

whence $$\mathrm{R} = \pi \mathrm{N}$$

3. *The Energy Density in a fully Radiating Enclosure.*—The energy density E can be expressed in terms of R or N. For let the whole sphere (Fig. 190) be sending out radiation; that passing through O comes normally from every element of the surface. If U is the velocity of travel of the radiation, the energy density at O, due to an element $d\mathrm{S}$, will be $\frac{\mathrm{N}d\mathrm{S}}{\mathrm{U}r^2}$, since $\frac{\mathrm{N}d\mathrm{S}}{r^2}$ passes per second through a sq. cm. at O, and it travels U cm. in the second. Then the energy per c.c. is $\frac{4\pi r^2\mathrm{N}}{\mathrm{U}r^2}$, or

$$\mathrm{E} = \frac{4\pi \mathrm{N}}{\mathrm{U}} = \frac{4\mathrm{R}}{\mathrm{U}}$$

The Pressure on a Fully Radiating Surface.—Put at O (Fig. 190), a sq. cm. of fully radiating, fully absorbing surface at the temperature of the enclosure. The energy density of a pencil falling on it from an element $d\mathrm{S}$ of the sphere is $\frac{\mathrm{N}d\mathrm{S}}{\mathrm{U}r^2}$, and if $d\mathrm{S}$ is at the end of a radius making θ with ON this produces pressure $\frac{\mathrm{N}\cos^2\theta d\mathrm{S}}{\mathrm{U}r^2}$.

If we divide the hemisphere into rings each of area $2\pi r^2\sin\theta d\theta$, then

$$\text{total pressure due to incident radiation} = \int_0^{\frac{\pi}{2}}\frac{2\pi r^2\mathrm{N}\cos^2\theta\sin\theta d\theta}{\mathrm{U}r^2}$$

$$= \frac{2\pi \mathrm{N}}{3\mathrm{U}}$$

But the radiation issuing from the area at O will be equal in every direction to that incident, so that this will be doubled, and we shall have

$$\text{Total pressure } \mathrm{P} = \frac{4\pi \mathrm{N}}{3\mathrm{U}} = \frac{4\mathrm{R}}{3\mathrm{U}} = \frac{\mathrm{E}}{3}$$

Relation between E and θ in full Radiation. The Fourth Power Law.—Let us imagine a sphere of which the inside surface can be made, at will, fully radiating or totally reflecting. Let us suppose that it can be surrounded at will by a perfectly conducting substance maintained at any desired temperature. Let us further suppose, and this is the essential feature of the method of working, that the sphere can be compressed against or extended by the pressure of the radiation $P=\frac{E}{3}$ without doing any work on or receiving any energy from the matter or ether or whatever we suppose to be the medium in which the radiant energy is localised. Let us suppose, in fact, that the surface of the sphere is a sort of semi-permeable membrane, permeable to the

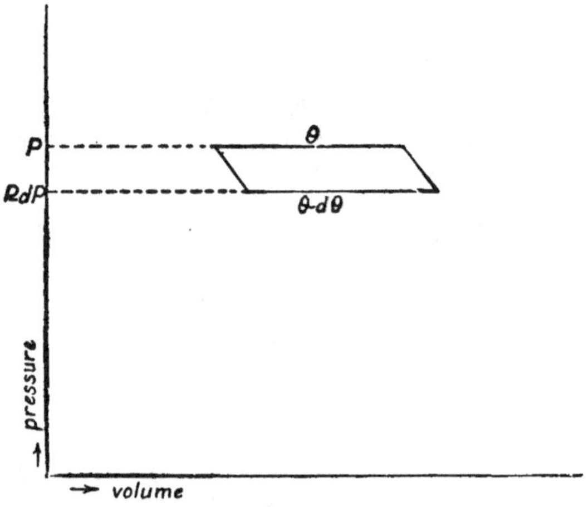

FIG. 191.

radiation medium, but impermeable, when we so choose, to the radiation and its energy.

Now let us take the radiation-filled space through the following cycle:—

1. Start at temperature θ with the inside surface fully radiating, and allow the surface to be pushed out by the pressure $P=\frac{E}{3}$ from radius r to radius dr, increasing the volume by $4\pi r^2 dr$. The surface is surrounded meanwhile by a conductor maintained at θ, so that not only is work $P \times 4\pi r^2 dr$ done, but radiant energy $E \times 4\pi r^2 dr$ is supplied to fill the extra volume. Hence the total supply of energy at θ is

$$dQ = 4\pi r^2 dr(P+E)$$

$$= 4\pi r^2 dr \times \frac{4E}{3}$$

2. Now make the inside surface of the sphere non-conducting and totally reflecting, and allow the radiation within to push out the surface a little further, no energy being received from outside. This is an adiabatic expansion. The energy density decreases, both through the

increased volume and the work done on the boundary, and the new energy density will correspond to a lower temperature $\theta - d\theta$. The pressure will fall say to $P - dP$. We shall assume that the radiation is so altered in quality that at every stage of the adiabatic expansion it remains full radiation for the temperature to which it is reduced. We shall seek to justify this assumption below.

3. When the adiabatic expansion is completed, and the temperature is $\theta - d\theta$ replace the reflecting surface by one fully radiating and surrounded by a conductor maintained at $\theta - d\theta$. Now compress the sphere at $\theta - d\theta$ to such an amount that

4. On replacing the fully radiating surface by a totally reflecting one, an adiabatic compression will restore the system to the original volume and temperature.

This cycle may be represented by Fig. 191, and evidently the net work done is $dP \cdot 4\pi r^2 dr$.

Every part of the process is reversible, and indeed we have a Carnot cycle. Hence

$$dQ \times \frac{d\theta}{\theta} = dP \cdot 4\pi r^2 dr$$

substitute for

$$dQ = 4\pi r^2 dr \times \frac{4E}{3}$$

and for

$$dP = \frac{dE}{3}$$

and we have

$$\frac{4d\theta}{\theta} = \frac{dE}{E}$$

Integrating

$$\log \theta^4 = \log E + \text{ constant}$$

then

$$E = a\theta^4$$

and since

$$E = \frac{4R}{U}$$

$$R = \frac{Ua\theta^4}{4}$$

Comparing this with the value of the radiation, page 250,

$$R = \sigma\theta^4$$

we see that

$$\frac{Ua}{4} = \sigma.$$

Full Radiation remains full Radiation in any Adiabatic Change.—Let us now examine this assumption, on which the reversibility of the cycle depends. Imagine an enclosure containing full radiation at θ, and let it expand adiabatically till the energy density corresponds to that in full radiation at $\theta - d\theta$. But let us suppose that the energy, though of the same density, is not of the same quality as the full radiation for $\theta - d\theta$. Suppose, for example, that the red is in excess and the blue in defect. Put into the enclosure substances at $\theta - d\theta$, absorbing and radiating respectively red and blue rays. The red body will be heated and the blue will be cooled, and by establishing a Carnot engine between them a little work can be obtained by equalising the temperatures and by gradually bringing back the radiation towards the quality of full radiation. When we have arrived at this quality the temperature will be a little below $\theta - d\theta$. Now remove the red and blue bodies and compress adiabatically till the original volume is reached. Since we started on this compression at a temperature below $\theta - d\theta$, when we arrive at the original volume we shall have a temperature below θ. Again, we shall have, by supposition, a quality differing from full radiation, and we can again insert different absorbing bodies, get out a little more work and lower the temperature still further in reducing to full radiation. Hence we have in the end the original volume, but colder, and we have obtained work in the process. That is, we have obtained work in a process the net result of which is merely the cooling of a body which may be, if we like, the coldest body in the system. This we must consider as contrary to experience, and so we reject the supposition on which it is based, that full radiation after an adiabatic change ceases to be full radiation.

Relation between Volume and Temperature in an Adiabatic Change of Volume.—Let a volume V contain full radiation at temperature θ and with energy density E. Then the total energy is

$$W = VE$$

Now let V change adiabatically to $V + dV$. If P is the radiation pressure the work done is

$$PdV = \frac{EdV}{3}$$

But this is equal to the diminution in internal energy, or

$$W - (W + dW) = \frac{EdV}{3}$$

or
$$VdE + EdV + \frac{EdV}{3} = 0$$

$$\therefore \quad VdE + \frac{4}{3}EdV = 0$$

Integrating $\quad EV^{\frac{4}{3}} = \text{constant}$

But $\quad E = a\theta^4$

Then $\quad \theta^4 V^{\frac{4}{3}} = \text{constant}$

or $\quad \theta V^{\frac{1}{3}} = \text{constant}$

If the space maintains the same shape $V^{\frac{1}{3}}$ is proportional to the linear dimension r. Then

$$\theta r = \text{constant}.$$

Entropy.—If ϕ is the entropy per volume 1, then in an adiabatic change ϕV is constant.

But $EV^{\frac{4}{3}}$ is constant also and so is $E^{\frac{3}{4}}V$. Hence

$$\phi \text{ is proportional to } E^{\frac{3}{4}} \text{ or to } \theta^3$$

Application of Doppler's Principle in an Adiabatic Expansion.—If a train of waves is reflected at a surface which is itself moving, then by

Doppler's principle the wave-length of the reflected train is altered. Suppose that the reflecting surface B is moving with velocity u away from the source A (Fig. 192). Consider first a normal pencil. It will appear, when reflected, to come from a source A′, as far behind B as A is in

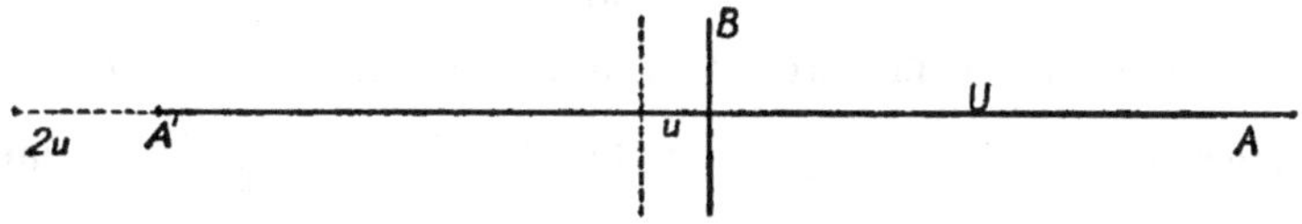

FIG. 192.

front, so that A′ is moving back with velocity $2u$ and the waves, which in the incident beam fill a length U equal to the velocity of radiation, fill in the reflected train a length $U+2u$. Hence if λ is the wave-length in the incident train, λ' that in the reflected,

$$\frac{\lambda'}{\lambda}=\frac{U+2u}{U}$$

or if $\lambda'=\lambda+d\lambda$

$$\frac{d\lambda}{\lambda}=\frac{2u}{U}$$

If the train is inclined at θ to the normal, then (Fig. 193) the reflected train may be supposed to come from A′, the image of the source A in the

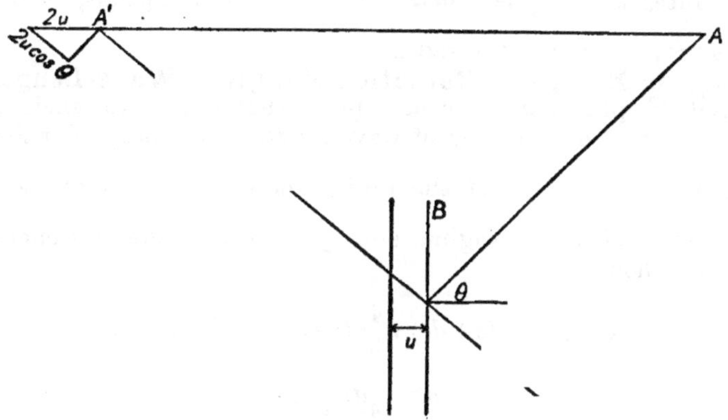

FIG. 193.

reflecting surface B, and again A′ is moving back with velocity $2u$. But this only increases the length of the reflected path by $2u\cos\theta$. Hence the waves originally in U will be now spread over $U+2u\cos\theta$ and

$$\frac{d\lambda}{\lambda}=\frac{2u\cos\theta}{U}$$

Now let us apply this to radiation originally of wave-length λ in a perfectly reflecting sphere, which is expanding with velocity

$$u=\frac{dr}{dt}$$

If a ray strikes the surface at θ to the normal, its path between successive reflections is $2r\cos\theta$, and it strikes the surface $\frac{U}{2r\cos\theta}$ times per second. At each reflection its wave-length is increased by

$$d\lambda=\lambda\,.\,\frac{2u\cos\theta}{U}$$

so that the increase per second is

$$\frac{d\lambda}{dt}=\lambda\,.\,\frac{2u\cos\theta}{U}\times\frac{U}{2r\cos\theta}$$

$$=\frac{\lambda u}{r}$$

or

$$=\frac{\lambda}{r}\frac{dr}{dt}$$

Then integrating, $\frac{\lambda}{r}$ is constant as the sphere expands, or the wave-length is proportional to the radius.

Change of Energy in Radiation of a given Wave-Length in an Adiabatic Expansion.—Let us suppose that a spherical enclosure only contains one kind of radiation of wave-length λ, of energy density e, and exerting pressure $p=\frac{e}{3}$. If the radius increases from r to $r+dr$ in an adiabatic expansion, the original energy is equal to the new energy + the work done; then

$$e\,.\,\frac{4}{3}\pi r^3=(e+de)\,.\,\frac{4}{3}\pi(r+dr)^3+\frac{e}{3}\,.\,4\pi r^2dr$$

whence

$$\frac{de}{e}+4\frac{dr}{r}=\theta$$

and integrating

$$er^4=\text{constant.}$$

Change of Energy of each Wave-Length in Adiabatic Expansion of full Radiation.—We shall assume that the above result applies to each wave-length separately when we have full radiation. If a sphere, then, containing full radiation expands from radius r to radius r', θ changing to θ' where $r\theta=r'\theta'$ (p. 338), then

$$\frac{\lambda}{\lambda'}=\frac{r}{r'}=\frac{\theta'}{\theta}$$

or $\lambda\theta$ is constant during the change.

Let the energy density of the radiation in the range of wave-length from λ to $\lambda+d\lambda$ be denoted by $ed\lambda$; e may be termed the energy per unit range about the wave-length λ. By the expansion let λ, $\lambda+d\lambda$ change to λ', $\lambda'+d\lambda'$, while $ed\lambda$ changes to $e'd\lambda'$.

As shown above

$$ed\lambda r^4 = e'd\lambda' r'^4.$$

But

$$d\lambda' = \frac{r'}{r}d\lambda$$

then

$$er^5 = e'r'^5$$

or since

$$\frac{r}{r'} = \frac{\theta'}{\theta}$$

$$\frac{e}{\theta^5} = \frac{e'}{\theta'^5}$$

or, comparing e, the energy per unit range, or in equal ranges $d\lambda$, in two full radiations at different temperatures and taking corresponding wave-lengths given by $\lambda\theta=$ constant we see that e is proportional to the fifth power of the temperature.

Suppose, then, at any one temperature we plot $\frac{e}{\theta^5}$ as ordinate against $\lambda\theta$ as abscissa, we shall obtain a curve which will be identical for full radiation at *all* temperatures.

Then we may put

$$\frac{e}{\theta^5} = f(\lambda\theta)$$

or

$$e = \theta^5 f(\lambda\theta),$$

where f is a function of the product $\lambda\theta$ only.

Maximum Value of Energy for Given Range of Wave-Length.

For a given temperature, when e is a maximum, $\frac{e}{\theta^5}$ is a maximum.

Then writing the maximum value of e as e_m, $\frac{e_m}{\theta^5}$ is the same for all temperatures, and therefore $\lambda_m\theta$ is the same for all temperatures, where λ_m is the wave-length for maximum energy per unit range.

Form of the Function expressing the Distribution of Energy in the Spectrum.—The methods used up to this point do not afford us any information as to the form of the curve expressing the distribution of the energy in the radiation spectrum, except that it must be represented by

$$e = \theta^5 f(\lambda\theta).$$

At present it appears to be necessary to introduce some hypothesis as to the way in which radiation is produced, in order to find the form of $f(\lambda\theta)$.

We shall not discuss this subject, but shall refer the reader to Wien's paper already cited for a particular hypothesis by which he obtained as the value of e

$$e = c_1 \lambda^{-5} e_1^{-\frac{c_2}{\lambda\theta}}$$

The work of Lummer and Pringsheim * and others shows that Wien's formula does not agree with experiment through a very wide range of wave-length.

Planck † by another hypothesis obtains a value

$$e = c_1 \lambda^{-5} \frac{1}{e^{\frac{c_2}{\lambda\theta}} - 1}$$

which agrees more closely with experiment. ‡

It may be noted that if $\lambda\theta$ is small compared with c_2, that is, if the radiation is well on the ultra-violet side of the maximum, $e^{\frac{c_2}{\lambda\theta}}$ is large and Planck's formula is practically the same as Wien's. If, on the other hand, $\lambda\theta$ is large compared with c_2, $e^{\frac{c_2}{\lambda\theta}} - 1$ may be put equal to $\frac{c_2}{\lambda\theta}$ and Planck's formula becomes

$$e = \frac{c_1 \lambda^{-4} \theta}{c_2}$$

or the radiation of a given wave-length well on the ultra-red side of the maximum is proportional to the absolute temperature.

* *Congrès Int. de Phys.*, ii. p. 41.

† *Deutsch Phys. Gesell.*, 1900, p. 202; *Science Abstracts*, iv., No. 507, p. 230.

‡ Rubens and Kurlbaum, *Preuss. Ak. Wiss. Berlin, Sitz. Ber.* xli. 1900, p. 929; *Science Abstracts*, iv., No. 371 p. 167.

INDEX

Made in United States
Troutdale, OR
11/03/2026

56847412R00209